COLLISION ACTIONS ON STRUCTURES

This textbook covers the collision of a moving, falling or flying object on a rigid barrier or a structural element, and the transmission of the transient action to the rest of the structural system. It is the only up-to-date book on this under-researched topic that confronts engineers on a day-to-day basis. The book deals with a range of real-life engineering problems and focuses on the application of knowledge and skillsets from structural analysis and structural dynamics. Fundamental principles and concepts on structural collision are first introduced, followed by their specific applications such as vehicular collision on bridge structures, boulder impact on rockfall barriers and collision by hail and windborne debris. Analytical solutions provided are in the form of closed-form expressions, which can be directly adopted in conventional manual calculations. The use of spreadsheets to simulate the dynamic response behaviour is also covered.

- The only standalone book covering the topic from a civil engineering perspective
- Practical guidance on real-life engineering problems, and use of computational and physical methods
- Conveys methodology validated experimentally

The book provides an excellent guide for practitioners and sets out fundamental principles for graduate students in civil, structural and mechanical engineering.

COLLISION ACTIONS ON STRUCTURES

Arnold C.Y. Yong, Nelson T.K. Lam, and
Scott J. Menegon

CRC Press
Taylor & Francis Group
Boca Raton London New York

CRC Press is an imprint of the
Taylor & Francis Group, an **informa** business

First edition published 2023
by CRC Press
6000 Broken Sound Parkway NW, Suite 300, Boca Raton, FL 33487-2742

and by CRC Press
4 Park Square, Milton Park, Abingdon, Oxon, OX14 4RN

CRC Press is an imprint of Taylor & Francis Group, LLC

Library of Congress Cataloguing-in-Publication Data
A catalog record for this title has been requested

ISBN: 978-0-367-67817-3 (hbk)
ISBN: 978-0-367-67830-2 (pbk)
ISBN: 978-1-003-13303-2 (ebk)

DOI: 10.1201/9781003133032

Typeset in Sabon
by MPS Limited, Dehradun

For Cindy Low and Joseph Yong and all the unconditional support you have provided; this book would not have been possible without you.

– Arnold Yong

Contents

Foreword

The practice of structural engineering is constantly evolving to meet the changing expectations and demands of those we serve. Whereas it was once sufficient to design a structure to support itself plus the anticipated loading from its expected use, as time passed it became an expectation that structures would also be designed to safely resist other actions such as wind loading, temperature effects and changes in material behaviour with time. Further expectations and demands of the public required that structures should safely respond to extreme events, such as earthquakes, typhoons and storm surges. The latest addition to the expectations and demands on structural engineers is to safely design for the collision actions of moving, flying or falling objects.

This evolution of structural design requires ongoing research and training to ensure that practising engineers are equipped with the tools to safely design structures to satisfy new requirements. There is always a challenge to provide tools that adequately represent the behaviour of a structure, while at the same time being sufficiently simple and usable to enable widespread adoption. To date, the structural engineering profession has been somewhat lacking design tools that adequately represent the behaviour of a structure subjected to collision loading while at the same time being sufficiently simple and usable. This book is an important step in addressing that.

Because the origins of structural design were in dealing with the "static" loads due to the weight of building and its contents, structural engineers are much more comfortable if "dynamic" loads such as wind, earthquake and collision can be simplified as equivalent static effects. There are, of course, limitations to the applicability of a simplified theory, and it is important that these are clearly presented so that the practising structural engineer is aware of the bounds within which a simplified theory can be safely applied. This book provides simplified methods for designing structures for collision loads, and sets out the basis and limitations for the simplified approach.

This book will provide an important reference for practising structural engineers when designing a structure to resist collision loads. The content would also be a valuable addition to undergraduate education in structural engineering, helping future generations of engineers to equip themselves with the knowledge and skills to meet the changing expectations and demands of the public we serve.

Geoff Taplin

Technical Director, AECOM (retired), Recipient of John Connell Gold Medal 2014

Preface

Few engineers have the skills to estimate, with a reasonable level of confidence, the level of damage sustained by structural elements from a collision scenario. For example, estimating the amount of deflection of a column when struck by a moving, or flying, object is a challenge to most structural engineers. Design guides for ensuring an adequate level of structural robustness from a rare event such as a collision are mainly about avoiding collapse of the structure, or part of the structure. They usually recommended that design checks are undertaken to ensure that if a key structural member (a column, for example) is removed by the collision, the damaged structure is able to make use of alternative load paths to avoid collapsing. As static analyses are employed in undertaking such checks, the need to analyse the "true" dynamic nature of the collision action and estimate the damage caused by the collision scenario is circumvented. The general lack of knowledge in regard to assessing structures for collision actions means the costs and benefits of retrofitting a facility to provide an optimised level of protection is often not known.

Education and training on structural analyses provided to structural engineers have been based on the principles of statics. There are a few textbooks on structural dynamics authored by well-respected experts in the field. Some elite tertiary institutions offer graduate courses on structural dynamics, but much of the emphasis is on predicting the behaviour of a structure responding to a predefined forcing function or ground motion time history generated by an earthquake. Few engineers have the skills to reliably predict the behaviour of a structure from a collision scenario, an explosion, an earthquake or a severe storm. Engineers who have been trained on structural dynamics would still need to consult regulatory documents, handbooks and the like to look for ways of representing the event scenario by parameters for input into a well-defined calculation procedure, or a forcing function for input into a structural analysis software. Earthquake and wind engineering are well researched, and many regulatory documents and design guides have been written for assisting engineers. More recently, the same is said of protective technology for countering explosion hazards. However, relatively little has been written on collision actions.

This book is concerned with collision actions of a moving, flying or falling object, which is an issue confronting structural engineers on a day-to-day basis. The topic in itself is significant even though many other types of hazards tend to generate more publicity. Research undertaken on collision actions on a structure is *adhoc*. Specialist software such as LS-DYNA is frequently used (as a "black box" tool) to make predictions of the outcome of a given collision scenario. Costly physical experimentation in the field, or in the laboratory (such as a crash test) is usually still required to validate the results that are generated by these specialist software packages. Much of what has been presented in the literature is about estimating, and modelling, the response of a structure to a collision by first undertaking a physical test and then matching (calibrating) the response in a computer simulation. Engineers often find it difficult to make use of the reported findings from such case studies, or observations made on certain experiments, to guide their own designs as they're often atypical in nature compared to the research studies. Fundamentals such as principles of equilibrium, kinematics compatibility and constitutive relationships have helped guide engineers in their reasoning. However, most engineers only have confidence in exercising those basic analytical skills when dealing with static load scenarios. The lack of understanding of the fundamentals that are associated with the transient action of a collision is still a major knowledge gap to be filled.

A key objective in the writing of this book is introducing fundamental principles on structural collision that are not widely known amongst practising structural engineers. The style of presentation of the early chapters of the book is likened to that of a textbook on structural mechanics in the sense that the underlying fundamental principles are presented in a logical fashion. Much of the techniques that can be learnt from the book can be implemented by conventional manual calculations. The use of spreadsheets to expedite the calculations is also covered in the later chapters. In addition to introducing fundamental concepts on impact dynamics weaving through the entire book, there are chapters emphasising applications dealing with specific types of construction. Case studies and worked examples are presented throughout the book alongside the theoretical material and are of different forms and scales.

This book was primarily written for self-learning by structural engineers; however, it can also be used as courseware for the structured tuition in a coursework program at either the undergraduate level (year 3 or above in a four-year bachelor degree) or at the graduate level (diploma or master's degree level). The book can also be used to assist the delivery of short courses forming part of a continuous professional development (CPD) program for practising civil and structural engineers.

Acknowledgements

The authors wish to first acknowledge Geoff Taplin, John Mander and the third (anonymous) book reviewer all of whom provided very positive feedback to the publisher over the proposal to publish the book. In addition, Geoff Taplin has also generously shared with the authors a great deal of his knowledge and experience on vehicular barrier collision. The writing of Chapter 9 and Chapter 10 was very much inspired by his input.

The authors also wish to acknowledge the following academic colleagues who have worked with the authors, and have made original contributions to knowledge, on the topic of structural collision as disseminated in the book: Yi Yang, Muneeb Ali, Jing Sun, Shihara Perera, Mahil Pathirana, Jude Shalitha Perera, Zireen Zanoofar Abdul Majeed, Shuangmin Shi and Yiwen Cui.

Assistance provided by Jude Shalitha Perera, Zireen Zanoofar Abdul Majeed, Shuangmin Shi, Prashidha Khatiwada and Tu Nguyen in advising the authors and in proofreading the book is most gratefully acknowledged. The acknowledgement of support is extended to other academics, all of whom have a depth of experience and expertise on impact and structural engineering. The following experts deserve special acknowledgement: Mahdi Miri Disfani, Lihai Zhang, Tuan Ngo, Emad Gad, Dong Ruan and Guoxing Lu.

The authors also wish to express their gratitude towards Julian Kwan, Harris Lam, Carlos Lam and Anthony Wong, at the Geotechnical Engineering Office of the Government of the Hong Kong Special Administrative Region, for their sharing of knowledge, and experience, during the time when the first and second authors were engaged as international advisors to the government over the protection of hill slopes from rockfalls and landslides.

A significant body of knowledge conveyed by the book was generated from research which took place in recent times involving the authors and their collaborators. The large-scale experimental programs forming part of the related research was made viable by financial support from the

Australian Research Council through Discovery Project DP170102858 and Linkage Project LP190100208, and also by Government of Hong Kong Special Administrative Region through numerous service contracts with the University of Melbourne. Scholarships in support of research higher-degree candidatures at the University of Melbourne also contributed to the success of the related research, so are the infrastructure and experimental facilities provided by the University of Melbourne and Swinburne University of Technology.

Authors

Arnold C. Y. Yong is a senior civil engineer at Sepakat Setia Perunding (SSP) in Malaysia. He earned his PhD degree at the University of Melbourne where he continued his research as a post-doctoral fellow. During his time at the university, his research primarily focused on collision actions on structures. He led a number of experimental programs involving large-scale dynamic testing funded by the Australian Research Council (ARC), and has provided expert advice to the Geotechnical Engineering Office (GEO) of the Government of the Hong Kong Special Administrative Region to assist developing design guides for protection against landslides and rockfalls on hillslopes in built-up areas.

Nelson T. K. Lam is a professor and leader of the Structures and Buildings Discipline in the Department of Infrastructure Engineering at the University of Melbourne. He is editor of the *Australian Journal of Structural Engineering*, member of the Structural College Board of Engineers Australia and a member of the committee responsible for the Australian standard for seismic actions. Nelson has earned many awards for his research, including the Chapman Medal (1999, 2010, 2019) and Warren Medal (2006) from Engineers Australia.

Scott J. Menegon is a consulting structural engineer and academic with expertise in reinforced concrete, earthquake engineering and collisions actions. Scott earned his PhD from the Swinburne University of Technology, where he was the joint inaugural recipient of the prestigious Dr William Piper Brown AM scholarship. Scott currently works in industry and maintains his affiliation with Swinburne as an Adjunct Industry Fellow. His achievement in research was recently recognised by the award of the Chapman Medal (2019) by Engineers Australia.

Authors

Notation

A	Area
A_E	Effective area of shear failure surface
A_G	Gross area of shear failure surface
A_s	Area of reinforcement
A_{st}	Area of tensile reinforcement
A_v	Area of shear failure surface
a_k	Acceleration of the k-th node
$a_i(t)$	Acceleration time-history of impactor in a two-degree-of-freedom system
$a_t(t)$	Acceleration time-history of target in a two-degree-of-freedom system
B	Width of a cross section
b	Width of a cross section
C	Depth of virtual fixity
C_D	Drag coefficient
$C_{F_{c,\max}}$	Correction factor for hybrid model
C_j	Generalised damping coefficient
C_n	Magnification factor for the n-th occurrence of collision
COR	Coefficient of restitution
$[C]$	Damping coefficient matrix
c	Damping coefficient
c_{cr}	Critical damping coefficient
D	Depth of a cross section
D_h	Diameter of hailstone
D_i	Diameter of impactor
D_n	Damping coefficient of a non-linear damper
D_p	Flexural rigidity of a plate
$[D]$	Dynamic matrix
d_c	Perpendicular distance of point of strike from the centre of a square plate
d_b	Diameter of reinforcement bar
d_d	Perpendicular distance from the centre to the edge of a plate

d_o	Depth to the centroid of the outermost layer of longitudinal reinforcement from the extreme compressive fibre of a cross section
d_{om}	Mean value of depth to the longitudinal reinforcement in each direction
E	Young's modulus
E_c	Young's modulus of concrete
E_s'	Post-elastic modulus of reinforcement
E_T	Transformed Young's modulus
E_1	Young's modulus of impactor
E_2	Young's modulus of target
EI	Flexural rigidity
EI_{combined}	Combined flexural rigidity of pole cast in cast into a concrete jacket
$E_c I_{\text{eff}}$	Effective flexural rigidity of concrete
EI_{eff}	Effective flexural rigidity
EI_{pole}	Flexural rigidity of pole
e	Thickness of cushion
F	Force
F_c	Contact force
F_{capacity}	Punching (shear) capacity
$F_{c,\max}$	Maximum contact force
F_f	Frictional force
F_{hard}	Base shear in a hard impact scenario
$F_{I,p}$	Inertial resistance of shear plug
$F_{\max}$	Maximum force
F_R	Reaction force
F_u	Ultimate force
F_y	Yield force
F^*	Design force
$[F]$	Flexibility matrix
f	Natural frequency of vibration
f_c	In-situ concrete compressive strength
f_c'	Specified characteristic concrete cylinder strength
f_{ct}	Direct tensile strength of concrete
f_{ctm}	Mean tensile strength of concrete
f_j	Natural frequency for the *j*-th mode of vibration
f_{su}	Ultimate strength of reinforcement
f_{sy}	Yield strength of reinforcement
f_y	Yield strength
g	Gravitational acceleration (9.81 m/s^2)
H	Height of pole above ground
H_B	*Brinell* hardness
H_w	Wall height
h	Barrier height

h_i	Initial height above a target that an impactor is released from
I	Second moment of area
I_c	Rotational inertia of a section about its centroid
I_θ	Rotational inertia of a barrier taken about the pivot point of overturning
$[I]$	Identity matrix
J	Impulse
J_i	The i-th short-duration impulse
J_t	Total impulse
j	Modal number
K_j	Generalised stiffness for the j-th mode of vibration
K_t	Dynamic increase factor
KE	Kinetic energy
KE_0	Kinetic energy delivered by impactor
KE_1	Kinetic energy carried by impactor on rebound
KE_2	Kinetic energy transferred to target
$[K]$	Stiffness matrix
k	Target stiffness
k_{eff}	Effective stiffness
k_R	Rotational stiffness at the base of a bridge pier
k_{R2}	Rotational stiffness at the base of a rigid rod
k_e	Lateral stiffness of soil-embedded pole
k_n	Contact stiffness defined by a non-linear spring
k_{n100}	Contact stiffness between granite impactor of 100 mm in diameter and concrete target
k^*	Generalised stiffness for the fundamental mode of vibration
L	Span length
L_e	Embedded length of pole
L_j	Excitation factor
L_p	Plastic hinge length
L_{sp}	Strain penetration in the supporting element
l	Length of base slab
M	Mass
M^*	Design moment
M_{base}	Mass of a base slab
M_c	Combinational mass
M_E	Modulus of elasticity of granular material
$M_{E,n}$	Value of M_E prior to the n-th occurrence of collision
M_{hard}	Base bending moment in a hard impact scenario
M_k	Mass of the k-th node
M_{side}	Mass of a side wall
$M_{simulation}$	Simulated bending moment
M_{stem}	Mass of a stem wall
M_t	Total mass of a member

M_j	Generalised mass for the *j*-th mode of vibration
M_y	Yield moment
$M(x)$	Bending moment
$[M]$	Mass matrix
m	Impactor mass
m_{add}	Additional lumped mass of a portion of a bridge that has not been included in its model explicitly
m_f	Front mass of a vehicle
m_g	Gabion mass
m_i	Tributary mass corresponding to v_i
m_{pole}	Pole mass
m_r	Rear mass of a vehicle
m_t	Target lumped mass
m_w	Wall mass
m^*	Generalised mass for the fundamental mode of vibration
N	Total number of impulses
N^*	Design axial load
N_{impact}	Total number of impact
N_n	Total number of nodes in a multi-degree-of-freedom lumped mass system
n	Total number of modes of vibration
n_s	Number of side walls or number of pairs of side walls
P	Point load
P_m	Maximum contact stress
PE	Potential energy
$P(\gamma)$	Probability density function of γ
p	Exponent characterising non-linearity of a contact force model, taken as 1.5 for *Hertzian* contact theory
p_{st}	Reinforcement ratio
R	Distance between point of contact and point of pivot
R_c	Radius of curvature
R_i	Radius of impactor
R_t	Radius of target
RSa	Response spectrum acceleration
RSd	Response spectrum displacement
r	Distance between centre of gravity and point of pivot
r_a	Radius of surface area of contact
r_c	Radial offset of point of strike from the centre of a square plate
r_d	Half the length of the diagonal of a square plate
r_{stem}	Distance between centroid of stem wall and point of pivot
SE	Strain energy
$[S]$	Sweeping matrix
s	Arc length
T	Natural period of vibration

T_j	Natural period for the j-th mode of vibration
T_M	Natural period of vibration of the rear spring in a 2DOF system
T_m	Natural period of vibration of the frontal spring in a 2DOF system
T_s	Soil-stiffness parameter
t	Time
t_d	Total duration of a forcing function
t_g	Time lag between the first and second collision
t_p	Plate thickness
t_t	Total duration of a simulated response
t_u	Flight time of hailstone
t_τ	Duration of a short-duration impulse
U_H	Horizontal velocity of hail
U_R	Resultant velocity of hail
U_V	Vertical velocity of hail
U_w	Horizontal wind velocity
u_{max}	Amplitude of displacement
u_{st}	Static displacement
$u(t)$	Displacement time-history
$\dot{u}_{max}$	Amplitude of velocity
$\dot{u}(t)$	Velocity time-history
$\dot{u}(\tau)$	Velocity of a target when subject to a short-duration impulse
$\ddot{u}(t)$	Acceleration time-history
V_b	Design force
v_i	Velocity at any point of a plate immediately following an impact
$v_i(t)$	Velocity time-history of impactor in a two-degree-of-freedom system
v_k	Velocity of the k-th node immediately following impact
v_n	Nodal velocity of a plate measured at the point of strike immediately following an impact
$v_t(t)$	Velocity time-history of target in a two-degree-of-freedom system
v_0	Incident velocity of impactor
v_1	Velocity of impactor on rebound in opposite direction
v_2	Velocity of target immediately following impact
v_θ	Incident velocity of impactor at an angle θ to target
W_f	Energy dissipated by frictional force
$W(r_c)$	Deformation profile of a plate
w	Barrier width
w_{base}	Width of base slab
w_o	Permanent indentation
w_{stem}	Width of stem wall

$x_i(t)$	Displacement time-history of impactor in a two-degree-of-freedom system
$x_t(t)$	Displacement time-history of target in a two-degree-of-freedom system
$\bar{x}$	Horizontal distance measured from the centre of gravity of a barrier to the point of pivot at the base of the barrier
Y	Compressive yield strength of material
$Y_j(t)$	Time dependent scaling factor of displacement time-history for the j-th mode of vibration
$\ddot{Y}(t)$	Time dependent scaling factor of acceleration time-history for the j-th mode of vibration
$\bar{y}$	Vertical distance measured from the centre of gravity of a barrier to the point of pivot at the base of the barrier
Z	Elastic section modulus
α_v	Angle of inclination of shear cracks in the shear plug
Δ	Maximum displacement
$\Delta_{C.G.}$	Vertical displacement of centre of gravity
$\Delta_{C.G.(crit)}$	Critical vertical displacement of centre of gravity
Δ_c	Cumulative sliding displacement
Δ_G	Pole deflection at ground level
Δ_H	Horizontal displacement of pole-slab assemblage
Δ_p	Plastic displacement
Δ_{pole}	Pole deflection
Δ_s	Static deflection mg/k
Δ_{top}	Total deflection at the top of soil-embedded pole
Δ_u	Ultimate displacement
Δ_y	Yield displacement
ΔE	Energy absorbed by localised deformation of materials
δ	Indentation
δ_b	Dynamic component of target deflection in a vertical collision scenario
δ_{max}	Maximum indentation
δ_s	Angle of skin friction
$\dot{\delta}$	Rate of indentation
$\dot{\delta}_0$	Initial rate of indentation
$\dot{\varepsilon}$	Strain rate
ε_s	Tensile strain
$\varepsilon_{s,\,max}$	Maximum tensile strain
$\varepsilon_s(r_c)$	Strain profile of a plate
ε_{su}	Ultimate stress of reinforcement
ε_{sy}	Yield stress of reinforcement
η_b	Coefficient of subgrade modulus variation
γ	Reduction factor to take into account delay in momentum transfer
κ	Dimensionless factor defined as I_0/mbR

λ	Target mass to impactor mass ratio
μ	Coefficient of friction
ν	Poisson's ratio
ν_1	Poisson's ratio of impactor
ν_2	Poisson's ratio of target
ω_D	Damped natural angular velocity
ω_j	Angular velocity for the *j*-th mode of vibration
ω_n	Natural angular velocity
ϕ	Capacity reduction factor
ϕ'	Effective angle of shear resistance
ϕ_k	Angle of shear resistance
ϕ_p	Phase angle
ϕ_p	Plastic curvature (in Chapter 8 only)
$\phi_{p,c}$	Plastic curvature governed by the compressive strain limit of 0.004 in the concrete being reached
$\phi_{p,t}$	Plastic curvature governed by the tensile strain limit of 0.04 in the reinforcement being reached
ϕ_u	Ultimate curvature
ϕ_y	Yield curvature
$\phi(x)$	Curvature
$\{\phi_j\}$	Mode shape vector for the *j*-th mode of vibration
$\psi(x)$, $\psi(x, y)$	Shape function
$\psi''(x)$	Unitless curvature
ρ	Density
ρ_a	Density of air
ρ_b	Density of boulder
ρ_c	Density of concrete
ρ_g	Density of gabion
ρ_h	Density of hailstone
σ_{vm}	*Von Mises* stress
τ	Time instance when a short-duration impulse occurs
θ	Angle of rotation
θ_G	Rotation of pole at ground level
θ_p	Plastic rotation
θ_{pole}	Rotation of pole
θ_{top}	Total rotation of soil-embedded pole
$\dot{\theta}$	Angular velocity
ξ	Damping ratio

Chapter 1

Introduction

1.1 EMPHASIS ON FUNDAMENTAL CONCEPTS

The analysis procedures and closed-form design expressions presented in this book for assessing structures to collision actions are founded on first principles and fundamental concepts. Empirical relationships (from experimental testing) are employed, where necessary, to constrain input parameters into the respective models, and to reaffirm the accuracy of the recommended expressions for predicting response behaviour and damage. Readers are introduced to threshold concepts that are transformative and irreversible in nature. Transformative refers to how the way of thinking is changed. Irreversible refers to the fact that once the concept is learnt, it will not be forgotten and the change in thinking is permanently engrained. By contrast, much of the impact engineering investigations that are reported in the literature on collision actions mainly convey observations from physical and/or numerical simulated experimentation. Curve-fitting approaches are commonly adopted to create empirical models, which are in turn used for designing a calculation procedure. Methodologies developed in this manner inherently fall short of delivering new threshold concepts. Whereas the emphasis in this book on fundamental concepts allows readers to holistically understand the problem at hand, which ultimately allows them to make more informed engineering decisions that result in better performing structures.

Closed-form expressions are mostly favoured by design engineers, when their basis is explained from first principles (from a reliable source), and when the accuracy of results derived from the expressions has been validated experimentally. A prominent example of such an algebraic expression is the one shown on the cover of this book. This expression for determining the elastic deflection generated by a collision is first introduced in Chapter 3 and a detailed derivation of the expression from first principles is presented. Expressions of a similar form can be found in many other chapters to help solve a range of problems. Citations to publications by the authors (and their co-authors) in high-ranking scholarly journals in the field for

DOI: 10.1201/9781003133032-1

supporting the methodologies introduced in the book are provided in the relevant chapters. A broad summary of the major experimental verifications that have been performed is provided towards the end of this chapter.

1.2 FORMAT AND PRE-REQUISITE KNOWLEDGE

Core concepts and calculation methodology relevant to each chapter of the book are presented initially by textual descriptions presented in plain English to explain the natural phenomena and concepts of modelling in physical terms. The presented analytical methods typically involve idealising a structure into a simple structural system such as a spring connected lumped mass system. The practical application of the calculation methodologies is illustrated by the use of worked examples in each chapter. Additional examples on each topic are presented in the final chapter of the book for consolidating the command of specific skills and knowledge, which the reader may well find useful when designing a new structure, or undertaking a robustness assessment of an existing structure, for any of the collision scenarios presented in this book.

Good working knowledge on static analyses involving application of virtual work principles and use of stiffness and flexible matrices (for analysing structural systems including indeterminate systems) are expected of the reader, so is knowledge on code compliant design of reinforced concrete (RC) and structural steel. No prior knowledge on structural dynamics nor contact mechanics is required of the readers in order that the book is suitable for reading by structural engineers at different stages of their career including students at graduate or undergraduate levels. Mathematical treatments presented in the book are at a level that should be comprehensible by engineers (and graduate students) who have received undergraduate education in civil or mechanical engineering. Thus, good working knowledge on basic calculus and linear algebra is expected.

1.3 CONTENTS

This book contains chapters that deal with collision scenarios of different scales. This includes scenarios where a moving, or flying, object strikes a structural element causing it to bend; the sliding and/or overturning of a free-standing installation (e.g. a barrier) from a moving, or flying, object; a heavy falling object landing on a floor; the localised action (contact force) of an impact scenario; the effects of providing cushioning between the impactor and the structure (e.g. gabions); different forms of damage caused to RC in a collision; a vehicle colliding on a bridge; a flying object (e.g. fallen boulder) impacting on a baffle, which is embedded into soil; and the impact of debris or hailstones on cladding. The final chapters of the book facilitate readers working through numerical examples to consolidate their learning of the analytical techniques presented in the book. With the case studies of fallen boulders on a rockfall barrier, the velocities of the collisions were up to 20 m/s. Vehicular collision scenarios had velocities of up to 30 m/s (100 km/h approximately). Hailstones under free fall and the effects of wind can reach a velocity close to 50 m/s. A more detailed chapter-by-chapter description of the book is presented in the following.

As this book does not require prerequisite knowledge on structural dynamics, Chapter 2 introduces relevant basic concepts in order to allow the average civil or structural engineer to use the book without the need to learn the dynamics of structures from other sources. Chapter 3 introduces the use of the principles of conservation of momentum and energy in estimating the amount of displacement generated by a collision. A method for estimating the deflection of a structural member when responding in flexure to a collision action is then introduced. These fundamental principles form the basis of calculation methodologies developed in the rest of the book. For example, the equal momentum and energy methodology is extended in Chapter 4 for estimating the overturning and sliding behaviour of a free-standing element. The overturning stability of a structure that is supported on a shallow foundation, and not free-standing, can also be checked by employing the calculation techniques introduced in the chapter. The analysis of the impact generated by a falling object on a structural member such as a floor beam (taking into account gravitational actions) is dealt with in Chapter 5. The measurement and modelling of contact force, which is required for assessing the effects of localised impact action at the point of contact with the surface of the target, is introduced in Chapter 6. The modelling methodologies introduced in Chapters 3 and 6 are further developed in Chapter 7 to incorporate the mechanism of cushioning; crushed rock-filled gabion cells that are placed in front of a rigid rockfall barrier, and the mitigating mechanism occurring in the gabion when impacted upon, are the main themes of the chapter. Recognising that RC is a mainstream construction material, Chapter 8 is devoted fully to modelling different forms of damage that is common to RC namely denting, spalling, punching and plastic hinging (i.e. non-linear flexural response).

Chapters 9 to 12 deal with specific engineering applications of the learnt modelling methodology in predicting damage caused from a range of collision scenarios. The modelling of the effects of vehicular impact on a bridge deck is treated in Chapters 9 and 10. Collision at a steel pole (baffle) that is embedded into the ground (soil) is covered in Chapter 11. Impact of debris on an aluminium panel is dealt with in Chapter 12. Chapter 13 illustrates the numerical simulations of the dynamic action of an impact aided by the use of Excel spreadsheets. Chapter 14, which is the final chapter of the book, presents a collection of case studies of impact experimentation to allow readers to benchmark predictions from analytical or numerical modelling, and to consolidate their learning by going through more worked examples on various topics covered in the book.

1.4 EXPERIMENTAL VERIFICATIONS

The authors had the benefits of making use of research funding provided by the Australian Research Council and industry sponsors, along with other sources of funding, to fill the knowledge gaps in the literature. Much of the methodology conveyed in the book has been verified experimentally as presented in publications co-authored by the authors.

The writing of Chapter 3, which was aimed at introducing the fundamental principles of utilising equal momentum and energy to predict the displacement demand of a collision, is based on original published materials that can be found in Lam et al. [1.1], Yong et al. [1.2] and Yong et al. [1.3]. The latter reference introduces and verifies a simple method of modelling static and collision actions that are applied concurrently. Modelling techniques that are introduced in Chapter 4, which deals with impact generated overturning and sliding of free-standing structures, have also been verified by dynamic experimentations as presented in [1.1, 1.4]. The modelling principles have been extended in Chapter 5 to address the considerations of vertical impact. The good accuracy of the model's predictions is demonstrated by the strong correlation with test results from 68 drop tests conducted on RC beam specimens [1.5].

The *Hunt and Crossley* (H&C) model is introduced in Chapter 6 for predicting the forcing function at the point of contact from a collision scenario (referred herein as the contact force). The good accuracy of the H&C model to predict contact force has been demonstrated by dynamic experiments conducted at different scales. Findings from verification studies based on impact experimentation were presented in Sun et al. [1.6], which deals with impact by hail, and in Perera et al. [1.7] and Perera et al. [1.8], which deal with impact by windborne debris. Original contributions to the model itself can be found in Sun et al. [1.9]. More recently, a verification study was conducted on larger impactor specimens that had been derived from rocks and boulders [1.10].

The mitigation action of the cushioning of a collision scenario can be modelled by combining equal momentum and energy principles with the use of the H&C model as presented in Chapter 7. Experimental verification of the accuracy of this newly derived predictive expression can be found in Perera et al. [1.11].

The modelling of different forms of impact generated damage to RC is presented in Chapter 8. The global response behaviour of the impacted RC member in flexure can be modelled using the calculation techniques presented, and verified, in Yong et al. [2], which incorporates RC modelling techniques presented in Menegon et al. [1.12] and Menegon et al. [1.13].

Calculation procedures for predicting localised damage in the form of denting and spalling have been derived, and verified by Majeed et al. [1.14]. Modelling methodologies and simulation techniques as presented in the book have been adapted in Chapters 9 and 10 for modelling the behaviour of a road bridge when subject to vehicular collision.

The calculation procedure for predicting the behaviour of a pole, or baffle, which is embedded into the ground, is presented in Chapter 11. The good accuracy of the predictions made has been demonstrated by impact testing as reported in Perera and Lam [1.15].

Methods of estimating the impact resistance of aluminium panels when subject to impact by hail and windborne debris are presented in Chapter 12. The good accuracy of the estimates has been demonstrated in dynamic tests as presented in Pathirana et al. [1.16] and Shi et al. [1.17]. The article by Shi et al. [1.17] is to be published in the special issue titled "Extreme Engineering" in the journal of *Thin-Walled Structures*.

Inexpensive techniques of simulating the dynamic response behaviour to a collision employing generic calculation tools (e.g. MATLAB and Excel) are presented in Chapter 13. The good accuracy of estimates made using the new techniques has been demonstrated in [1.18–1.21].

This book also introduces methodologies that were derived from investigations undertaken during its preparation. For example, there are detailed illustrations of unpublished techniques that are concerned with predicting sliding generated by multiple impact on a barrier (Chapter 4), determining maximum contact force by a hybrid model (Chapter 6), modelling flexural actions of a reinforced concrete barrier (Chapter 8), modelling vehicular collision on a bridge deck (Chapter 9 and 10) and time-history simulations of the response of a beam, or pole, with different boundary conditions to a collision action (Chapter 13). Research into impact actions of hail on claddings and glazing panels is ongoing at the time of writing of the book. The contents presented in Chapter 12 are the latest findings at press time.

1.5 EDUCATION AND TRAINING

Some contents presented in the book can be used as part of the syllabus of a degree program in civil engineering at either the undergraduate or graduate level. When used in undergraduate teaching, fundamental concepts as conveyed in Chapters 2 to 5 can be included in the core curriculum in the third, or final, year of a four-year bachelor of engineering degree program in civil or mechanical engineering. It is preferred that students are introduced to the book after completing core subjects in structural analysis and design and have acquired the skills and knowledge, as described earlier.

A graduate-level subject, which is fully devoted to topics covered in the book, would occupy approximately 50 percent of the total student workload in a semester. Alternatively, topics covered by the textbook may be blended with other topics namely seismic, blast, wind and waves actions. In the graduate-level elective subject titled *Extreme Loading of Structures,* which is taught by the second author at the University of Melbourne, lectures on contents delivered in Chapters 2 to 7 are blended with lectures on seismic, blast, ballistic and wave actions. That subject accounts for 25 percent of the total workload in a semester. In another elective subject titled *Structural Dynamics and Modelling*, Chapters 2 and 13 are blended with other topics on structural dynamics as applied to building structures.

A professional short course, which is aimed at introducing all the topics delivered in this book, to professional engineers would take at least two full days. It would also be viable to deliver short courses that blend with other topics in a manner as described previously in graduate teaching.

1.6 CLOSING REMARKS

The overarching philosophy of the authors is to present design models and assessment procedures that are founded on first principles and fundamental concepts. The book introduces novel methods for assessing structures to various different collisions scenarios, which for many readers, will be threshold concepts that permanently change the way the reader thinks about collision actions. A rundown of the topics covered in the book is presented in this chapter. References to the literature reporting the experimental work performed to validate the methodologies presented in the book are then given to establish the credibility and accuracy of the procedures. This chapter concludes with a discussion on the potential utility of the book for structured learning.

REFERENCES

1.1 Lam, N.T.K., Yong, A.C.Y., Lam, C., Kwan, J.S.H., Perera, J.S., Disfani, M.M., and Gad, E., 2018, "Displacement-based approach for the assessment of overturning stability of rectangular rigid barriers subjected to point impact", *Journal of Engineering Mechanics*, Vol. 144(2), pp. 04017161-1–04017161-15. doi: 10.1061/(ASCE)EM.1943-7889.0001383

1.2 Yong, A.C.Y., Lam, N.T.K., Menegon, S.J., and Gad, E.F., 2020, "Experimental and analytical assessment of the flexural behaviour of cantilevered RC walls subjected to impact actions", *Journal of Structural Engineering*, Vol. 146(4), p. 04020034, doi:10.1061/(ASCE)ST.1943-541X.0002578

1.3 Yong, A.C.Y., Lam, N.T.K., Menegon, S.J., and Gad, E.F., 2020, "Cantilevered RC wall subjected to combined static and impact actions", *International Journal of Impact Engineering*, Vol. 143, p. 103596, doi: 10.1016/j.ijimpeng.2020.103596

1.4 Yong, A.C.Y., Lam, C., Lam, N.T.K., Perera, J.S., and Kwan, J.S.H., 2019, "Analytical solution for estimating sliding displacement of rigid barriers subjected to boulder impact", *Journal of Engineering Mechanics*, Vol. 145(3), p. 04019006, doi:10.1061/(ASCE)EM.1943-7889.0001576
1.5 Yong, A.C.Y., Lam, N.T.K., and Menegon, S.J., 2021, "Closed-form expressions for improved impact resistant design of reinforced concrete beams", *Structures*, Vol. 29, pp. 1828–1836, doi:10.1016/j.istruc.2020.12.041
1.6 Sun, J., Lam, N., Zhang, L., Ruan, D., and Gad, E., 2015, "Contact forces generated by hailstone impact", *International Journal of Impact Engineering*, Vol. 84, pp. 145–158, doi:10.1016/j.ijimpeng.2015.05.015
1.7 Perera, S., Lam, N., Pathirana, M., Zhang, L., Ruan, D., and Gad, E., 2016, "Deterministic solutions for contact force generated by impact of windborne debris", *International Journal of Impact Engineering*, Vol. 91, pp. 126–141, doi:10.1016/j.ijimpeng.2016.01.002
1.8 Perera, S., Lam, N., Pathirana, M., Zhang, L., Ruan, D., and Gad, E., 2017, "Use of static tests for predicting damage to cladding panels caused by storm debris", *Journal of Building Engineering*, Vol. 12, pp. 109–117, doi:10.1016/j.jobe.2017.05.012
1.9 Sun, J., Lam, N., Zhang, L., Ruan, D., and Gad, E., 2018, "A note on Hunt and Crossley model with generalized visco-elastic damping", *International Journal of Impact Engineering*, Vol. 121, pp. 151–156, doi:10.1016/j.ijimpeng.2018.07.007
1.10 Majeed, Z.Z.A., Lam, N.T.K., Lam, C., Gad, E., and Kwan, J.S.H., 2019, "Contact force generated by impact of boulder on concrete surface", *International Journal of Impact Engineering*, Vol. 132, p. 103324, doi:10.1016/j.ijimpeng.2019.103324
1.11 Perera, J.S., Lam, N., Disfani, M.M., and Gad, E., 2021, "Experimental and analytical investigation of a RC wall with a Gabion cushion subjected to boulder impact", *International Journal of Impact Engineering*, Vol. 151, p. 103823, doi:10.1016/j.ijimpeng.2021.103823
1.12 Menegon, S.J., Wilson, J.L., Lam, N.T.K., and Gad, E.F., 2020, "Development of a user-friendly and transparent non-linear analysis program for RC walls", *Computers and Concrete*, Vol. 25(4), pp. 327–341, doi:10.12989/cac.2020.25.4.327
1.13 Menegon, S.J., Wilson, J.L., Lam, N.T.K., and Gad, E.F., 2021, "Tension stiffening model for limited ductile reinforced concrete walls", *Magazine of Concrete Research*, Vol. 73(7), pp. 366–378, doi:10.1680/jmacr.20.00211
1.14 Majeed, Z.Z.A., Lam, N.T.K., and Gad, E.F., 2021, "Predictions of localised damage to concrete caused by a low-velocity impact", *International Journal of Impact Engineering*, Vol. 149, p. 103799, doi:10.1016/j.ijimpeng.2020.103799
1.15 Perera, J.S. and Lam, N., 2022, "Soil-embedded steel baffle with concrete footing responding to collision by a fallen or flying object", *International Journal of Geomechanics*, Vol. 22(3), p. 04021311, doi:10.1061/(ASCE)GM.1943-5622.0002299
1.16 Pathirana, M., Lam, N., Perera, S., Zhang, L., Ruan, D., and Gad, E., 2017, "Damage modelling of aluminium panels impacted by windborne debris", *Journal of Wind Engineering and Industrial Aerodynamics*, Vol. 165, pp. 1–12, doi:10.1016/j.jweia.2017.02.014

1.17 Shi, S., Lam, N.T.K., Cui, Y., Zhang, L., Lu, G., and Gad, E.F., (in press), "Indentation into an aluminium panel by a rigid spherical object", *Thin-Walled Structures*.
1.18 Lam, N., Tsang, H., and Gad, E., 2010, "Simulations of response to low velocity impact by spreadsheet", *International Journal of Structural Stability and Dynamics*, Vol. 10(03), pp. 483–499, doi:10.1142/S0219455410003580
1.19 Yang, Y., Lam, N.T.K., and Zhang, L., 2012, "Evaluation of simplified methods of estimating beam responses to impact", *International Journal of Structural Stability and Dynamics*, Vol. 12(3), p. 1250016, doi:10.1142/S0219455412500162
1.20 Yang, Y., Lam, N., and Zhang, L., 2012, "Estimation of response of plate structure subject to low veloctiy impact by a solid object", *International Journal of Structural Stability and Dynamics*, Vol. 12(06), p. 1250053, doi:10.1142/S0219455412500538
1.21 Sun, J., Lam, N., Zhang, L., Ruan, D., and Gad, E., 2016, "Computer Simulation of Contact Forces Generated by Impact", *International Journal of Structural Stability and Dynamics*, Vol. 17(01), p. 1750005. doi: 10.1142/S0219455417500055

Chapter 2

Structural Dynamics Basics

2.1 INTRODUCTION

The purpose of this chapter is to provide engineering readers who do not have any prior knowledge of structural dynamics with sufficient grounding in the subject matter such that they can comprehend the contents presented in the book without the need of first reading through a textbook on structural dynamics. The contents presented are limited to natural vibration. Thus, it is not the intention of this chapter to provide comprehensive training on structural dynamics. Readers who had education and training in structural dynamics may skip this chapter. Popular textbooks on structural dynamics include Clough and Penzien [2.1] and Chopra [2.2].

Introducing the basic concept of representing a structural system as a single-degree-of-freedom (SDOF) lumped mass system (Section 2.2), multi-degree-of-freedom (MDOF) lumped mass systems (Section 2.3) and distributed mass systems (Section 2.4) is a key aim of the chapter. Readers should at least be familiar with the concept of representing a structure by a SDOF, or MDOF, lumped mass systems, and the meaning of the natural period of vibration and its significance, before reading into the later chapters of the book. The reading of Section 2.4 and after may be postponed until reading into the part of the book which deals with dynamic computations (Chapter 13) or when it is necessary to learn how to derive a lumped mass model to represent a structural element such as a beam, or a slab. Much of the presented modelling methodology (in the later chapters of the book) is based on considering the interaction of the impactor with the lumped mass of a SDOF system. When reading the book, it is essential to have the knowledge of transforming a MDOF lumped mass system, or a distributed mass system, into an equivalent SDOF system. Basic principles of estimating the natural period of vibration (T) of a structural system in conditions of natural vibration is first introduced. Some modelling methodology introduced in the book involves the use of T as a modelling parameter.

DOI: 10.1201/9781003133032-2

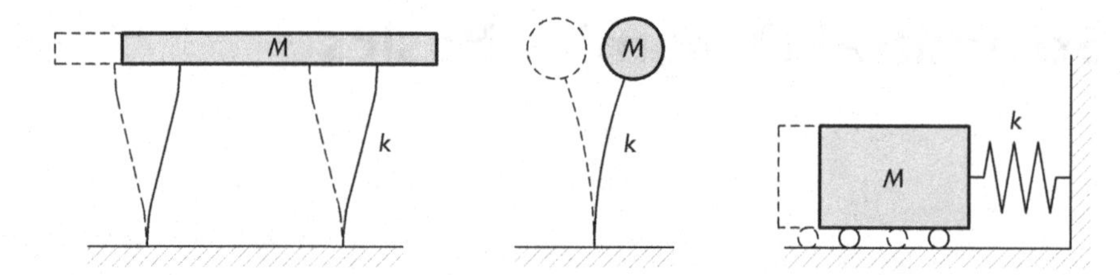

Figure 2.1 Single-degree-of-freedom lumped mass system.

2.2 DYNAMICS OF SINGLE-DEGREE-OF-FREEDOM LUMPED MASS SYSTEMS

SDOF lumped mass systems can be presented in different forms, as shown in Figure 2.1. Principles of equilibrium may be applied at any snapshot in dynamic conditions as in static conditions. What is exclusive of dynamic equilibrium (which is distinguished from static equilibrium) is having the inertia force included in the equation of equilibrium in the direction of the collision. Inertia force is the product of mass and acceleration and is applied in a direction opposite to that of acceleration. In a linear elastic analysis, the elastic force is the second force to be considered in the dynamic equilibrium of the lumped mass. The elastic force is the product of lateral stiffness and displacement, and is applied in a direction opposite to that of displacement. Damping force is the third force to be considered. The most common form of damping that is employed in structural dynamic analysis is viscous damping, which will be covered later in this section. In summary, the inertia force and elastic force are the two forces that need be considered in the dynamic equilibrium of the lumped mass system when undergoing un-damped natural vibration (Figure 2.2).

Parameters characterising the SDOF lumped mass model for the idealised condition of un-damped natural vibration are namely: mass (M) and the elastic spring stiffness (k). Equation (2.1a) is the differential equation which shows the inertia force (the first term) to be in equilibrium with the elastic force (the second term). The solution to the differential equation gives the time history of displacement, $u(t)$, which is expressed as a sinusoidal function of time, as shown by Equation (2.1b). Expressions for determining

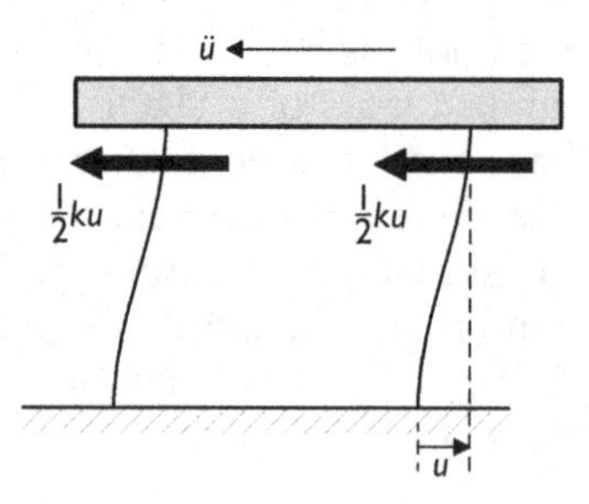

Figure 2.2 Dynamic equilibrium in un-damped natural vibration.

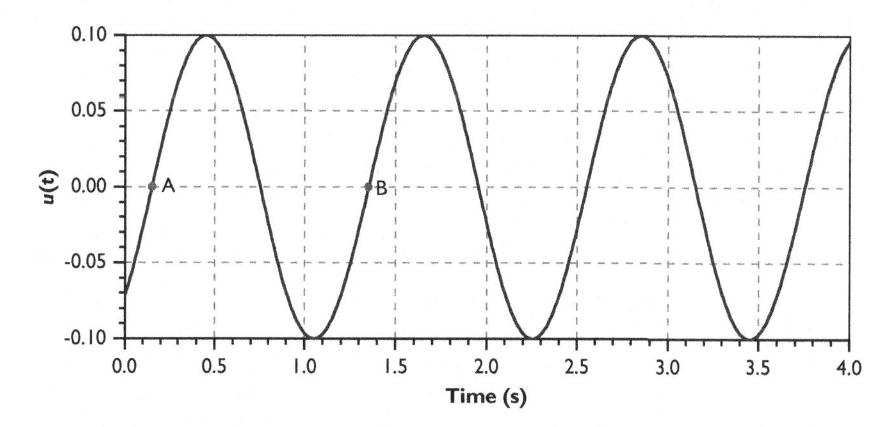

Figure 2.3 Displacement time-history of system in un-damped natural vibration.

the velocity and acceleration time-histories are accordingly obtained by differentiating $u(t)$ with respect to time (either once or twice), as shown by Equations (2.1c) and (2.1d), respectively. Note that the variables, $u(t)$, $\dot{u}(t)$ and $\ddot{u}(t)$, which represent the time-histories are written as u, $\dot{u}$ and $\ddot{u}$, respectively, in some of the equations hereafter for better readability. Once the time-histories of displacement, velocity and acceleration are known, the magnitude of the inertia force and elastic force are also known. An example displacement time-history of un-damped natural vibration, as defined by Equation (2.1b), is shown in Figure 2.3.

$$M\ddot{u} + ku = 0 \tag{2.1a}$$

$$u(t) = \Delta \sin\left(\omega_n t - \phi_p\right) \tag{2.1b}$$

$$\dot{u}(t) = \omega_n \Delta \cos\left(\omega_n t - \phi_p\right) \tag{2.1c}$$

$$\ddot{u}(t) = -\omega_n^2 \Delta \sin\left(\omega_n t - \phi_p\right) \tag{2.1d}$$

In structural dynamics, the state of the dynamic response of the structure is characterised by an angle. For example, point A on Figure 2.3 refers to an initial at rest condition of zero displacement whereas point B refers to the return to zero-displacement following a cycle of vibration. Thus, moving from A to B corresponds to a change in angle of 2π radian. The natural angular velocity of the vibration (ω_n) controls the pace of change. If A is at t

= 0, the phase angle (ϕ_p) is equal to zero. The natural frequency of vibration (f) is the number of such cycles per unit time, whereas the natural period of vibration (T) is the amount of time taken for one vibration cycle to elapse and hence its value is equal to the reciprocal of f (i.e. $T = 1/f$). Since the change in angle across one vibration cycle is 2π, the natural angular velocity of the vibration is equal to 2π divided by the natural period of vibration, i.e. $\omega_n = 2\pi/T$. Δ is the amplitude of displacement, i.e. $u_{max} = \Delta$. Relationships between the three parameters, (i) Δ, (ii) ω_n and (iii) ϕ_p, are defined in the expressions of Equations (2.1b) – (2.1d).

The maximum displacement of the system enables the maximum amount of strain energy absorption (SE) to be found, as shown by Equation (2.2a). Meanwhile, the maximum velocity of the system, which is equal to $\omega_n\Delta$, enables the maximum amount of kinetic energy (KE) to be found as shown by Equation (2.2b). By equal energy principles, SE can be equated to KE, as shown by Equation (2.2c), which can be rearranged into Equation (2.2d) for finding the value of ω_n. Finally, the expression for determining the value of T as function of M and k is shown by Equation (2.2e).

$$SE = \frac{1}{2}k\Delta^2 \tag{2.2a}$$

$$KE = \frac{1}{2}M\dot{u}_{max}^2 = \frac{1}{2}M(\omega_n\Delta)^2 \tag{2.2b}$$

$$\frac{1}{2}k\Delta^2 = \frac{1}{2}M(\omega_n\Delta)^2 \tag{2.2c}$$

$$\omega_n = \sqrt{\frac{k}{M}} \tag{2.2d}$$

$$T = 2\pi\sqrt{\frac{M}{k}} \tag{2.2e}$$

In adopting the more realistic model that incorporates energy dissipation during the course of natural vibration, the additional (third) modelling parameter is the damping coefficient (c). The dynamic equilibrium of the lumped mass considering inertia force, damping force and elastic force is shown in the schematic diagram (Figure 2.4).

If viscous damping is assumed, the value of the damping force is taken as product of the damping coefficient and velocity of motion of the lumped masses (which represents the masses carried by the structural system). Dynamic equilibrium of the system of Figure 2.4 is written in the form of Equation (2.3).

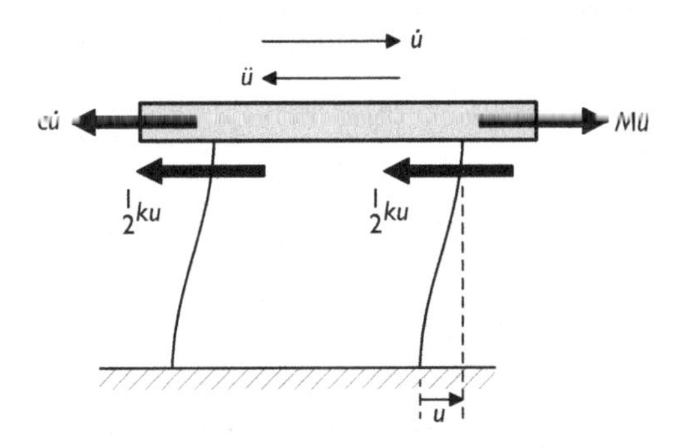

Figure 2.4 Dynamic equilibrium in viscously damped natural vibration.

$$M\ddot{u} + c\dot{u} + ku = 0 \tag{2.3}$$

As c reaches the critical value, the system is so heavily damped that no oscillation takes place. This critical damping coefficient (c_{cr}) is defined by Equation (2.4a). Damping ratio (ξ) is introduced to express c as a fraction of the critical damping, c_{cr}, by Equation (2.4b). Combining Equations (2.4a) and (2.4b) gives Equation (2.4c), which defines the parameter c.

$$c_{cr} = 2M\omega_n \tag{2.4a}$$

$$\xi = \frac{c}{c_{cr}} \tag{2.4b}$$

$$c = 2\xi M\omega_n \tag{2.4c}$$

Equation (2.3) features three controlling parameters: M, c and k. The equation can be rearranged by substituting $k = M\omega_n^2$ (as inferred from Equation (2.2d)) and Equation (2.4c) into Equation (2.3) and then dividing every term in the expression by M. The new form of expression so obtained from the substitution has the number of controlling parameters reduced from three to two (ω_n, ξ), as shown by Equation (2.5).

$$\ddot{u} + 2\xi\omega_n\dot{u} + \omega_n^2 u = 0 \tag{2.5}$$

The solution to the differential equation gives the displacement time-history, $u(t)$, for viscously damped natural vibration as shown by Equation (2.6). Note in Equation (2.6) the natural angular velocity is replaced by the term ω_D, the damped natural angular velocity, which is calculated using Equation (2.7). For ξ values not exceeding 0.2, ω_D approximately equals ω_n.

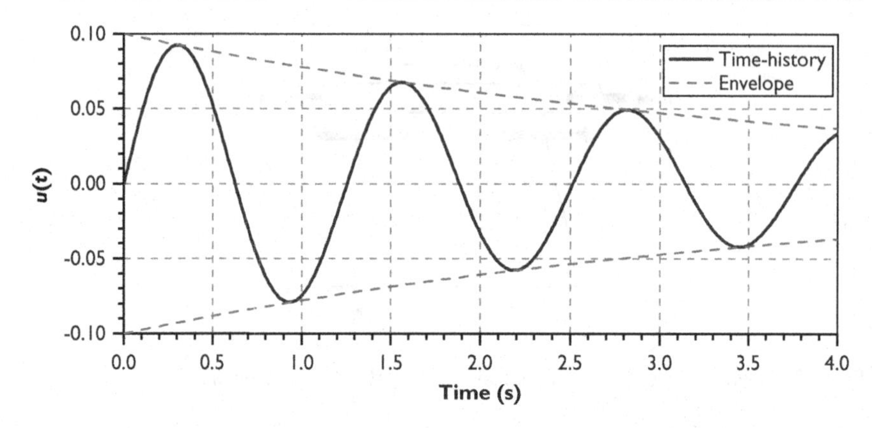

Figure 2.5 Displacement time-history of system in viscously damped natural vibration.

$$u(t) = \Delta e^{-\xi \omega_D t} \sin(\omega_D t - \phi_p) \tag{2.6}$$

$$\omega_D = \omega_n \sqrt{1 - \xi^2} \tag{2.7}$$

An example time-history of viscously damped natural vibration as obtained by applying Equation (2.6) is shown in Figure 2.5.

2.3 DYNAMICS OF MULTI-DEGREE-OF-FREEDOM LUMPED MASS SYSTEMS

MDOF lumped mass systems can be used to represent multi-storey buildings (Figure 2.6). As the building becomes taller, each lumped mass can be used to represent the mass of more than one storey. A cantilever structure like a tall concrete chimney may also be modelled by discrete lumped masses as a simplified representation of the actual structure.

Take a two-degree-of-freedom (2DOF) lumped mass system as example for illustration purposes (Figure 2.7). An equation of dynamic equilibrium can be written for each of the two lumped masses in the same manner as for the SDOF lumped mass system, assuming un-damped vibration. The two equations of equilibrium can be presented in the format of a matrix, as shown by Equations (2.8a) and (2.8b).

$$\begin{bmatrix} M_1 & 0 \\ 0 & M_2 \end{bmatrix} \begin{Bmatrix} \ddot{u}_1(t) \\ \ddot{u}_2(t) \end{Bmatrix} + \begin{bmatrix} K_{11} & K_{12} \\ K_{21} & K_{22} \end{bmatrix} \begin{Bmatrix} u_1(t) \\ u_2(t) \end{Bmatrix} = \begin{Bmatrix} 0 \\ 0 \end{Bmatrix} \tag{2.8a}$$

$$[M]\{\ddot{u}(t)\} + [K]\{u(t)\} = \{0\} \tag{2.8b}$$

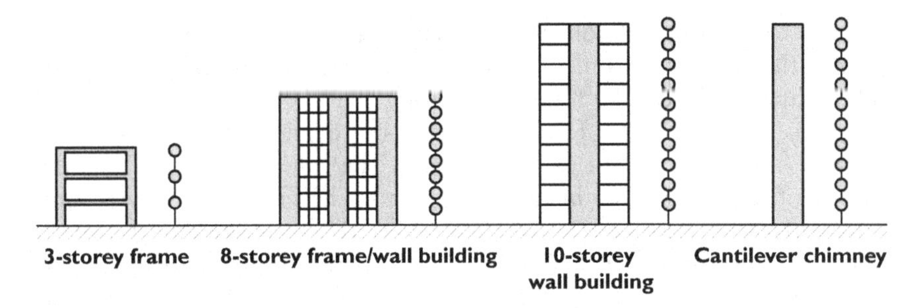

Figure 2.6 Multiple lumped masses systems.

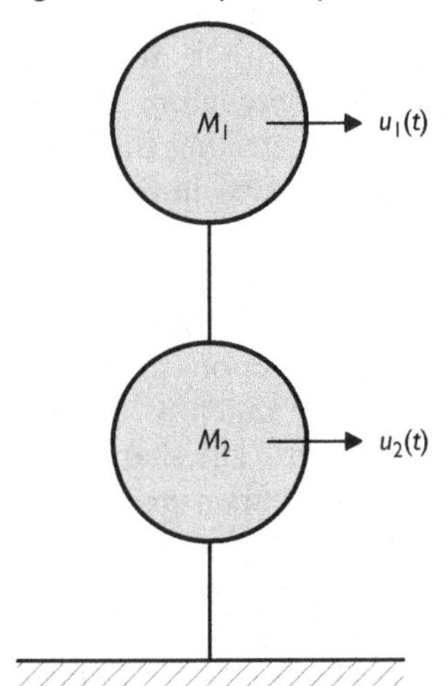

Figure 2.7 Two-degree-of-freedom lumped masses system.

The column vectors shown in { } each represents motion behaviour (i.e. displacement and acceleration) of the two lumped masses and can be expressed as product of a time-dependent scaling factor, $Y(t)$, along with the mode shape vector $\{\phi\}$, which is time independent (as shown by Equation (2.9a)). The total dynamic response behaviour of the system comprises a number of modes of vibration. Subscript j in Equation (2.9a) denotes the modal number. Differentiating Equation (2.9a) twice with respect to t gives Equation (2.9b).

$$\{u(t)\} = \sum_{j=1}^{n} Y_j(t)\{\phi_j\} \tag{2.9a}$$

$$\{\ddot{u}(t)\} = \sum_{j=1}^{n} \ddot{Y}_j(t)\{\phi_j\} \tag{2.9b}$$

Methods for finding solutions to the mode shape vectors are not within the scope of this chapter. Detailed description of a method for determining the mode shape vector can be found in Chapter 13. There can be n modes of vibration (i.e. j = 1 to n) for representing the dynamic behaviour of the structure in response to the collision action. The value of n depends on the number of lumped masses in the stick model. The first mode of vibration (j = 1) of the lowest natural frequency (i.e. the highest natural period) contributes most to the deflection of the structure. Contributions from the higher modes ($j > 1$) diminish with increasing natural period of vibration. It is common to only take into account two or three significant modes of vibration depending on the nature of the contact. For example, a hard impact generated by the impact of a fallen metal sphere (or a machined spherical specimen of fresh granitic sphere) would excite more higher modes than a bag of sand carrying the same weight and striking the floor with the same velocity. The key difference between the two impact scenarios is the duration of contact. The lower the duration of contact, the higher the frequency of excitations generated by the impact. Thus, higher mode participations are controlled by factors other than the amount of energy delivered by the impact. When higher mode contributions are taken into account, the total response of the structure may be taken as the arithmetic sum of the modal contributions in each time step. The shape of a higher mode is characterised by the change in sense of curvature up the height of the '"stick" (refer to Figure 2.8 as typical mode shapes of a stick model). The higher the modal number, the higher the rate of change. Thus, although higher modes contributions to deflection are usually minor, there can be implications on the intensity of shear and localised actions. Details of modal superposition considering higher modes effects can be found in Chapter 13, which deals with simulations of impact dynamics.

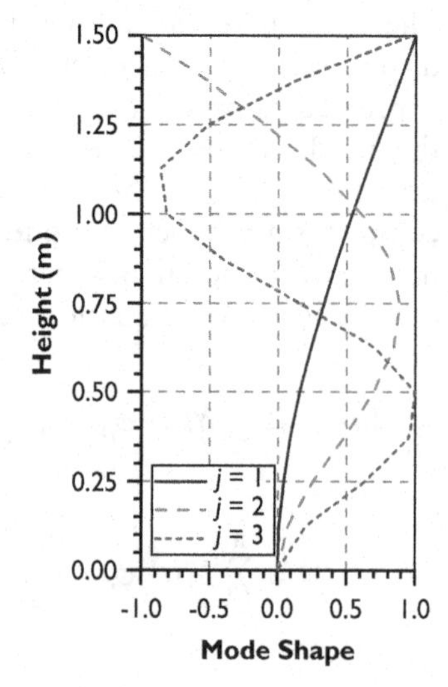

Figure 2.8 Mode shapes of vibration.

In this chapter only the first fundamental mode of vibration is considered, assuming that the response of the targeted structural element to the collision action is flexural dominant. Substituting Equations (2.9a) and (2.9b) into Equation (2.8b) and pre-multiplying every term in the expression by $\{\phi_j^T\}$ would result in Equation (2.10a) for un-damped vibration of the MDOF system, which can be presented in the format of Equation (2.10b) and is to be read in conjunction with Equations (2.11a) and (2.11b).

$$\{\phi_j^T\}[M]\{\phi_j\}\ddot{Y}_j(t) + \{\phi_j^T\}[K]\{\phi_j\}Y_j(t) = \{0\} \tag{2.10a}$$

$$M_j \ddot{Y}_j(t) + K_j Y_j(t) = 0 \tag{2.10b}$$

$$M_j = \{\phi_j^T\}[M]\{\phi_j\} \tag{2.11a}$$

$$K_j = \{\phi_j^T\}[K]\{\phi_j\} \tag{2.11b}$$

Note that the triple product of a row vector { }, a square matrix [] and a column vector { } of the same dimension gives a scalar quantity. The two scalar quantities so resulted from the algebraic manipulation are namely the generalised mass (M_j) and the generalised stiffness (K_j) for modal number j. Operating on the triple product involving $[K]$ is in analogy to multiplying force by displacement to obtain the strain energy. Thus, the expression for determining the strain energy and kinetic energy can be written in the form as shown by Equations (2.12a) and (2.12b), respectively. Equating SE to KE as shown by Equation (2.12c) results in the expression for finding the value of ω_j and T_j, as shown in Equations (2.12d) and (2.12e), respectively. In perspective, Equations (2.12a) to (2.12e) for dealing with a MDOF system is in analogy to Equations (2.2a) to (2.2e), which deal with a SDOF system.

$$\text{SE} = \frac{1}{2}\{\phi_j^T\}[K]\{\phi_j\}Y_{j,\max}^2 \tag{2.12a}$$

$$\text{KE} = \frac{1}{2}\{\phi_j^T\}[M]\{\phi_j\}\omega_j^2 Y_{j,\max}^2 \tag{2.12b}$$

$$\frac{1}{2}\{\phi_j^T\}[K]\{\phi_j\}Y_{j,\max}^2 = \frac{1}{2}\{\phi_j^T\}[M]\{\phi_j\}\omega_j^2 Y_{j,\max}^2 \tag{2.12c}$$

$$\omega_j = \sqrt{\frac{\{\phi_j^T\}[K]\{\phi_j\}}{\{\phi_j^T\}[M]\{\phi_j\}}} = \sqrt{\frac{K_j}{M_j}} \tag{2.12d}$$

$$T_j = 2\pi\sqrt{\frac{M_j}{K_j}} \tag{2.12e}$$

2.4 DYNAMICS OF DISTRIBUTED MASS SYSTEMS

An element of continuously distributed mass across its length, e.g. a tall concrete chimney, can be modelled as a stick with multiple discrete lumped masses to model its dynamic response behaviour (e.g. Figure 2.6). The more finely divided the lumped masses, the more representative is the stick model idealisation. However, the increase in the number of lumped masses in the model is at the expense of increasing the computational costs as the dimension of the mass and stiffness matrices are both increased. A more expedient solution is to treat the structure as a distributed mass system and to derive the classical solution to the generalised mass and stiffness by use of calculus. With a distributed mass model, the mode shape vector is replaced with a shape function, $\psi(x)$, which is usually normalised to unity at the point of maximum displacement.

Readers may skip the rest of this chapter and proceed to Chapter 3. Refer to Table 2.1 for a list of definitions for generalised mass and stiffness for different types of structural elements. The remaining sections of this chapter are for readers who are interested in the derivation of the values/equations shown in Table 2.1.

2.4.1 Beams

A simple example of such a shape function purely by intuition and curve-fitting is shown in Figure 2.9 for a simply supported beam which is dominated by flexural actions. With the analysis of a distributed mass system, only the first (fundamental) mode of vibration is considered. Thus, implicit in the analysis is a flexural dominant response behaviour of the structural element. The generalised mass and stiffness are denoted as m^* and k^*, respectively.

The strain energy of absorption derived from flexural actions of the beam may therefore be taken as the integral of the product of the bending moment and unitless curvature ($\psi''(x)$) along the entire length of the beam (where $\psi''(x)$ denotes the curvature at a distance of x from the start of the beam with reference to Figure 2.9). The curvature is the second derivative of the shape function, i.e. $\psi''(x) = (d^2/dx^2)[\psi(x)]$. With the beam responding within the elastic limit, the bending moment of any given beam segment is equal to the product of flexural rigidity (EI) and $\psi''(x)$ of the segment, i.e. $M(x) = EI\psi''(x)$. The amount of strain energy absorbed by the beam segment is accordingly equal to $0.5M\psi''(x)dx$ or $0.5EI\psi''(x)^2dx$ for linear elastic behaviour. Equation (2.13a) is for determining the value of SE.

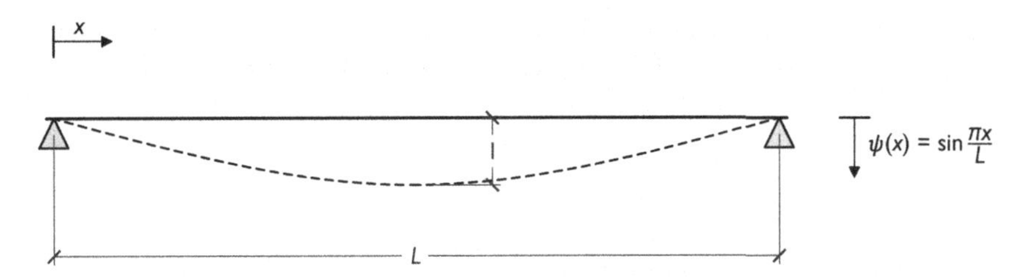

Figure 2.9 Example shape function of a simply supported beam.

In a similar manner, the amount of KE to be absorbed by the beam may be calculated using Equation (2.13b). To calculate the amount of KE to be absorbed, it is necessary to obtain the product of mass of every beam segment (which is density times sectional area: ρA) and the square of the velocity of the respective segment. The displacement of the segment as per the given shape function is: $\psi(x)$, and hence $\omega_n^2 \psi(x)^2$ represents the squaring of the velocity of the beam segments. The amount of kinetic energy to be absorbed by any given beam segment is accordingly equal to $0.5\omega_n^2 \rho A \psi(x)^2$. Equating Equation (2.13a) to Equation (2.13b), as shown by Equation (2.13c), gives Equations (2.13d) and (2.13e) for finding the value of ω_n and T, respectively. In perspective, Equations (2.13a) to (2.13e) are in analogy with Equations (2.12a) to (2.12e).

$$\mathrm{SE} = \frac{1}{2}\int_0^L EI\psi''(x)^2 dx \tag{2.13a}$$

$$\mathrm{KE} = \frac{1}{2}\omega_n^2 \int_0^L \rho A\psi(x)^2 dx \tag{2.13b}$$

$$\frac{1}{2}\int_0^L EI\psi''(x)^2 dx = \frac{1}{2}\omega_n^2 \int_0^L \rho A\psi(x)^2 dx \tag{2.13c}$$

$$\omega_n = \sqrt{\frac{\int_0^L EI\psi''(x)^2 dx}{\int_0^L \rho A\psi(x)^2 dx}} \tag{2.13d}$$

$$T = 2\pi\sqrt{\frac{\int_0^L \rho A\psi(x)^2 dx}{\int_0^L EI\psi''(x)^2 dx}} \tag{2.13e}$$

Expressions for finding the value of the generalised stiffness (k^*) and mass (m^*) of the beam are accordingly given by Equations (2.14a) and (2.14b), respectively. Detailed derivations of Equations (2.13) to (2.14) are presented in an appendix to this chapter (Section 2.7).

$$k^* = \int_0^L EI\psi''(x)^2 dx \tag{2.14a}$$

$$m^* = \int_0^L \rho A\psi(x)^2 dx \tag{2.14b}$$

Equations (2.13) and (2.14) are applicable for beam structures that only span in one direction.

Details of calculation for determining the value of k^* and m^* are illustrated by working through some numerical examples in a step-by-step fashion in Section 2.5. Separate examples based on different positioning of the collision along a beam will be presented. Graphs showing changes to the value of these parameters are used to illustrate the trends and their sensitivities.

2.4.2 Plates

For plate structures, dimensions in two directions need to be taken into account. Consider a plate in bending with lengths a and b in the x and y directions, respectively, and thickness t_p. The amount of strain energy absorbed by the plate can be determined using Equation (2.15a) [2.3, 2.4]. Similar to Equation (2.14a), the shape function, $\psi(x, y)$, is shown in place of the actual deflection. The value of the KE to be absorbed by a plate structure is as defined by Equation (2.15b). Going by the same line of reasoning as in deriving Equations (2.13a) to (2.13e) when dealing with a beam, equating Equations (2.15a) to (2.15e) are derived in a similar manner for the analysis of a plate element, as shown by 2.15c. Equations (2.15d) and (2.15e) can be used for finding the values ω_n and T, respectively. Note that $\psi(x, y)$ is written as ψ in the following equations for sake of brevity (and improved readability).

$$\begin{aligned} \text{SE} = \tfrac{1}{2}\int_0^b \int_0^a D_p \Bigg[& \left(\frac{\partial^2\psi}{\partial x^2} + \frac{\partial^2\psi}{\partial y^2}\right)^2 \\ & - 2(1-\nu)\left(\frac{\partial^2\psi}{\partial x^2}\frac{\partial^2\psi}{\partial y^2} - \left(\frac{\partial^2\psi}{\partial x\partial y}\right)^2\right)\Bigg] dxdy \end{aligned} \tag{2.15a}$$

$$\text{KE} = \frac{1}{2}\omega_n^2 \int_0^b \int_0^a \rho t\psi^2 dxdy \tag{2.15b}$$

$$\frac{1}{2}\int_0^b\int_0^a D_p\left[\left(\frac{\partial^2\psi}{\partial x^2}+\frac{\partial^2\psi}{\partial y^2}\right)^2-2(1-\nu)\left(\frac{\partial^2\psi}{\partial x^2}\frac{\partial^2\psi}{\partial y^2}-\left(\frac{\partial^2\psi}{\partial x\partial y}\right)^2\right)\right]dxdy = \frac{1}{2}\omega_n^2\int_0^b\int_0^a \rho t\psi^2 dxdy \quad (2.15c)$$

$$\omega_n = \sqrt{\frac{\int_0^b\int_0^a D_p\left[\left(\frac{\partial^2\psi}{\partial x^2}+\frac{\partial^2\psi}{\partial y^2}\right)^2-2(1-\nu)\left(\frac{\partial^2\psi}{\partial x^2}\frac{\partial^2\psi}{\partial y^2}-\left(\frac{\partial^2\psi}{\partial x\partial y}\right)^2\right)\right]dxdy}{\int_0^b\int_0^a \rho t\psi^2 dxdy}} \quad (2.15d)$$

$$T = 2\pi\sqrt{\frac{\int_0^b\int_0^a \rho t\psi^2 dxdy}{\int_0^b\int_0^a D_p\left[\left(\frac{\partial^2\psi}{\partial x^2}+\frac{\partial^2\psi}{\partial y^2}\right)^2-2(1-\nu)\left(\frac{\partial^2\psi}{\partial x^2}\frac{\partial^2\psi}{\partial y^2}-\left(\frac{\partial^2\psi}{\partial x\partial y}\right)^2\right)\right]dxdy}} \quad (2.15e)$$

where D_p is the plate flexural rigidity as defined by Equation (2.16).

$$D_p = \frac{Et_p^3}{12(1-\nu^2)} \quad (2.16)$$

$k*$ and $m*$ can then be given by Equations (2.17a) and (2.17b), respectively.

$$k* = \int_0^b\int_0^a D_p\left[\left(\frac{\partial^2\psi}{\partial x^2}+\frac{\partial^2\psi}{\partial y^2}\right)^2 - 2(1-\nu)\left(\frac{\partial^2\psi}{\partial x^2}\frac{\partial^2\psi}{\partial y^2}-\left(\frac{\partial^2\psi}{\partial x\partial y}\right)^2\right)\right]dxdy \quad (2.17a)$$

$$m* = \int_0^b\int_0^a \rho t\psi^2 dxdy \quad (2.17b)$$

2.5 WORKED EXAMPLES

2.5.1 Simply Supported Beam with Sinusoidal Shape Function

Estimate the generalised mass and stiffness of a simply supported beam when subject to a collision at mid-span, assuming a sinusoidal shape function (refer Figure 2.10).

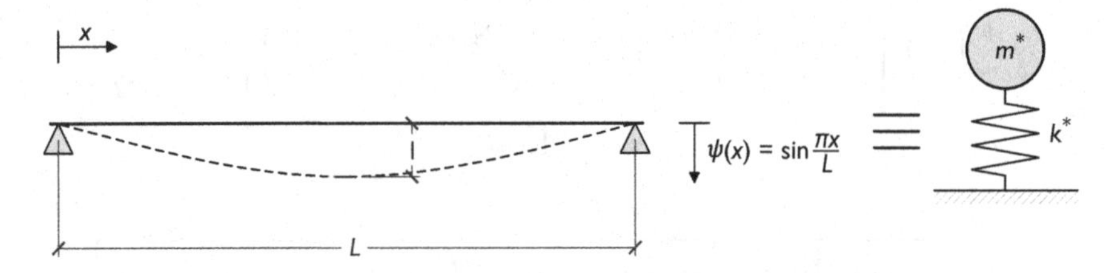

Figure 2.10 Simply supported beam with sinusoidal shape function.

Solution

$$\psi(x) = \sin\frac{\pi x}{L}$$

Derivation of generalised mass:

$$m^* = \int_0^L \rho A\psi(x)^2 dx = \int_0^L \rho A \sin^2\frac{\pi x}{L}dx = \int_0^L \frac{\rho A}{2}\left(1 - \cos\frac{2\pi x}{L}\right)dx$$
$$= \frac{\rho A}{2}\left[x - \frac{L}{2\pi}\sin\frac{2\pi x}{L}\right]_0^L = \frac{\rho AL}{2}$$

Note that ρAL represents the total mass of the beam.

The first and second derivatives of the shape function, $\psi(x)$, with respect to x are as follows:

$$\psi'(x) = \frac{\pi}{L}\cos\frac{\pi x}{L}$$

$$\psi''(x) = -\left(\frac{\pi}{L}\right)^2 \sin\frac{\pi x}{L}$$

Derivation of generalised stiffness:

$$k^* = \int_0^L EI\psi''(x)^2 dx = \int_0^L EI\left(\left(\frac{\pi}{L}\right)^2 \sin\frac{\pi x}{L}\right)^2 dx = \left(\frac{\pi}{L}\right)^4 \int_0^L EI\sin^2\frac{\pi x}{L}dx$$
$$= \frac{EI}{2}\left(\frac{\pi}{L}\right)^4 \int_0^L \left(1 - \cos\frac{2\pi x}{L}\right)dx = \frac{\pi^4 EI}{2L^4}\left[x - \frac{L}{2\pi}\sin\frac{2\pi x}{L}\right]_0^L = \frac{\pi^4 EI}{2L^3} = \frac{48.7EI}{L^3}$$

In summary,

$$m^* = 50\% \text{ of total mass}$$

$$k^* = \frac{48.7EI}{L^3}$$

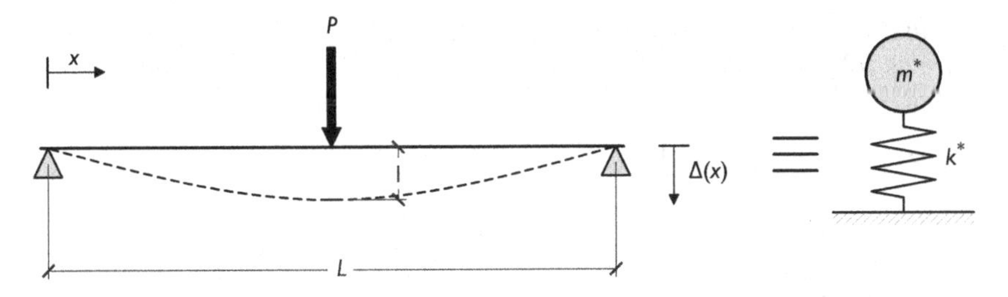

Figure 2.11 A simply supported beam with a point load at mid-span.

2.5.2 Simply Supported Beam with a Point Load at Mid-Span

Estimate the generalised mass and stiffness of the same simply supported beam when subject to collision at mid-span of the beam, assuming a shape function consistent with the deflection of a pointed loaded beam (refer Figure 2.11).

Solution

The deflection profile of a beam as shown in Figure 2.11 is written as:

$$\Delta(x) = \frac{Px}{48EI}(3L^2 - 4x^2) \text{ for } x \leq \frac{L}{2}$$

The maximum deflection of the beam at mid-span is defined as:

$$\Delta_{\max} = \frac{PL^3}{48EI}$$

The deflection of the beam may be normalised to unity at the point of maximum deflection to obtain the shape function as shown in the following:

$$\psi(x) = \frac{\Delta(x)}{\Delta_{\max}} = \frac{\frac{Px}{48EI}(3L^2 - 4x^2)}{\frac{PL^3}{48EI}} = 3\left(\frac{x}{L}\right) - 4\left(\frac{x}{L}\right)^3 \text{ for } x \leq \frac{L}{2}$$

Derivation of generalised mass:

$$m* = 2\int_0^{\frac{L}{2}} \rho A\psi(x)^2 dx = 2\int_0^{\frac{L}{2}} \rho A\left[3\left(\frac{x}{L}\right) - 4\left(\frac{x}{L}\right)^3\right]^2 dx$$

$$= 2\rho AL\left[\frac{9}{3}\left(\frac{x}{L}\right)^3 - \frac{24}{5}\left(\frac{x}{L}\right)^5 + \frac{16}{7}\left(\frac{x}{L}\right)^7\right]_0^{\frac{L}{2}}$$

$$= 0.486\rho AL$$

The first and second derivatives of the shape function, $\psi(x)$, with respect to x are as follows:

$$\psi'(x) = \frac{3}{L} - 12\left(\frac{x^2}{L^3}\right)$$

$$\psi''(x) = -24\left(\frac{x}{L^3}\right)$$

Derivation of the generalised stiffness:

$$k^* = 2\int_0^{\frac{L}{2}} EI\psi''(x)^2 dx = 2\int_0^{\frac{L}{2}} EI\left(-24\left(\frac{x}{L^3}\right)\right)^2 dx$$

$$= 2EI\left(\frac{24^2}{L^3}\right)\left[\frac{1}{3}\left(\frac{x}{L}\right)^3\right]_0^{\frac{L}{2}} = \frac{48EI}{L^3}$$

In summary,

$$m^* = 48.6\% \text{ of total mass}$$

$$k^* = \frac{48EI}{L^3}$$

By replacing point load P in Figure 2.11 by a uniformly distributed load, the generalised mass and stiffness are estimated to be:

$$m^* = 50.4\% \text{ of total mass}$$

$$k^* = \frac{49.2EI}{L^3}$$

2.5.3 Simply Supported Beam with a Point Load at Varying Positions

Estimate the generalised mass and stiffness of the same simply supported beam when subject to collision at the position shown in Figure 2.11, assuming a shape function consistent with the deflection of a pointed loaded beam.

Solution

The deflection profile of the beam shown in Figure 2.12 is written as follows:

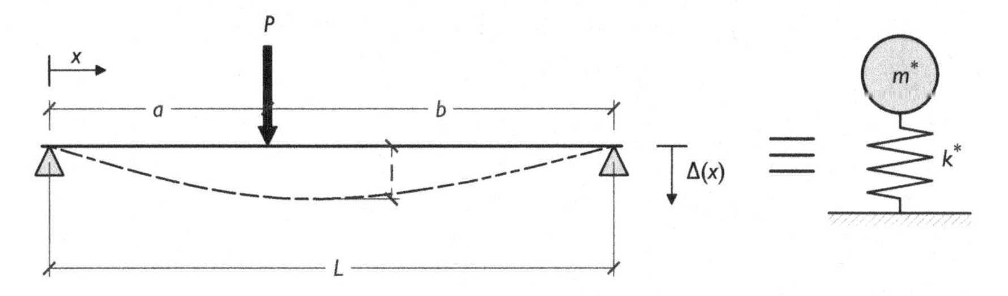

Figure 2.12 A simply supported beam with a point load at varying positions.

$$\Delta_1(x) = \frac{Pbx}{6EIL}(L^2 - b^2 - x^2) \quad \text{for } x < a$$

$$\Delta_2(x) = \frac{Pa(L-x)}{6EIL}(2Lx - x^2 - a^2) \quad \text{for } x > a$$

The maximum deflection at mid-span is defined as follows:

$$\Delta_{max} = \frac{Pab(a+2b)\sqrt{3a(a+2b)}}{27EIL}$$

The deflection profile may be normalised to unity at the point of maximum deflection to give the shape function.

$$\psi_1(x) = \frac{\Delta_1(x)}{\Delta_{max}} = \frac{\frac{Pbx}{6EIL}(L^2 - b^2 - x^2)}{\frac{Pab(a+2b)\sqrt{3a(a+2b)}}{27EIL}} = \frac{3\sqrt{3}x(L^2 - b^2 - x^2)}{2a(a+2b)\sqrt{a(a+2b)}}$$

$$\psi_2(x) = \frac{\Delta_2(x)}{\Delta_{max}} = \frac{\frac{Pa(L-x)}{6EIL}(2Lx - x^2 - a^2)}{\frac{Pab(a+2b)\sqrt{3a(a+2b)}}{27EIL}} = \frac{3\sqrt{3}(L-x)(2Lx - x^2 - a^2)}{2b(a+2b)\sqrt{a(a+2b)}}$$

Derivation of the generalised mass:

$$\begin{aligned} m^* &= \int_0^a \rho A\psi_1(x)^2 dx + \int_a^L \rho A\psi_2(x)^2 dx \\ &= \int_0^a \rho A\left[\frac{3\sqrt{3}x(L^2 - b^2 - x^2)}{2a(a+2b)\sqrt{a(a+2b)}}\right]^2 dx + \int_a^L \rho A\left[\frac{3\sqrt{3}(L-x)(2Lx - x^2 - a^2)}{2b(a+2b)\sqrt{a(a+2b)}}\right]^2 dx \\ &= \left[\frac{18(a^4 + 6a^3b + 11.5a^2b^2 + 6ab^3 + b^4)}{35a(a+2b)^3}\right]\rho AL = \text{Mass Factor} \times \rho AL \end{aligned}$$

where

$$\text{Mass Factor} = \frac{18(a^4 + 6a^3b + 11.5a^2b^2 + 6ab^3 + b^4)}{35a(a+2b)^3}$$

The second derivatives of the shape functions with respect to x are as follows:

$$\psi''_1(x) = -\frac{9\sqrt{3}x}{(a^2 + 2ab)^{\frac{3}{2}}}$$

$$\psi''_2(x) = -\frac{9\sqrt{3}(a + b - x)}{b\sqrt{a}(a + 2b)^{\frac{3}{2}}}$$

Derivation of the generalised stiffness:

$$\begin{aligned}
k^* &= \int_0^a EI\psi''_1(x)^2 dx + \int_a^L EI\psi''_2(x)^2 dx \\
&= \int_0^a EI\left[-\frac{9\sqrt{3}x}{(a^2 + 2ab)^{\frac{3}{2}}}\right]^2 dx + \int_a^L EI\left[-\frac{9\sqrt{3}(a + b - x)}{b\sqrt{a}(a + 2b)^{\frac{3}{2}}}\right]^2 dx \\
&= \left[\frac{81(a + b)^4}{a(a + 2b)^3}\right]\frac{EI}{L^3} = \text{Stiffness Factor} \times \frac{EI}{L^3}
\end{aligned}$$

where

$$\text{Stiffness Factor} = \frac{81(a + b)^4}{a(a + 2b)^3}$$

Plotting mass factor and stiffness factor versus a: L ratio gives Figures 2.13(a) and 2.13(b), respectively.

There is little change in the equivalent lumped mass system representing the dynamic response behaviour of a simply supported beam as the position of load application moves away from the mid-span position, and is within the middle third of the beam. As shown in Figure 2.13(a), the mass factor is insensitive to changes in the position of collision along the beam and is close to 0.5, while the stiffness factor is more sensitive to changes in the position along the beam but is not as sensitive as the static stiffness, as shown in Figure 2.13(b).

2.5.4 Cantilever Beam with a Point Load at the End

Estimate the generalised mass and stiffness of a cantilever beam when subject to a collision at the free end of the beam, assuming a shape function consistent with a pointed loaded beam (refer Figure 2.14).

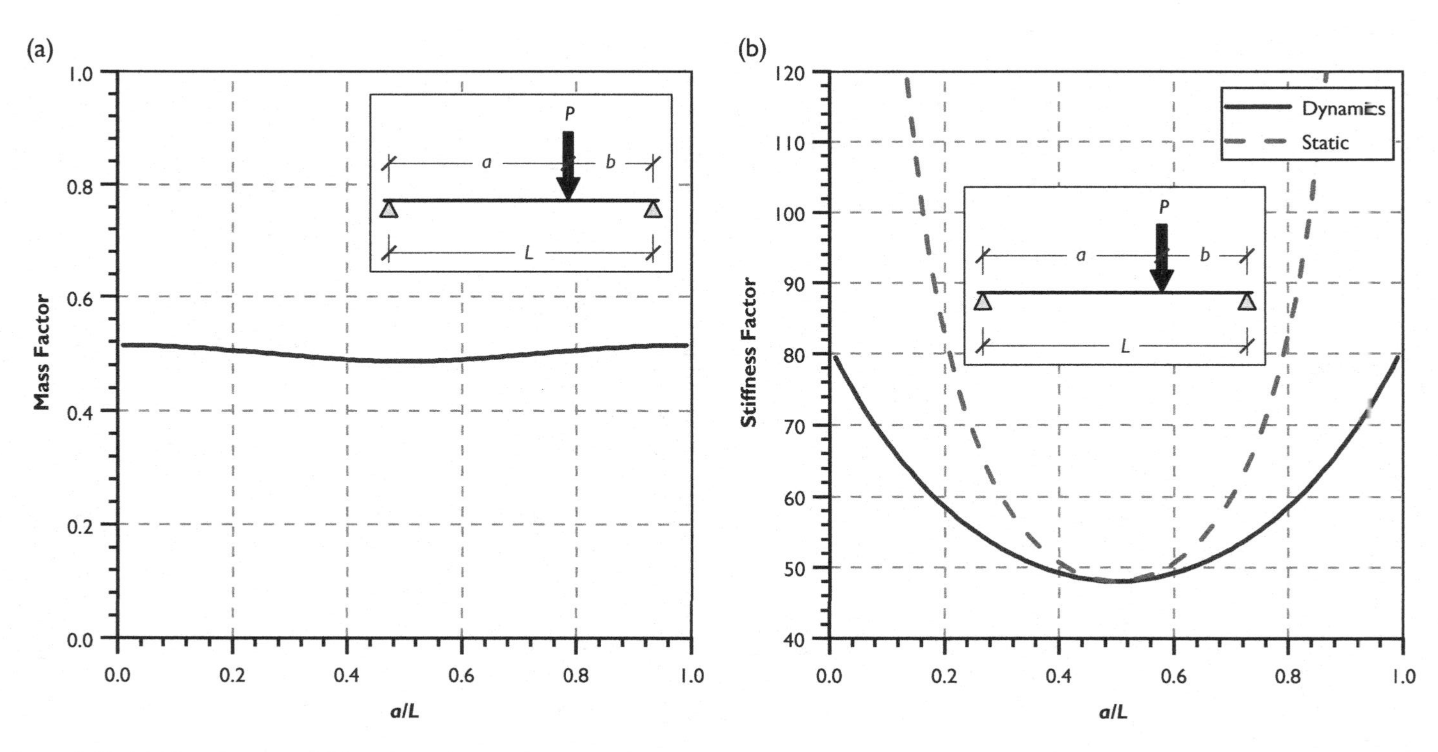

Figure 2.13 (a) Mass factor and (b) stiffness factor of a simply supported beam with a point load at varying locations.

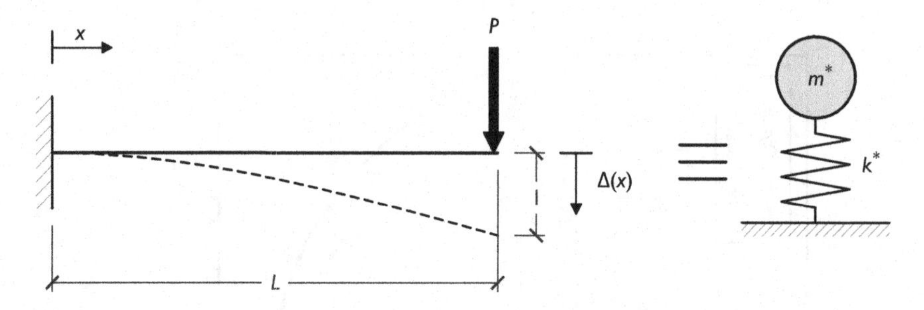

Figure 2.14 A cantilever beam with a point load at free end.

Solution

Deflection profile of a beam shown in Figure 2.14 is written as follows:

$$\Delta(x) = \frac{P}{6EI}(3Lx^2 - x^3)$$

The maximum deflection at the free end of the beam is defined as follows:

$$\Delta_{max} = \frac{PL^3}{3EI}$$

The deflection profile may be normalised to unity at the point of maximum deflection to give the shape function.

$$\psi(x) = \frac{\Delta(x)}{\Delta_{max}} = \frac{\frac{P}{6EI}(3Lx^2 - x^3)}{\frac{PL^3}{3EI}} = \frac{1}{2}\left[3\left(\frac{x}{L}\right)^2 - \left(\frac{x}{L}\right)^3\right]$$

Derivation of the generalised mass:

$$m^* = \int_0^L \rho A\psi(x)^2 dx = \int_0^L \frac{\rho A}{2^2}\left[3\left(\frac{x}{\ell}\right)^2 - \left(\frac{x}{\ell}\right)^3\right]^2 dx$$

$$= \frac{\rho AL}{4}\left[\frac{9}{5}\left(\frac{x}{L}\right)^5 - \left(\frac{x}{L}\right)^6 + \frac{1}{7}\left(\frac{x}{L}\right)^7\right]_0^L = 0.236\rho AL$$

The first and second derivatives of the shape function, $\psi(x)$, with respect to x are as follows:

$$\psi'(x) = \frac{1}{2}\left[6\left(\frac{x}{L^2}\right) - 3\left(\frac{x^2}{L^3}\right)\right]$$

$$\psi''(x) = \frac{3}{L^2}\left(1 - \frac{x}{L}\right)$$

Derivation of the generalised stiffness:

$$k* = \int_0^L EI\psi''(x)^2 dx = \int_0^L \left(\frac{3}{L^2}\right)^2 EI\left(1 - \frac{x}{L}\right)^2 dx$$

$$= \frac{9EI}{L^4}\left[L\left(\frac{x}{L} - \left(\frac{x}{L}\right)^2 + \frac{1}{3}\left(\frac{x}{L}\right)^3\right]_0^L = \frac{3EI}{L^3}$$

In summary,

$m* = 23.6\%$ of total mass

$$k* = \frac{3EI}{L^3}$$

By replacing point load, P, in Figure 2.14 by a uniformly distributed load, the generalised mass and stiffness are estimated to be:

$m* = 25.7\%$ of total mass

$$k* = \frac{3.2EI}{L^3}$$

2.5.5 Cantilever Beam with a Point Load at Varying Positions

Estimate the generalised mass and stiffness of the same cantilever beam as for the previous example except that the adopted shape function is consistent with a uniformly loaded beam (refer Figure 2.15).

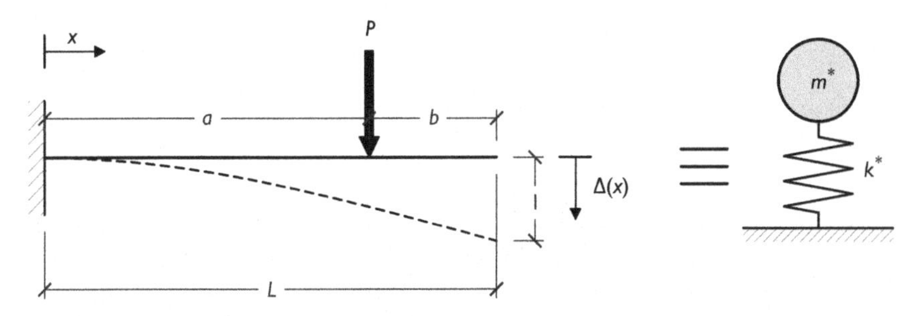

Figure 2.15 A cantilever beam with point load at varying positions.

Solution

The deflection profile of the beam as shown in Figure 2.14 is written as follows:

$$\Delta_1(x) = \frac{Px^2}{6EI}(3a - x) \quad \text{for } x < a$$

$$\Delta_2(x) = \frac{Pa^2}{6EI}(3x - a) \quad \text{for } x < a$$

The maximum deflection at the free end of the beam is defined as follows:

$$\Delta_{\max} = \frac{Pa^2}{6EI}(3L - a)$$

The deflection profile may be normalised to unity at the point of maximum deflection to give the shape function, as shown in the following:

$$\psi_1(x) = \frac{\Delta_1(x)}{\Delta_{\max}} = \frac{\frac{Px^2}{6EI}(3a - x)}{\frac{Pa^2}{6EI}(3L - a)} = \left(\frac{x}{a}\right)^2\left(\frac{3a - x}{3L - a}\right)$$

$$\psi_2(x) = \frac{\Delta_2(x)}{\Delta_{\max}} = \frac{\frac{Pa^2}{6EI}(3x - a)}{\frac{Pa^2}{6EI}(3L - a)} = \frac{3x - a}{3L - a}$$

Derivation of the generalised mass:

$$\begin{aligned} m^* &= \int_0^a \rho A\psi_1(x)^2 dx + \int_a^L \rho A\psi_2(x)^2 dx \\ &= \int_0^a \rho A\left[\left(\frac{x}{a}\right)^2\left(\frac{3a - x}{3L - a}\right)\right]^2 dx + \int_a^L \rho A\left[\frac{3x - a}{3L - a}\right]^2 dx \\ &= \left[\frac{33a^3 + 140a^2b + 210ab^2 + 105b^3}{35(a + b)(2a + 3b)^2}\right]\rho AL = \text{Mass Factor} \times \rho AL \end{aligned}$$

where

$$\text{Mass Factor} = \frac{33a^3 + 140a^2b + 210ab^2 + 105b^3}{35(a + b)(2a + 3b)^2}$$

The second derivatives of the shape functions with respect to x are as follows:

$$\psi''_1(x) = \frac{6(a - x)}{a^2(2a + 3b)}$$

$$\psi''_2(x) = 0$$

Derivation of the generalised stiffness:

$$\begin{aligned} k* &= \int_0^a EI\psi''_1(x)^2 dx + \int_a^L EI\psi''_2(x)^2 dx \\ &= \int_0^a EI\left[\frac{6(a - x)}{a^2(2a + 3b)}\right]^2 dx + \int_a^L EI\,[0]^2 dx \\ &= \left[\frac{12(a + b)^3}{a(2a + 3b)^2}\right]\frac{EI}{L^3} = \text{Stiffness Factor} \times \frac{EI}{L^3} \end{aligned}$$

where

$$\text{Stiffness Factor} = \frac{12(a + b)^3}{a(2a + 3b)^2}$$

Plotting mass factor and stiffness factor versus a: L ratio gives Figures 2.16(a) and 2.16(b), respectively.

As for a simply supported beam, there is also little change in the equivalent lumped mass system representing the dynamic response behaviour of a cantilevered beam as the position of load application moves away from the end of the beam by up to half the beam length. As shown in Figure 2.16(a), the mass factor is insensitive to changes in the position of collision along the beam and is close to 0.25, while the stiffness factor is more sensitive to changes in the position along the beam but is not as sensitive as the static stiffness, as shown in Figure 2.16(b).

2.5.6 Plate Simply Supported on All Edges

Estimate the generalised mass and stiffness of a plate which is simply supported on all the edges when subject to a collision at the centre of the plate.

Solution

The following shape function is taken:

$$\psi(x, y) = \sin\frac{\pi x}{a}\sin\frac{\pi y}{b}$$

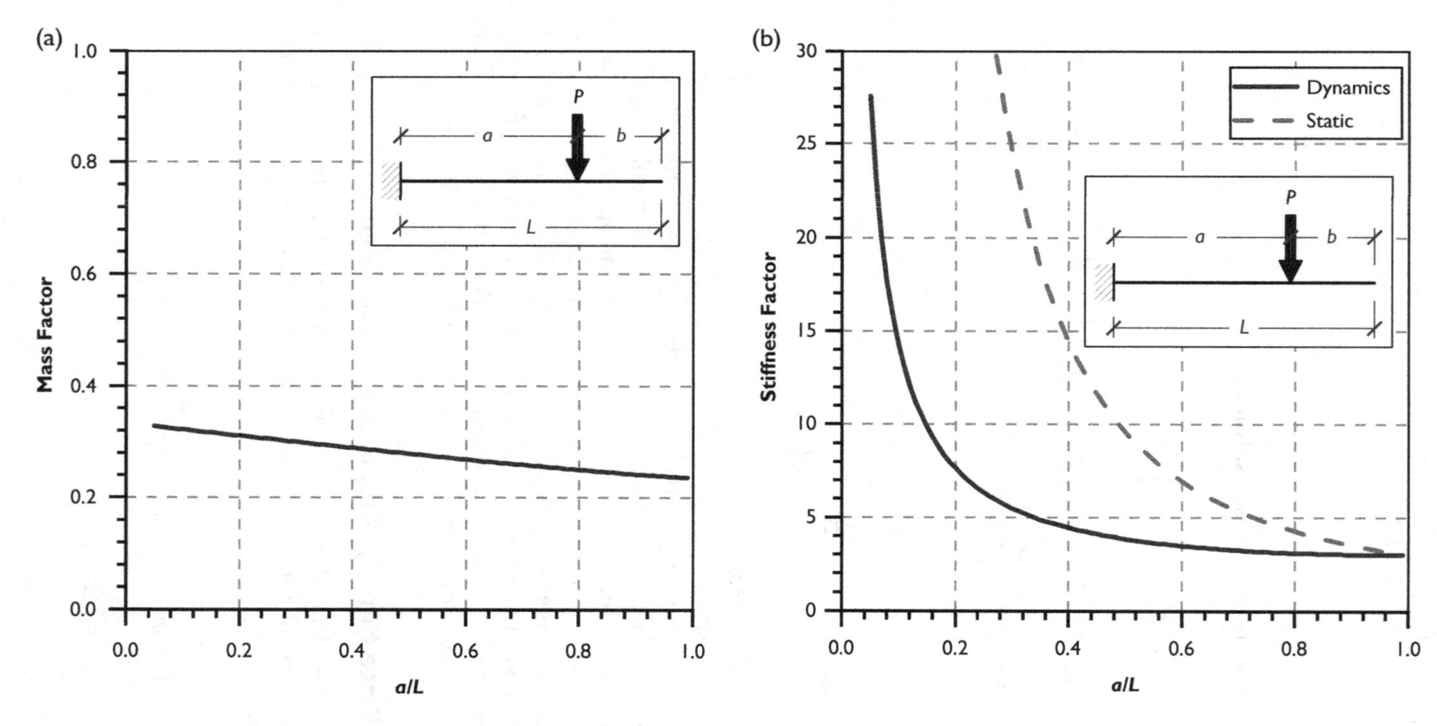

Figure 2.16 (a) Mass factor and (b) stiffness factor of a cantilever beam with a point load at varying locations.

The first and second derivatives of the shape function with respect to x and y are as follows:

$$\frac{\partial \psi}{\partial x} = \frac{\pi}{a} \cos \frac{\pi x}{a} \sin \frac{\pi y}{b}$$

$$\frac{\partial^2 \psi}{\partial x^2} = -\left(\frac{\pi}{a}\right)^2 \sin \frac{\pi x}{a} \sin \frac{\pi y}{b}$$

$$\frac{\partial \psi}{\partial y} = \frac{\pi}{b} \sin \frac{\pi x}{a} \cos \frac{\pi y}{b}$$

$$\frac{\partial^2 \psi}{\partial y^2} = -\left(\frac{\pi}{b}\right)^2 \sin \frac{\pi x}{a} \sin \frac{\pi y}{b}$$

Check boundary conditions of the shape function.

Recognising that there must be zero displacement and zero bending moment at all the edges, the following expressions are written:

$$\text{At } x = 0 \text{ and } x = a, \ \psi = \frac{\partial^2 \psi}{\partial x^2} = 0$$

$$\text{At } y = 0 \text{ and } y = b, \ \psi = \frac{\partial^2 \psi}{\partial y^2} = 0$$

Maximum displacement at the centre of the plate is as follows:

$$\text{At } x = \frac{a}{2} \text{ and } y = \frac{b}{2}, \ \psi = 1$$

Expressions that have been written in the above satisfy all the boundary conditions.

Derivation of the generalised mass:

$$m* = \int_0^b \int_0^a \rho t \psi^2 dx dy = \int_0^b \int_0^a \rho t \left(\sin \frac{\pi x}{a} \sin \frac{\pi y}{b}\right)^2 dx dy = \frac{1}{4} ab\rho t_p$$

Derivation of the generalised stiffness:

$$\frac{\partial^2 \psi}{\partial x \partial y} = \frac{\pi^2}{ab} \cos \frac{\pi x}{a} \cos \frac{\pi y}{b}$$

$$k^* = \int_0^b \int_0^a D_p \left[\left(\frac{\partial^2 \psi}{\partial x^2} + \frac{\partial^2 \psi}{\partial y^2} \right)^2 - 2(1-\nu) \left(\frac{\partial^2 \psi}{\partial x^2} \frac{\partial^2 \psi}{\partial y^2} - \left(\frac{\partial^2 \psi}{\partial x \partial y} \right)^2 \right) \right] dxdy$$

$$= \int_0^b \int_0^a D_p \left[\left(-\left(\frac{\pi}{a}\right)^2 \sin\frac{\pi x}{a} \sin\frac{\pi y}{b} - \left(\frac{\pi}{b}\right)^2 \sin\frac{\pi x}{a} \sin\frac{\pi y}{b} \right)^2 \right.$$

$$\left. - 2(1-\nu) \left(\left(\frac{\pi}{a}\right)^2 \left(\frac{\pi}{b}\right)^2 \sin^2\frac{\pi x}{a} \sin^2\frac{\pi y}{b} - \left(\frac{\pi^2}{ab} \cos\frac{\pi x}{a} \cos\frac{\pi y}{b} \right)^2 \right) \right] dxdy$$

$$= \frac{1}{4} D_p \pi^4 ab \left(\frac{1}{a^2} + \frac{1}{b^2} \right)^2$$

For a square plate where $a = b$, solutions to the generalised mass and stiffness can be simplified into:

$$m^* = \frac{1}{4} a^2 \rho t_p$$

$$k^* = \frac{D_p \pi^4}{a^2}$$

2.5.7 Plate Fully Fixed on All Edges

Estimate the generalised mass and stiffness of a rectangular plate which is fully fixed on all the edges when subject to collision at the centre of the plate.

Solution

The following shape function is taken:

$$\psi(x, y) = \frac{1}{4} \left(1 - \cos\frac{2\pi x}{a} \right) \left(1 - \cos\frac{2\pi y}{b} \right)$$

The first and second derivatives of the shape function with respect to x and y are as follows:

$$\frac{\partial \psi}{\partial x} = \frac{1}{2} \left(\frac{\pi}{a} \sin\frac{2\pi x}{a} \right) \left(1 - \cos\frac{2\pi y}{b} \right)$$

$$\frac{\partial^2 \psi}{\partial x^2} = \left(\frac{\pi}{a} \right)^2 \left(\cos\frac{2\pi x}{a} \right) \left(1 - \cos\frac{2\pi y}{b} \right)$$

$$\frac{\partial \psi}{\partial y} = \frac{1}{2}\left(1 - \cos\frac{2\pi x}{a}\right)\left(\frac{\pi}{b}\sin\frac{2\pi y}{b}\right)$$

$$\frac{\partial^2 \psi}{\partial y^2} = \left(1 - \cos\frac{2\pi x}{a}\right)\left(\frac{\pi}{b}\right)^2\left(\cos\frac{2\pi y}{b}\right)$$

Check boundary conditions of shape function.

Recognising that there must be zero displacement and zero rotation at all the edges, the following expressions are written:

$$\text{At } x = 0 \text{ and } x = a,\ \psi = \frac{\partial \psi}{\partial x} = 0$$

$$\text{At } y = 0 \text{ and } y = b,\ \psi = \frac{\partial \psi}{\partial y} = 0$$

Maximum displacement at the centre of the plate is as follows:

$$\text{At } x = \frac{a}{2} \text{ and } y = \frac{b}{2},\ \psi = 1$$

Expressions that have been written in the above satisfy all the boundary conditions.

Derivation of the generalised mass:

$$m^* = \int_0^b \int_0^a \rho t \psi^2 dxdy = \int_0^b \int_0^a \rho t \left[\frac{1}{4}\left(1 - \cos\frac{2\pi x}{a}\right)\left(1 - \cos\frac{2\pi y}{b}\right)\right]^2 dxdy$$

$$= \frac{9}{64} ab\rho t_p$$

Derivation of the generalised stiffness:

$$\frac{\partial^2 \psi}{\partial x \partial y} = \left(\frac{\pi^2}{ab}\right)\left(\sin\frac{2\pi x}{a}\right)\left(\sin\frac{2\pi y}{b}\right)$$

$$k^* = \int_0^b \int_0^a D_p\left[\left(\frac{\partial^2 \psi}{\partial x^2} + \frac{\partial^2 \psi}{\partial y^2}\right)^2 - 2(1 - \nu)\left(\frac{\partial^2 \psi}{\partial x^2}\frac{\partial^2 \psi}{\partial y^2} - \left(\frac{\partial^2 \psi}{\partial x \partial y}\right)^2\right)\right] dxdy$$

$$= D_p \pi^4 ab\left[\frac{3}{4}\left(\frac{1}{a^2} + \frac{1}{b^2}\right)^2 - \left(\frac{1}{ab}\right)^2\right]$$

For a square plate where $a = b$, solutions to the generalised mass and stiffness can be simplified into:

$$m^* = \frac{9}{64}a^2\rho t_p$$

$$k^* = \frac{2D_p\pi^4}{a^2}$$

2.6 CLOSING REMARKS

The dynamic equilibrium of SDOF and MDOF lumped mass systems have both been examined in this chapter. The natural angular velocity and period of vibration for both types of systems can be found by equating the elastic strain energy of absorption with the kinetic energy delivered by the colliding object. It was found that the strain energy of absorption of the MDOF system can be calculated by taking the triple product of the transpose of the mode shape (row) vector, the stiffness matrix and the mode shape (column) vector. This triple product gives the generalised stiffness (K) of the MDOF system. A similar triple product formed around the mass matrix gives the generalised mass (M). This formulation has been extended to a system of distributed mass in order that both the generalised stiffness and the generalised mass can be calculated by use of standard calculus techniques based on an assumed shape function of deflection. The calculation methodology has been illustrated with examples of a simply supported beam and a cantilever beam of uniform cross sections assuming different shape functions. Results have been shown to be insensitive to change in the shape function, thereby demonstrating the robustness of the results. The generalised mass and generalised stiffness of beam and rectangular plate (with plan dimensions a and b) with different boundary conditions are summarised in Table 2.1. Worked examples showing the derivation of the presented expressions can be found in the appendix to this chapter (Section 2.7).

2.7 APPENDIX

Derivations of Equations (2.13) to (2.14) are presented in this section.

The derivation is based on equating the amount of strain energy (SE) and kinetic energy (KE) to be absorbed by a member, as shown by Equation (2.18):

$$\text{SE} = \text{KE} \tag{2.18}$$

Table 2.1 Generalised mass and stiffness

Type of Structure	*Generalised Mass (% of total mass)*	*Generalised Stiffness*
Simply Supported Beam (Collision at Mid-Span)	50	$\frac{48EI}{L^3}$
Cantilever Beam (Collision at Free End)	25	$\frac{3EI}{L^3}$
Beam Fixed at Both Ends (Collision at Mid-Span)	40	$\frac{192EI}{L^3}$
Plate Simply Supported on All Edges (Collision at Centre)	25	$\frac{1}{4}D_p\pi^4ab\left(\frac{1}{a^2}+\frac{1}{b^2}\right)^2$
Plate Full Fixed on All Edges (Collision at Centre)	14	$D_p\pi^4ab\left[\frac{3}{4}\left(\frac{1}{a^2}+\frac{1}{b^2}\right)^2-\left(\frac{1}{ab}\right)^2\right]$

Consider a vibrating structural member segment that deforms in the z-direction, as shown in Figure 2.17.

The segment has length of dx and deforms at an angle of $d\theta$ when subjected to bending moment $M(x)$.

Equations (2.1b) and (2.1c) define the displacement and velocity time-histories of an un-damped system. Amplitude of the displacement Δ in the equations is replaced by shape function, $\psi(x)$, to incorporate the variation of displacement along the member, as shown by Equations (2.19a) and (2.19b) for displacement and velocity, respectively.

$$u(t, x) = \sin(\omega_n t - \phi_p)\psi(x) \tag{2.19a}$$

$$\dot{u}(t, x) = \omega_n \cos(\omega_n t - \phi_p)\psi(x) \tag{2.19b}$$

Strain energy experienced by the segment can thus be written in the form of Equation (2.20):

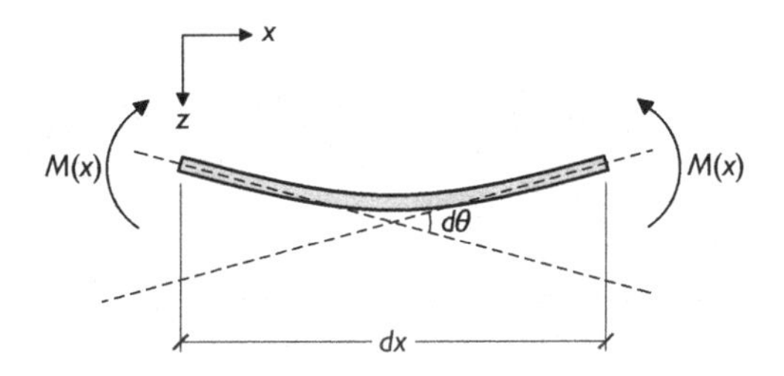

Figure 2.17 Member segment deforming in z-direction.

$$d(\text{SE}) = \frac{1}{2} M(x)\, d\theta \tag{2.20}$$

where $M(x)$ and $d\theta$ are defined by Equations (2.21a) and (2.21b), respectively.

$$M(x) = EI\phi(x) \tag{2.21a}$$

$$d\theta = \phi(x)\, dx \tag{2.21b}$$

Substituting Equations (2.21a) and (2.21b) into Equation (2.20) gives Equation (2.22):

$$d(\text{SE}) = \frac{1}{2}[EI\phi(x)^2]\, dx \tag{2.22}$$

$\phi(x)$ is the curvature of the member, which can be determined by differentiating displacement defined by Equation (2.19a) twice with respect to x, as shown by Equation (2.23):

$$\phi(x) = \frac{\partial^2 u(t, x)}{\partial x^2} = \sin(\omega_n t - \phi_p)\psi''(x) \tag{2.23}$$

Substituting Equation (2.23) into Equation (2.22) gives Equation (2.24):

$$d(\text{SE}) = \frac{\sin^2(\omega_n t - \phi_p)}{2}[EI\psi''(x)^2]\, dx \tag{2.24}$$

To determine the maximum value of SE, $\sin(\omega_n t - \phi_p) = 1$ is adopted and Equation (2.24) is re-written in the integral form of Equation (2.25) by assuming member length of L, which is identical to Equation (2.13a).

$$\text{SE} = \frac{1}{2}\int_0^L EI\psi''(x)^2\, dx \tag{2.25}$$

Mass per unit length along the member is defined as ρA; thus, the mass of the segment shown in Figure 2.17 is equal to $\rho A dx$. Segmental kinetic energy can then be defined by Equation (2.26):

$$d(\text{KE}) = \frac{1}{2}(\rho A dx)(\dot{u}(t, x))^2 \tag{2.26}$$

Substituting Equation (2.19b) into Equation (2.26) gives Equation (2.27):

$$d(\text{KE}) = \frac{1}{2}(\rho A dx)[\omega_n \cos(\omega_n t - \phi_p)\psi(x)]^2 \tag{2.27}$$

By adopting $\cos(\omega_n t - \phi) = 1$, which corresponds to maximum kinetic energy, Equation (2.27) may be re-arranged into the integral form of Equation (2.28) by assuming member length of L, which is identical to Equation (2.13b).

$$\text{KE} = \frac{1}{2}\omega_n^2 \int_0^L \rho A \psi(x)^2 dx \tag{2.28}$$

Substituting Equations (2.25) and (2.28) into Equation (2.18) gives Equation (2.29), which can be re-arranged into Equation (2.30). Equations (2.29) and (2.30) are identical to Equations (2.13c) and (2.13d), respectively.

$$\frac{1}{2}\int_0^L EI\psi''(x)^2 dx = \frac{1}{2}\omega_n^2 \int_0^L \rho A \psi(x)^2 dx \tag{2.29}$$

$$\omega_n = \sqrt{\frac{\int_0^L EI\psi''(x)^2 dx}{\int_0^L \rho A \psi(x)^2 dx}} \tag{2.30}$$

ω_n is also commonly defined by Equation (2.31):

$$\omega_n = \sqrt{\frac{m^*}{k^*}} \tag{2.31}$$

By equating Equations (2.30) and (2.31), generalised mass and stiffness can be determined to be Equations (2.32a) and (2.32b), respectively.

$$k^* = \int_0^L EI\psi''(x)^2 dx \tag{2.32a}$$

$$m^* = \int_0^L \rho A \psi(x)^2 dx \tag{2.32b}$$

Equations (2.32a) and (2.32b) are identical to Equations (2.14a) and (2.14b), respectively.

REFERENCES

2.1 Clough, R.W. and Penzien, J.P., 2003, *Dynamics of Structures*, 3rd ed., Computers & Structures, Inc., Berkeley, CA.
2.2 Chopra, A.K., 2012, *Dynamics of Structures - Theory and Applications to Earthquake Engineering*, 4th ed., Pentice Hall, Upper Saddle River, NJ.
2.3 Ventsel, E. and Krauthammer, T., 2001, *Thin Plates and Shells: theory, Analysis, and Applications*, Marcel Dekker, United States.
2.4 Timoshenko, S. and Woinowsky-Krieger, S., 1959, *Theory of Plates and Shells*, McGraw-Hill Book Company, U.S.

Chapter 3

Fundamentals of Collision Actions on a Structure

3.1 INTRODUCTION

The principles of statics are very much a common denominator underlying the skill sets of all structural engineers. These basic principles are used by designers when exercising their judgement in all stages of the structural design process from conceptual design to performing sanity checks throughout the design process to ensure safety of the newly constructed structure. The distinction between impulsive and localised action of a collision is first explained in this chapter. Some example observations on the response of a simple structural element to an impulsive action of a collision are presented, which reveal anomalies that contradict the common beliefs held by engineers (that are based on the principles of statics). A novel method for determining the displacement demand of a collision, which makes use of equal momentum and equal energy principles, is then presented. This *displacement-based* approach is essentially a calibrated use of statics to make realistic predictions of the equivalent static design forces to represent the dynamic effects of a collision action. The potential adaptation of the newly introduced methodology for regulating structural design is presented at the end of the chapter.

3.2 IMPULSIVE AND LOCALISED ACTIONS OF A COLLISION

As the moving impactor collides with the structure (i.e. target) in a collision event, a force is developed at the point of contact. Like other types of load scenarios, the magnitude, position of application, direction and sense of the applied force need to be input into the analysis. However, the duration of the force application (i.e. duration of contact) is not taken into account using traditional force-based approaches where the collision action is considered as an equivalent static force. If the impact action is to be analysed by dynamic analysis a forcing function (i.e. force expressed as a function of time over the duration of contact) needs to be specified. The

DOI: 10.1201/9781003133032-3

transmitted impulse, which is the integral of force over time, is responsible for putting the structure, or part of it, into motion. The amount of kinetic energy transmitted to the target is controlled by its velocity when responding to the impact action. It is precisely this kinetic energy that causes damage to the structure. Given that velocity takes time to develop, the timing of the force application is an important factor controlling the likelihood of damage in a collision scenario.

In the case of a collision on a free-standing object, which can be a portable barrier, the kinetic energy delivered by the impactor can be dissipated in the form of sliding or overturning, or both, depending on the geometry of the barrier and the ground conditions (Figures 3.1(a) and (b)). In the case of a structure that is restrained at its base to prevent sliding or overturning, the kinetic energy has to be absorbed by the structure in the form of strain energy. Slender structures, such as a pole or a barrier wall, develop strain energy associated with bending/flexural actions (Figure 3.1(c)). For a low-intensity collision where no damage occurs, all the energy can be dissipated through damping over the duration that the structure is left to vibrate freely. This chapter deals with the bending of a structural element responding to a collision. The sliding and overturning behaviour of a free-standing structure is presented in Chapter 4.

Collision on a structure can cause it to respond in different ways causing different forms of damage. The transmitted impulsive action has the potential of causing damage to the '"weakest link" in the structure. Thus, the

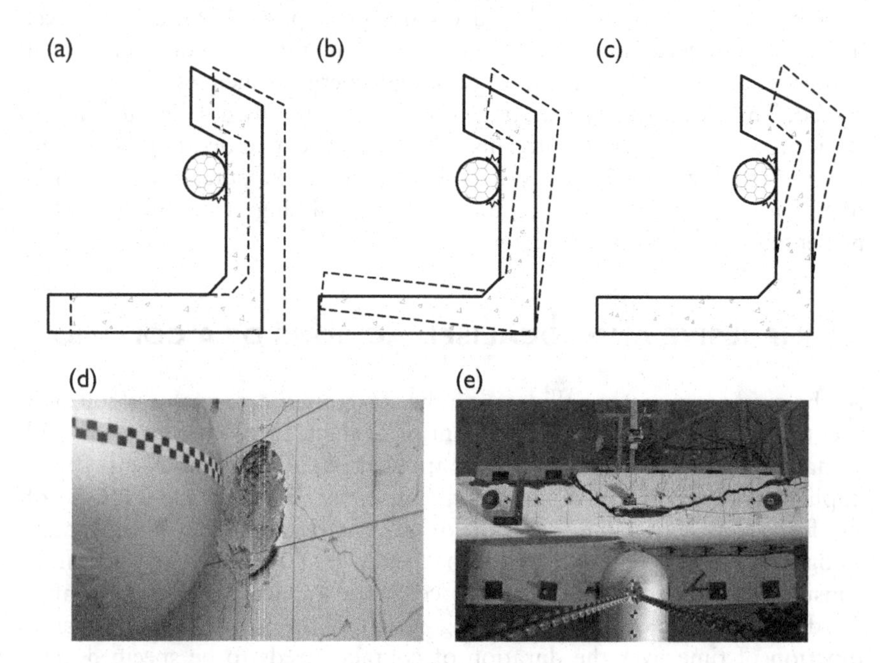

Figure 3.1 Failure mechanism in a collision.

part of the structure sustaining damage by the impulsive action of the collision can be remote from the position of contact. Damage in the form of indentation, or spalling, is instead localised at the point of contact (Figure 3.1(d)). Localised damage is controlled principally by the magnitude of the contact force and the geometry of the part of the impactor that makes direct contact with the surface of the target. Punching failure can also occur close to the point of contact and is controlled by both the magnitude of the contact force and the amount of impulse transmitted (Figure 3.1(e)). Localised actions, with respect to reinforced concrete (RC) structures (targets), are presented and discussed in detail in Chapter 8.

3.3 ANOMALIES WITH COLLISION ACTIONS

An inherent limitation of using a static force to emulate a collision action is the omission of the time dimension in the analysis. This limitation is evident in the description of impulsive actions and is an important factor controlling damage to the structure. Anomalies observed from the behaviour of a simply supported beam responding to a collision, as presented herein, will reveal further limitations.

Consider a series of impact tests conducted on a simply supported beam of varying span lengths (Figure 3.2). The bending moment experienced at mid-span of the beam can be shown to decrease with increasing span length when the mass of the impactor and the drop height are both held constant. This observed decrease in the bending moment contradicts the principles of

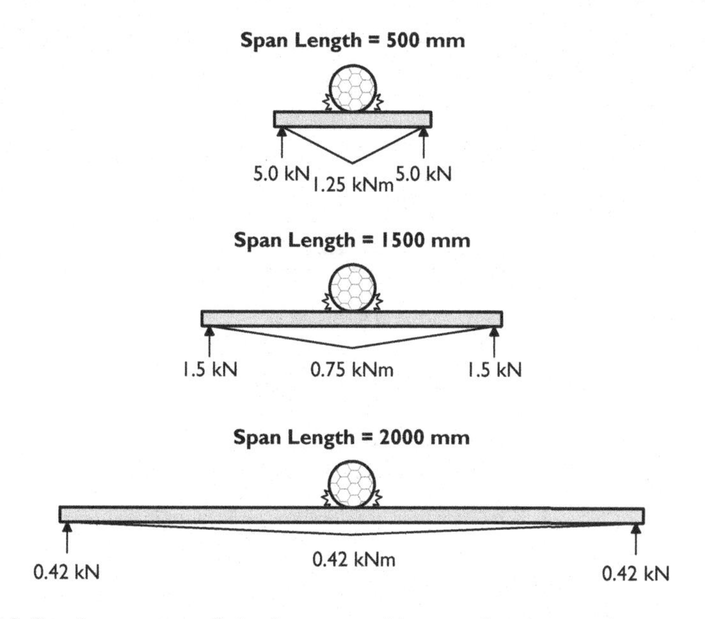

Figure 3.2 Bending moment of simply supported beams when impacted.

statics, for the scenario where an equivalent static force is adopted. Note, the linear form of the bending moment diagrams as presented in the figures is a simplification, as the actual bending moment diagrams caused from the impact are not identical in shape to that of static conditions (i.e. a vertical point load at mid-span). This simplification is considered reasonable for simple systems such as cantilever barrier walls and single span beams in view of observations from experimental investigations that have been undertaken (as described in the literature review presented in Chapter 1).

The maximum deflection of the beam at mid-span is also shown to reveal further anomalies (Figure 3.3). The rate of increase in the value of the deflection with increasing span length is shown to be much lower than the ratio of the span lengths raised to a power of 3 (as would be expected from static analyses of a beam subject to a point force applied at mid-span). These anomalies are due to the dynamic response behaviour of the beam, wherein the forces (e.g. bending moments) and displacements are under the influence of inertia forces within the beam that are generated by the collision.

Other anomalies with collision actions are failure mechanisms that are only seen in collision scenarios but otherwise not seen in quasi-static load scenarios. The punching failure on a stem wall of a RC barrier as revealed in the photo of Figure 3.1(e) is a failure mechanism which is distinctive of impact conditions and would not be seen in a static experiment in which the stem wall is subject to a quasi-static force.

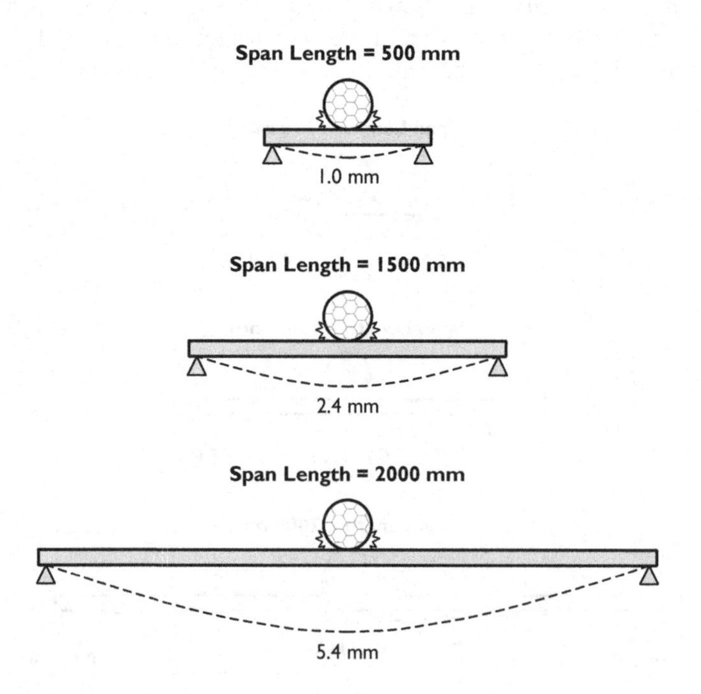

Figure 3.3 Deflection of simply supported beams when impacted.

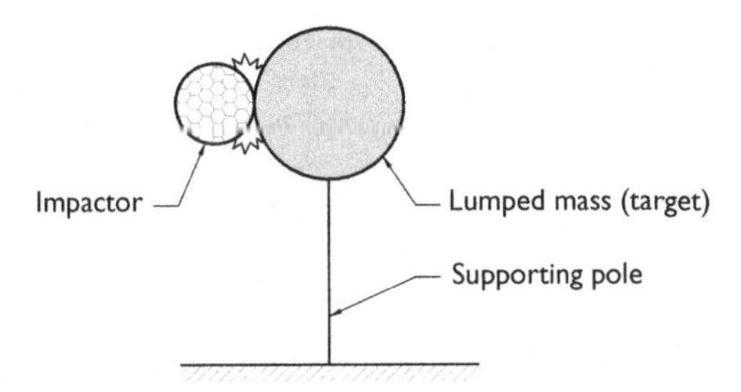

Figure 3.4 Impact on a lumped mass supported on a pole.

3.4 EQUAL MOMENTUM PRINCIPLES

The anomalies, as described previously, reveal limitations with the use of a quasi-static force for estimating the response of a structure to a collision. A simple target structure to be considered herein is a lumped mass, which is supported on a pole. The target is subject to impact by another lumped mass (the impactor), as shown in Figure 3.4. The recommended approach of modelling the impact is to first estimate the displacement demand on the targeted lumped mass employing the established principles of conservation of momentum.

Consider two lumped masses colliding in space. In the initial stage of the analysis, the pole in support of the lumped mass need not be considered given that the duration of contact in a collision is assumed to last for an infinitesimal amount of time (that can be no more than a few milliseconds). The time of contact is therefore too short for the pole to deflect and develop any significant amount of bending moment nor shear forces within it. Interferences of the pole on the lumped mass in the course of the (very rapid) transfer of momentum and energy may therefore be neglected.

In the illustration presented herein, the impactor (of mass = m) moves in a straight-line trajectory at a velocity of v_0 to come in contact with the target (of mass = λm). It is noted that the mass of the target, which is expressed as λ times the mass of the impactor, is initially stationary. Immediately following the collision, the impactor would rebound from the target at a velocity v_1 (the rebound velocity) if the mass of the target is equal to, or exceeds, the mass of the impactor (i.e. $\lambda \geq 1$). Such a scenario is defined herein as Scenario Type (1). Otherwise, the impactor would travel with the target in the direction of the collision after making the initial contact should one of the following situations apply: (i) the mass of the target is less than that of the impactor (i.e. $\lambda < 1$) and (ii) the collision is so intense that the impactor coalesces with the target forming a combined mass. Either of these scenarios is defined herein as Scenario Type (2). Refer the schematic diagram of Figure 3.5, which describes the two types of scenarios.

By equal momentum principles, Equations (3.1a) and (3.2a) can be written for Scenario Type (1) and Type (2), respectively, to define the relationship

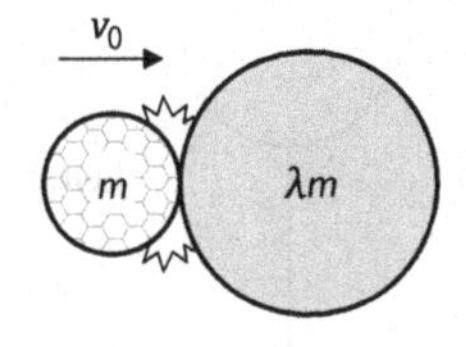

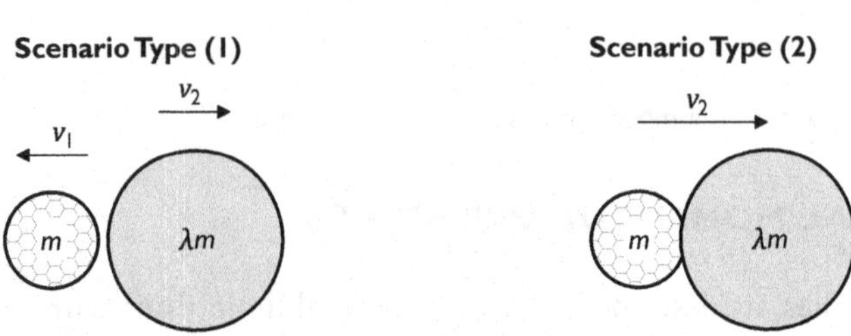

Figure 3.5 Lumped masses colliding in space.

between the "incident velocity," (v_0), the rebound velocity of the impactor (v_1) and the velocity of the target or the combined mass (v_2), immediately following the impact. Note that v_1 is only relevant for Scenario Type (1).

Scenario Type (1)

$$mv_0 = -mv_1 + \lambda mv_2 \tag{3.1a}$$

$$\frac{v_2}{v_0} = \frac{1 + \mathrm{COR}}{1 + \lambda} \tag{3.1b}$$

$$\mathrm{COR} = \frac{v_1 + v_2}{v_0} \tag{3.1c}$$

Scenario Type (2)

$$mv_0 = (1 + \lambda)mv_2 \tag{3.2a}$$

$$\frac{v_2}{v_0} = \frac{1}{1 + \lambda} \tag{3.2b}$$

The left-hand side of Equations (3.1a) and (3.2a) represents the momentum carried by the moving impactor prior to making contact with the target. The two terms on the right-hand side of Equation (3.1a) represent the momentum of the impactor on rebound and the momentum of the target, which has been put into motion following the collision. The single term on the right-hand side of Equation (3.2a) represents the momentum of the combined mass (i.e. impactor plus target). Re-arrangement of the terms in Equations (3.1a) and (3.2a) results in Equations (3.1b) and (3.2b), respectively. Clearly, the velocity developed in the target in the course of the collision is highly dependent on its size relative to that of the impactor (as reflected in the value of λ).

The coefficient of restitution (COR), which varies from 0 to 1, only appears in the expression associated with Scenario Type (1). The COR expression as defined by Equation (3.1c) is to characterise energy dissipation as contact occurs. It is found from the experimental works of the authors that the values of COR vary with the material properties of the impactor as well as the target, impact velocity and size of impactor. The value of COR is best obtained by physical impact testing using representative specimens of both the impactor and target. Table 3.1 which shows typical ranges of values for different impactor-target combinations, is intended to provide intial estimates when no representative information from impact testing is available. Given the same material combination, the value of COR is shown to decrease with the increase in impact velocity and impactor size. It should be noted that having a higher COR value would give conservative estimates on global response of a target structure, whereas a lower COR value would result in conservative estimates of the localised actions.

Consider a hypothetical Scenario Type (1) in which the target is initially stationary, and the value of COR and λ are both equal to unity. As soon as contact occurs the impactor would come to a standstill (i.e. $v_1 = 0$), and the

Table 3.1 Typical COR values for different materials

Impactor Material	*Target Material*	*Range of COR*	*References*
Steel	Steel	0.15–0.5	[3.1,3.2]
Cast Iron		0.35–0.75	[3.3–3.5]
Hailstone		0.02–0.04	[3.6]
Steel	Concrete	0.27–0.32	[3.1,3.7]
Large Granite		0.18–0.45	[3.7,3.8]
Small Granite		0.2–0.8	[3.9,3.10]
Volcanic Rock		0.07–0.25	Experiments by Authors
Timber	Aluminium Alloy	0.2–0.35	[3.11]
Cementitious Debris		0.05–0.1	[3.11]
Glass	Brass	0.72–0.8	[3.12]
	Granite	0.87–0.97	[3.12]
	Rubber	0.37–0.46	[3.12]
Granite	Granite and Sandstone	0.82–0.86	[3.13]

target mass is put into motion with an initial velocity equal to the incident velocity (i.e. $v_2 = v_0$). This idealised scenario bears resemblance with a common scenario seen on the snooker table where a snooker ball moving in a straight-line hits the centre point (projected from the straight-line) of a stationary snooker ball.

3.5 ENERGY TRANSFER IN A COLLISION

Initially, the amount of kinetic energy that is transferred to the target (KE_2) can be written as Equations (3.3a) and (3.4a) for Scenario Type (1) and Type (2), respectively. Velocity of the target, as determined from Equations (3.1b) and (3.2b), can be substituted into Equations (3.3a) and (3.4a) to give Equations (3.3b) and (3.4b), respectively. Rearranging Equations (3.3b) and (3.4b) gives Equations (3.3c) and (3.4c), respectively, which represents the ratio of the kinetic energy transferred to the target and the initial kinetic energy delivered by the impactor, ($KE_0 = 0.5mv_0^2$).

Scenario Type (1)

$$KE_2 = \frac{1}{2}\lambda m v_2^2 \quad (3.3a)$$

$$KE_2 = \frac{1}{2}mv_0^2\lambda\left(\frac{1 + COR}{1 + \lambda}\right)^2 \quad (3.3b)$$

$$\frac{KE_2}{KE_0} = \lambda\left(\frac{1 + COR}{1 + \lambda}\right)^2 \quad (3.3c)$$

Scenario Type (2)

$$KE_2 = \frac{1}{2}(1 + \lambda)mv_2^2 \quad (3.4a)$$

$$KE_2 = \frac{1}{2}mv_0^2\left(\frac{1}{1 + \lambda}\right) \quad (3.4b)$$

$$\frac{KE_2}{KE_0} = \frac{1}{1 + \lambda} \quad (3.4c)$$

In the idealised (snooker ball) collision scenario as described at the end of Section 3.4, the energy ratio is equal to unity meaning that much of the kinetic energy carried by the impactor is transferred totally to the target. If the target mass is doubled (i.e. $\lambda = 2$), the energy ratio is decreased slightly to 8/9 indicating a small amount of energy loss. If the value of λ is increased to 9, the energy ratio is decreased further to 0.36 inferring significant energy loss of 64%. This "energy loss" is being carried away by the impactor on rebound. Further, the energy ratio is reduced to 0.09 if the value of COR tends to zero. This happens when a significant amount of energy is dissipated when contact occurs and represents conditions where the impactor does not rebound from the target. A similar trend of decrease in the energy ratio resulted from an increase in the value of λ can also be seen from Equation (3.4c), which applies to Scenario Type (2).

As deflection of the pole takes place (Figure 3.4), the kinetic energy as defined by Equations (3.3b) and (3.4b) would be absorbed by the supporting pole in the form of strain energy (SE), as presented by Equation (3.5):

$$KE_2 = SE \tag{3.5}$$

The energy absorption by the pole assuming linear elastic behaviour, which can also be represented by a spring connected lumped mass model, is depicted in Figure 3.6 and represented by Equation (3.6):

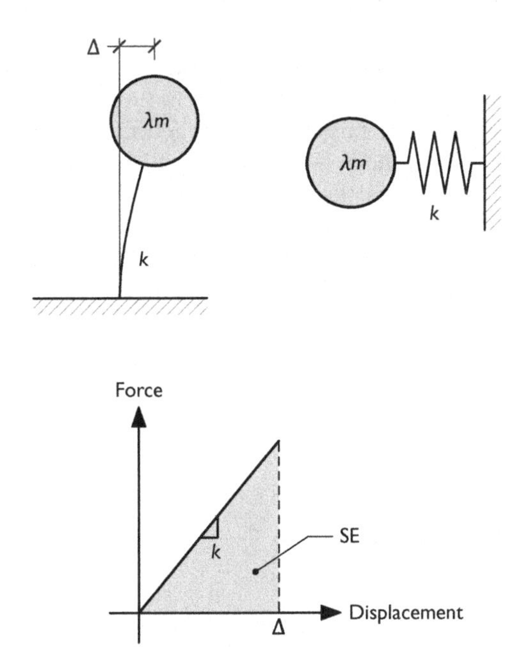

Figure 3.6 Strain energy of absorption.

$$SE = \frac{1}{2}k\Delta^2 \tag{3.6}$$

Substituting Equations (3.3b) and (3.6) into Equation (3.5) and rearranging the terms to solve for Δ gives Equation (3.7a). Similarly, substituting Equations (3.4b) and (3.6) into Equation (3.5) and rearranging the terms gives Equation (3.7b). Equations (3.7a) and (3.7b) can be used for estimating the displacement demand (Δ) on the pole for Scenario Type (1) and Type (2), respectively.

Scenario Type (1)

$$\Delta = \frac{mv_0}{\sqrt{mk}}\sqrt{\lambda\left(\frac{1 + COR}{1 + \lambda}\right)^2} \tag{3.7a}$$

Scenario Type (2)

$$\Delta = \frac{mv_0}{\sqrt{mk}}\sqrt{\frac{1}{1 + \lambda}} \tag{3.7b}$$

The square root term on the right of both expressions is the reduction factor, which takes into account energy losses, as described. Inertial resistance of the target is a main contributor to the reduction in the displacement demand of the collision. This phenomenon is reflected in the decrease in the value of the square root term in Equations (3.7a) and (3.7b) with increasing mass of the target (which is reflected in the value of λ).

The value of the base shear (V_b) experienced by the pole in a collision scenario can be obtained as the product of the deflection (Δ) and lateral stiffness of the pole (k), as shown by Equation (3.8). Substituting the deflection value as obtained from Equations (3.7a) and (3.7b) into Equation (3.8) gives Equations (3.9a) and (3.9b), respectively.

$$V_b = k\Delta \tag{3.8}$$

Scenario Type (1)

$$V_b = v_0\sqrt{mk}\sqrt{\lambda\left(\frac{1 + COR}{1 + \lambda}\right)^2} \tag{3.9a}$$

Scenario Type (2)

$$V_b = v_0\sqrt{mk}\sqrt{\frac{1}{1 + \lambda}} \tag{3.9b}$$

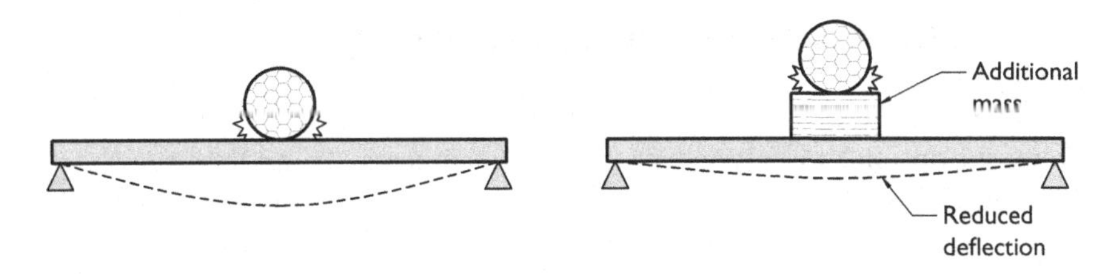

Figure 3.7 Impact on a beam without and with a dummy mass.

3.6 THE EFFECTS OF INERTIAL RESISTANCE

The square root term on the right of Equations (3.7) and (3.9) is the reduction factor, which is dependent on parameters λ and COR. The reduction factor represents the significant contributions of the inertial resistance of the target in mitigating the displacement demand of the impact. This reduction phenomenon implies that a structural element, such as a beam, which is targeted by a collision can be protected by simply placing a dummy mass on the target to receive (i.e. dampen) the impact. This phenomenon has been well demonstrated by collision experiments reported in Ali et al. [3.14]. The tests involved dropping an object to land at the mid-span position of a timber, or a mild steel, beam specimen. A dummy mass was placed on the beam to receive the collision in some of the experiments conducted (Figure 3.7). When all the parameters were kept unchanged, the deflection of the test specimen, which was fitted with a dummy mass, was reduced by up to 40%. This amount of reduction in deflection was achieved by simply drawing upon the inertial resistance developed in the dummy mass (and without the need of strengthening the beam nor shortening the span length).

The mass of the element also generates inertial resistance. The difference is that only a fraction of the mass of the beam (i.e. the generalised mass) is activated to develop the inertial resistance. As demonstrated in Chapter 2, the generalised mass of a simply supported beam is only half the total mass of the beam. It was observed in a series of collision experiments that (when the same object was dropped on beams at mid-span) the longer the span length of the beam, the greater the generalised mass and hence inertial resistance generated from within the beam. This offers explanation to the phenomenon of reduction in bending moment with increasing span length as presented earlier in Figure 3.2. The deflection anomaly as revealed in Figure 3.3 was attributed to the same cause. The same argument can be used to explain why a simply supported beam built of lightweight material would deflect more than a beam cast of concrete when the collision parameters, span length and flexural rigidity of the beam are all kept the same.

3.7 THE DISPLACEMENT-BASED APPROACH

Calculations based on the use of principles of statics remain to be the preferred option of analysis by structural engineers. To address the shortcomings of using an equivalent quasi-static force to represent a collision action, the magnitude of the static force must first be calibrated to give a realistic estimate of the displacement demand on the targeted structure, or structural element. This method of analysis is also known as the *displacement-based* method of analysis, which combines the use of the principles of conservation of momentum and energy, along with the established principles of statics.

The procedure as described is resolved into the following steps:

1. Estimate the deflection of the targeted structure by employing equal momentum and equal energy principles (Figure 3.8(a)), this involves idealising the structural element into a lumped mass system which is characterised by parameters m^* and k^* as introduced in Chapter 2;
2. Apply an equivalent static force to achieve the same amount of deflection in the structure (Figure 3.8(b)); and

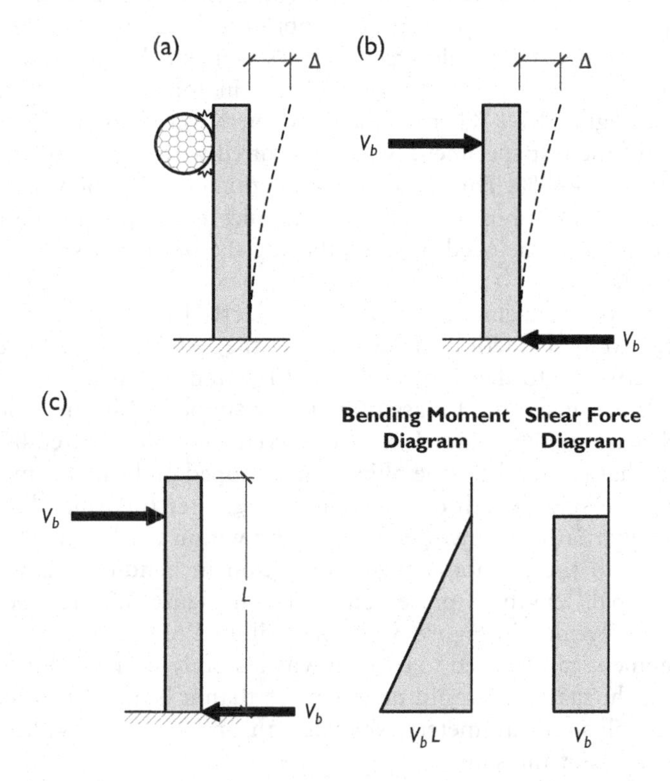

Figure 3.8 Schematic diagram showing the concept of a calibrated static model.

3. Analyse the structure to determine the bending moment and shear forces developed in the targeted element by equivalent static analyses (Figure 3.8(c)).

3.8 LIMITATIONS OF THE PROPOSED DISPLACEMENT-BASED APPROACH

An important limitation of the predictive model introduced in this chapter (as represented by Equations (3.7) and (3.9)) is that the momentum delivered by the impactor must be transferred to the target in such a short period of time that interferences by the supporting structure is negligible or else the deflection demand of the collision could be overstated. The duration of contact between the impactor and target is an important parameter controlling the outcome of the collision. Cushioning the impact is a way to prolong the contact. Take the collision scenario of Figure 3.4 as an example; it can be shown that prolonging the duration of contact could allow internal forces (shear and bending) to be developed in the pole during the course of the momentum transfer. Internal forces developed in the pole in a cushioned impact would then interfere with its motion. Deflection would be suppressed as a result. The degree of suppression is a function of the ratio of the duration of contact and the natural period of vibration of the pole (target structure). The delaying mechanism as described explains how deflection can be suppressed by cushioning, which is a topic to be treated in detail in Chapter 7.

Another limitation of the predictive methodology for Scenario Type (1) is that the collision scenario must be such that the mass of the impactor is exceeded by the mass (or generalised mass) of the target. This condition is necessary to ensure that the impactor rebounds from the target on impact (i.e. $v_1 > 0$). As the impactor rebounds and moves away from the target following the impact a second occurrence of contact can be avoided. There are two types of scenarios where this requirement is not met. The first type of scenario is where the impactor mass exceeds the target mass (i.e. $\lambda < 1$). Another condition is collision in the vertical direction under the influence of gravity which results in the impactor settling on the target permanently. With both types of scenarios multiple occurrences of contact can occur.

Multiple contacts (collisions) can be very complex to analyse. Making an attempt to derive a theoretical solution for multiple collisions is not recommended. An alternative heuristic model based on the assumption that the impactor is attached to the targeted structural element (a beam) following the first contact was adopted by the authors for predicting the deflection of a beam when struck by a fallen object, i.e. Scenario Type (2) as defined in Section 3.4. The assumption that the impactor becomes attached to the target (as depicted in Figure 3.5) is strictly not valid in theory but nonetheless the model predictions have been found to be in agreement with experimental measurements across a number of tests [3.14, 3.15]. The analysis for vertical impact is presented in detail in Chapter 5.

Another possible source of error with the three-step calculation procedure as presented in Section 3.7 is the neglect of the higher mode contributions, which are responsible for dynamic anomalies (modifications to the deflection profile) that might dominate the response of a slender structure when impacted by a hard moving object. Such anomalies were found to be minor from impact testing of cantilevered wall specimens (with an aspect ratio of 6.5) [3.1] and on simply supported beam specimens (with aspect ratios varying between 5 and 10) [3.15]. As a general rule, the extent of higher mode interferences on a structural member such as a barrier wall, a pole or a beam increases with its span to depth (*L*/*D*) ratio and span length. The collision of a stiff solid object featuring a hard instantaneous one-off contact with the target prior to rebounding is expected to generate more higher modes contributions than collision by a bag of sand, for example. The stem wall of a rockfall barrier, crash barrier or other forms of cantilevered structural elements having an *L*/*D* ratio of up to 6 to 7, and with height (or length) of up to 3 m, should not be subject to much higher mode interferences in a critical collision scenario. Beams other than cantilevered beams tend to attract more higher modes interferences because of their higher *L*/*D* ratio.

The displacement-based method can be used for estimating the maximum bending moment and curvature at the critical cross section and in modelling the deformation of the member regardless of higher mode interferences. Anomalies associated with higher modes dominance tend to shorten the effective span length of the member, and hence exacerbating the risk of premature failure by shear (punching), which can be localised around the point of impact. Structural design for safeguarding against punching failure of a beam or wall (slab) can be found in Chapter 8 (for a RC element). Higher mode effects can also cause short duration and localised high curvature demand that can only be predicted by undertaking dynamic simulations. To cover for these uncertainties, it is desirable to introduce some ductility into the design of the structure by incorporating recommendations as outlined in Chapter 8 for RC members, and to use compact sections for structural steel members.

If a detailed investigation is to be conducted to ascertain the extent of higher mode contributions, the duration of contact and the natural period of the fundamental mode of vibration of the targeted element would need to be predicted. Higher mode interference is expected to be minor if the contact duration comes close to, or exceed, the fundamental natural period of vibration of the affected structural element. Note that the duration of contact increases with the size of the impactor and can be sensitive to the compressive stiffness of the impactor and the surface of the target. Details of this type of analyses can be found in Chapter 13.

3.9 RIGID PLASTIC AND ELASTO-PLASTIC BEHAVIOUR

Algebraic expressions for determining the displacement demand of a collision as presented in Section 3.5 are only applicable to linear elastic behaviour of the targeted structure. In this section, the predictive model is extended for incorporating the effects of ductile yielding in the targeted structure. Two simplified models for ductile yielding are introduced herein: (1) a rigid perfectly plastic behaviour, which is suited to modelling highly ductile systems, and (2) elasto-plastic behaviour, which is suited to modelling systems with limited ductility. Refer to Figure 3.9 for their schematic illustration. Notations F_y and Δ_y, as presented in Figure 3.9, are the yield force and yield displacement, respectively.

The amount of energy absorbed by the target in the form of strain energy is indicated by the area under the force-displacement response curve of the targeted structure, as shown in Figure 3.9. The amount of strain energy of absorption by the system is as defined by Equations (3.10a) and (3.10b) for the rigid perfectly plastic and the elasto-plastic model, respectively.

$$SE = F_y \Delta \tag{3.10a}$$

$$SE = F_y \left(\Delta - \frac{\Delta_y}{2} \right) \tag{3.10b}$$

(a)

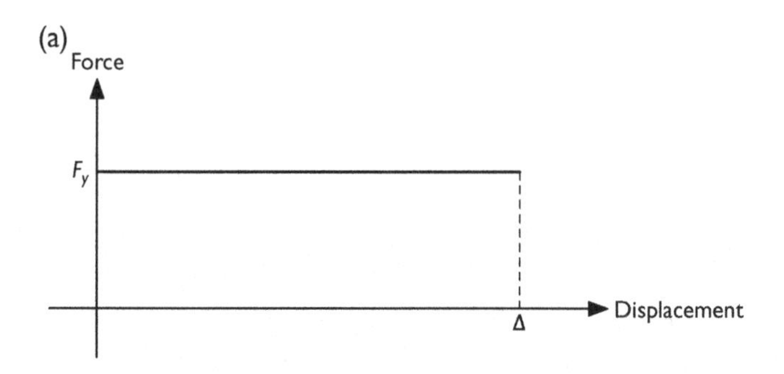

(b)

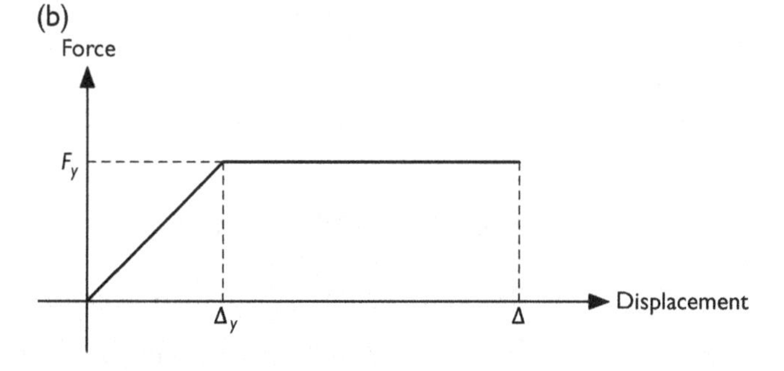

Figure 3.9 Collision on a ductile element assuming embedment of impactor: (a) rigid perfectly plastic behaviour, (b) elasto-plastic behavior.

As presented previously, the kinetic energy is absorbed by the deflection of the target structure in the form of strain energy and therefore it can be assumed that KE_2 = SE, i.e. Equation (3.5). Consider Scenario Type (2), as defined in Section 3.4. Equating KE_2 from Equation (3.4b) to SE from Equations (3.10a) and (3.10b) gives Equations (3.11a) and (3.11b), respectively, which can be used to determine the displacement demand from the collision for the structural system experiencing yielding. The predictions are based on the assumption that the impactor becomes attached to the target following the first contact.

$$\Delta = \frac{mv_0^2}{2F_y}\left(\frac{1}{1+\lambda}\right) \tag{3.11a}$$

$$\Delta = \frac{mv_0^2}{2F_y}\left(\frac{1}{1+\lambda}\right) + \frac{\Delta_y}{2} \tag{3.11b}$$

In situations where the impactor rebounds, and moves away from the target following the first contact, Equation (3.3b) should be used in place of (3.4b). Equating the energy demand on the target, as determined using Equation (3.3b), with the strain energy of absorption of Equations (3.10a) and (3.10b) gives Equations (3.12a) and (3.12b), respectively.

$$\Delta = \frac{mv_0^2}{2F_y}\lambda\left(\frac{1+\text{COR}}{1+\lambda}\right)^2 \tag{3.12a}$$

$$\Delta = \frac{mv_0^2}{2F_y}\lambda\left(\frac{1+\text{COR}}{1+\lambda}\right)^2 + \frac{\Delta_y}{2} \tag{3.12b}$$

The presented expressions for predicting the displacement demand can be used for estimating damage caused by the collision to a structural element experiencing ductile, or limited ductile, yielding. The calculated displacement demand can be used to infer the extent of plastic hinge rotation as experienced by the affected element (refer Chapter 8 for details with respect to RC elements).

3.10 IMPLICATIONS ON DESIGN PRACTICES

3.10.1 Existing Code Provisions for Collision Actions

Current codes of practices that have provisions for collision actions are mostly associated with the design of highway structures, railway bridges, vehicular crash barriers or rockfall barriers. The three main design approaches of these code provisions include: simply stipulating a prescriptive static design force that has a constant value to represent the collision action [3.16–3.19]; providing empirical/semi-empirical equations for estimating the design force [3.20–3.22]; or utilising an equal energy method for calculating the design force (V_b), as given by Equation (3.13) [3.19]:

$$V_b = v_0\sqrt{mk} \tag{3.13}$$

Note that Equation (3.13) can be derived easily from Equation (3.14a), which shows the kinetic energy equating with the elastic energy of absorption. Equation (3.14a) can also be rearranged into Equation (3.14b) for finding deflection demand of the collision. Note that multiplying Equation (3.14b) by k (as $V_b = k\Delta$) reverts back to Equation (3.13).

$$\frac{1}{2}mv_0^2 = \frac{1}{2}k\Delta^2 \tag{3.14a}$$

$$\Delta = \frac{mv_0}{\sqrt{mk}} \tag{3.14b}$$

None of the listed expressions considers energy losses occurring on impact. Note that the kinetic energy delivered by the impactor (e.g. a moving vehicle) can be resolved into different components as described in Section 3.5. As explained earlier, a portion of the delivered energy is dissipated at the point of contact; some energy is carried away by the impactor on rebound; and the remaining energy is transferred to the target (e.g. barrier) and the supporting structure. These stipulated design provisions have not taken into account energy partitioning at all and are therefore overly conservative. The comparison of Equations (3.9a) and (3.9b) with Equation (3.13) is evident of the conservatism as the square root term on the right of Equations (3.7) and (3.9), representing the mitigation effects of inertial resistance, has been omitted.

The aforementioned code provisions are however justified in situations where the kinetic energy delivered by the impactor (e.g. moving vehicle) is absorbed totally by the target (e.g. crash barrier). This is the case when the mass of the barrier is much smaller than that of the vehicle, which crashes into the barrier. The collision of a heavy truck on a metal barrier matches with this description but not necessarily so with collision of a truck colliding with a bridge pier. The code provisions can also be justified in a contrasting situation where a solid barrier is so rigid that all the deformations, and energy dissipation, may be assumed to occur in the moving (and crumbling) vehicle.

It is recommended herein that expressions introduced in this chapter (except for Equation (3.13)) be stipulated in codes of practice to guide structural design for accommodating the impulsive action of a collision.

3.10.2 Analogy with Representation for Seismic Actions

Expressions for determining the deflection or force demand generated by a projected collision scenario can be presented in the form of a response spectrum in the same manner as for seismic actions to facilitate their implementation in the design of built facilities including buildings and bridges (given that seismic actions have become better known amongst civil and structural engineers residing in regions of varying seismic activity levels around the globe). This analogy applies only to the impulsive effects of a collision as covered in Chapters 3 to 5, and does not apply to localised actions which is the subject matter of Chapter 6.

The acceleration response spectrum of the flat-hyperbolic form, which is commonly used in earthquake engineering, can be adapted for representing other forms of transient actions such as collision actions. The algebraic expression for defining the hyperbolic function of the response spectrum of Figure 3.10 is essentially the estimated velocity demand of the targeted structure (v_2) multiplied by $2\pi/T$, as shown by Equation (3.15). The expression has the gravitational acceleration g in the denominator meaning that RSa is in unit of g. The value of v_2 for input into Equation (3.15) can be obtained using Equations (3.1b) or (3.2b) for Scenario Types (1) and (2), respectively.

$$\text{RSa} = \frac{v_2}{g}\left(\frac{2\pi}{T}\right) \tag{3.15}$$

The velocity term in Equation (3.15) can be associated with the impulsive action of a collision, or similarly, the "throw effects" of earthquake ground shaking on a structure or that of a blast wave. The "flat part" of the response spectrum represents the limiting force demand of the collision action. The estimated amount of force generated on contact between the impactor and the surface of the target may be used to define this limit. Determination of the contact force is presented in detail in Chapter 6.

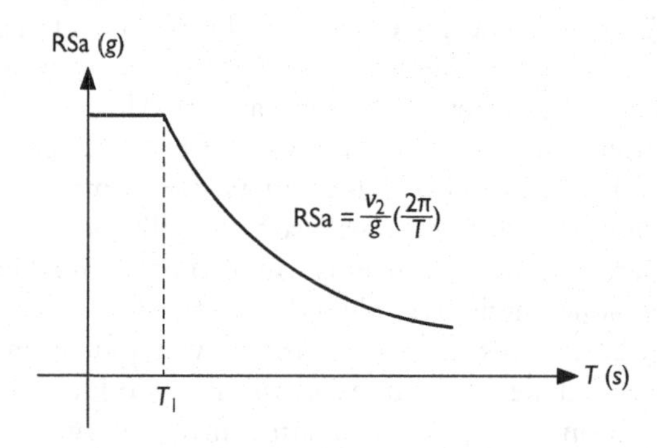

Figure 3.10 Acceleration response spectrum of a collision action.

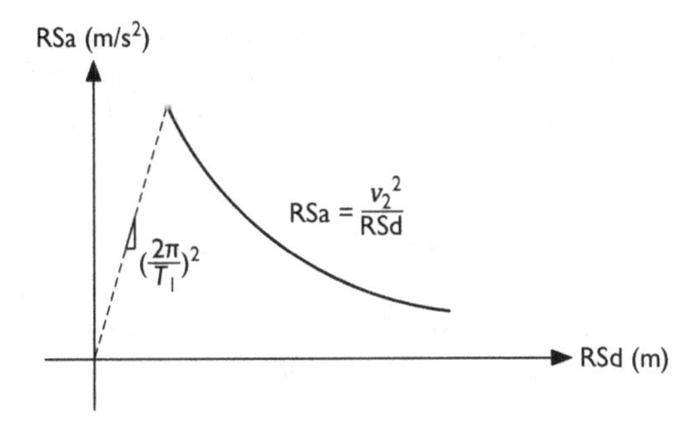

Figure 3.11 Acceleration-displacement response spectrum diagram of a collision action.

Should there be doubts as to what the value of the contact force is, the acceleration response spectrum may only feature the hyperbolic function only and omit the "flat part" of the spectrum (i.e. taking T_1 = 0s). The response spectrum of a fully hyperbolic form provides conservative estimates for the force, or acceleration, demand of the collision. In a similar manner, the impulsive action of a collision can also be presented in the form of an acceleration-displacement response spectrum (ADRS) diagram. which is commonly used to represent seismic actions. The algebraic expression for defining the hyperbolic function of the response spectrum of Figure 3.11 is essentially the square of the estimated velocity demand on the targeted structure divided by the drift demand (RSd), as shown by Equation (3.16). Again, the velocity term, v_2, can be estimated by the use of Equations (3.1b) and (3.2b) for Scenario Types (1) and (2), respectively.

$$\mathrm{RSa} = \frac{v_2^2}{\mathrm{RSd}} \tag{3.16}$$

The ADRS diagram has advantages over the conventional acceleration response spectrum as the drift demand is shown explicitly on the diagram and with good resolution. The use of the ADRS diagram is also favoured in situations where the designer prefers not to use the natural period of vibration as the parameter for characterising the structure. The ADRS diagram may contain the hyperbolic function only (i.e. without the "flat part") to provide conservative estimates of the force (or acceleration) and drift demand of the collision.

3.10.3 Limit State Design for Collision Actions

In limit state structural design, the transient action generated by a projected collision scenario should be considered to co-exist with dead and live load in the same manner as for seismic actions. The partial safety load factor for collision action should be determined in accordance with uncertainties in relation to the collision scenario as characterised by the impactor mass and the incident velocity of impact. There can be enhancement in strength and stiffness of the structure because of strain rate effects in transient conditions, but such enhancement may be very minor depending on the collision scenario. For limited ductile system such as lightly reinforced concrete and structural steel of compact cross section, material factors for designing against the impulsive action of a collision may be taken to be the same for seismic actions. Non-ductile structural elements such as steel of non-compact cross-sections, or RC not complying with the requirements specified in Section 8.4 (refer Chapter 8) should not be exposed to any significant collision hazards. This is so regardless of whether the structural element is designed to respond up to the limit of yield in the projected collision scenario. Existing installations of this type, that are exposed to collision hazards, are recommended to be retrofitted, or subject to detailed evaluations. Timber elements or elements built of other lightweight materials also warrant special considerations because of their lower level of protection by inertial resistance.

3.11 CLOSING REMARKS

This chapter is aimed at introducing the fundamentals of collision actions on structures. The chapter predominantly deals with the impulsive action of a collision and makes use of simple structural elements like a simply supported beam or a pole supporting a lumped mass to illustrate fundamental principles. Expressions for determining the displacement demand of a lumped mass system have been derived by combining the principles of equal momentum and equal energy. Two types of scenarios are presented; the first is when the impactor rebounds from target (Scenario Type 1) and the second is when the impactor "attaches" to the target (Scenario Type 2). Type 1 is generally applicable to horizontal collision scenarios featuring rebound of the impactor, which is the case when the mass of the target significantly exceeds the mass of the impactor. Type 2 is generally applicable to scenarios where the impactor is attached to the target following the collision, or where the impactor strikes and rebounds from the target multiple times. The latter type of behaviour is expected in a vertical collision of a heavy fallen impactor, or in a horizontal collision where the mass of the impactor exceeds that of the target.

In the displacement-based approach of analysis, the calculated deflection demand is used to calibrate a static model to give realistic predictions of the internal forces that are generated from the collision action. The phenomenon of inertial resistance, which can result in a significant reduction in the displacement and force demand of the collision, is highlighted in this chapter. Apparent anomalies presented in the early part of the chapter have accordingly been resolved and largely apportioned to the concept of inertial resistance. Predictive expressions based on linear elastic behaviour of the supporting structure are first presented. Further expressions have been presented to address structural elements that rely on ductility and respond within their inelastic limit to resist collision actions. In the final part of the chapter, current code provisions are reviewed; use of response spectra of different formats, which were originally developed for representing seismic actions, have been adapted for representing collision actions; and recommendations for design in alignment with limit state principles are given. References are also made to later chapters, which give detailed treatment in relation to overturning and sliding (Chapter 4), vertical impact under the influence of gravity (Chapter 5), contact force generated by the collision (Chapter 6) and the cushioning of the impact (Chapter 7).

REFERENCES

3.1 Yong, A.C.Y., Lam, N.T.K., Menegon, S.J., and Gad, E.F., 2020, "Experimental and analytical assessment of the flexural behaviour of cantilevered RC walls subjected to impact actions", *Journal of Structural Engineering*, Vol. 146(4), p. 04020034, doi:10.1061/(ASCE)ST.1943-541X.0002578

3.2 Yong, A.C.Y., Lam, N.T.K., Menegon, S.J., and Gad, E.F., 2020, "Cantilevered RC wall subjected to combined static and impact actions", *International Journal of Impact Engineering*, Vol. 143, p. 103596, doi:10.1016/j.ijimpeng.2020.103596

3.3 Lam, N.T.K., Yong, A.C.Y., Lam, C., Kwan, J.S.H., Perera, J.S., Disfani, M.M., and Gad, E., 2018, "Displacement-based approach for the assessment of overturning stability of rectangular rigid barriers subjected to point impact", *Journal of Engineering Mechanics*, Vol. 144(2), pp. 04017161-1 –04017161-15. doi: 10.1061/(ASCE)EM.1943-7889.0001383

3.4 Lam, C., Yong, A.C.Y., Kwan, J.S.H., and Lam, N.T.K., 2018, "Overturning stability of L-shaped rigid barriers subjected to rockfall impacts", *Landslides*, Vol. 149, pp. 1347–1357. doi:10.1007/s10346-018-0957-5

3.5 Yong, A.C.Y., Lam, C., Lam, N.T.K., Perera, J.S., and Kwan, J.S.H., 2019, "Analytical solution for estimating sliding displacement of rigid barriers subjected to boulder impact", *Journal of Engineering Mechanics*, Vol. 145(3), p. 04019006, doi:10.1061/(ASCE)EM.1943-7889.0001576

3.6 Sun, J., Lam, N., Zhang, L., Ruan, D., and Gad, E., 2015, "Contact forces generated by hailstone impact", *International Journal of Impact Engineering*, Vol. 84, pp. 145–158, doi:10.1016/j.ijimpeng.2015.05.015

3.7 Majeed, Z.Z.A., Lam, N.T.K., and Gad, E.F., 2021, "Predictions of localised damage to concrete caused by a low-velocity impact", *International Journal of Impact Engineering*, Vol. 149, p. 103799, doi:10.1016/j.ijimpeng.2020.103799

3.8 Wyllie, D.C., 2014, *Rock Fall Engineering*, CRC Press, Boca Raton, Florida.

3.9 Majeed, Z.Z.A., Lam, N.T.K., Lam, C., Gad, E., and Kwan, J.S.H., 2019, "Contact force generated by impact of boulder on concrete surface", *International Journal of Impact Engineering*, Vol. 132, p. 103324, doi: 10.1016/j.ijimpeng.2019.103324

3.10 Chau, K.T., Wong, R.H.C., and Lee, C.F., 1998, "Rockfall Problems in Hong Kong and some new experimental results for coefficients of restitution", *International Journal of Rock Mechanics and Mining Sciences*, Vol. 35(4), pp. 662–663, doi:10.1016/S0148-9062(98)00023-0
3.11 Perera, S., Lam, N., Pathirana, M., Zhang, L., Ruan, D., and Gad, E., 2016, "Deterministic solutions for contact force generated by impact of windborne debris", *International Journal of Impact Engineering*, Vol. 91, pp. 126–141, doi:10.1016/j.ijimpeng.2016.01.002
3.12 Sandeep, C.S., Luo, L., and Senetakis, K., 2020, "Effect of grain size and surface roughness on the normal coefficient of restitution of single grains", *Materials*, Vol. 13(4), p. 814, doi:10.3390/ma13040814
3.13 Imre, B., Räbsamen, S., and Springman, S.M., 2008, "A coefficient of restitution of rock materials", *Computers & Geosciences*, Vol. 34(4), pp. 339–350, doi:10.1016/j.cageo.2007.04.004
3.14 Ali, M., Sun, J., Lam, N., Zhang, L., and Gad, E., 2014, "Simple hand calculation method for estimating deflection generated by the low velocity impact of a solid object", *Australian Journal of Structural Engineering*, Vol. 15(3), pp. 243–259, doi:10.7158/13287982.2014.11465162
3.15 Yong, A.C.Y., Lam, N.T.K., and Menegon, S.J., 2021, "Closed-form expressions for improved impact resistant design of reinforced concrete beams", *Structures*, Vol. 29, pp. 1828–1836, doi:10.1016/j.istruc.2020.12.041
3.16 Standards Australia, 2017, *AS 5100.2 Bridge Design Part 2: Design Loads*, Standard Australia Limited, New South Wales, Australia.
3.17 Austroads, 2013, *Standardised Bridge Barrier Design*, Austroads Ltd, Sydney.
3.18 American Association of State Highway and Transportation Officials, 2012, *AASHTO LRFD Bridge Design Specifications, 6th edition*, American Association of State Highway and Transportation Officials, Washington DC, U.S.
3.19 European Committee for Standardization (CEN), 2006, *Eurocode 1: Actions on structures – Part 1–7: General actions – Accidental actions*, European Committee for Standardization (CEN), Brussels, Belgium.
3.20 Japan Road Association, 2000, "Manual for anti-impact structures against falling rock (in Japanese)". Japan Road Association, Tokyo, Japan.
3.21 ASTRA, 2008, "Einwirkungen infolge Steinschlags auf Schutzgalerien (in German)", *Richtlinie, Bundesamt für Strassen, Baudirektion SBB, Eidgenössische Drucksachen-und Materialzentrale*, Bern.
3.22 Kwan, J.S.H., *Supplementary technical guidance on design of rigid debris-resisting barriers (GEO Report No. 270)*. 2012: Geotechnical Engineering Office, the Government of the Hong Kong Special Administrative Region.

Chapter 4

Collision with a Free-Standing Barrier

4.1 INTRODUCTION

The overturning and sliding of a free-standing barrier can have dire consequences depending on its location and application. The methodology presented in this chapter can be used to assess portable barriers for collisions actions, such as temporary vehicular barriers or hillside barriers, which are for providing protection from vehicle collisions and landslides or rockfalls, respectively. Some hillside barriers are found on piles and are not free-standing. In this instance, the methodology introduced in this chapter may too be used for assessing their ultimate performance in an extreme scenario, which has surpassed the resistant capacity of the piles. The stability of the barrier would then resort to its own weight and geometry. Thus, the potential utility of the method of assessment presented in this chapter is not limited to free-standing barriers and can also be used for checking the overturning resistance of barriers, and structures, generally.

Conventional methods of checking overturning stability based on considering static equilibrium is based on the assumption that all the forces applied to the barrier are "permanent forces". This is only true of the gravitational load which stabilises the barrier, whereas the impact force generated by the collision of a moving (or fallen) object is transient in nature. Thus, the use of statics to check for overturning or sliding can give misleading predictions. A barrier that is deemed to have overturned, using a static equilibrium approach, in reality might only experience some limited rocking motion without necessarily being overturned. Calculations based on statics will only predict if some sliding would occur, when subject to the projected impact force; however, it would not predict the amount of sliding movement.

The method of calculation stipulated by codes of practices and design guidelines typically represents an impact action by what is known as an "equivalent static force" [4.1–4.6]. Barriers are then designed to provide sufficient sliding and overturning stability against such force. The "equivalent static force" modelling approach can only be used to estimate the limiting force to cause sliding and overturning motion (referred herein as sliding and overturning capacity) of traffic barriers in a collision scenario [4.7].

DOI: 10.1201/9781003133032-4

The methodology presented in this chapter gives predictions of the overturning and/or sliding capacity, and not just the limiting forces. Thus, a much more realistic assessment of the outcome of the collision (e.g. rockfall) scenario is provided. The methodology for assessing the risk of overturning is illustrated initially by considering a barrier of rectangular cross section. Further illustrations are based on considering free-standing barriers of different cross sections, such as an L-shaped cross-section that is consistent with the geometry of impact resistant barriers that are commonly employed for the protection of hill slopes. The later sections in the chapter deal with predictions of sliding displacements.

4.2 OVERTURNING OF A RECTANGULAR BLOCK

The principle of conservation of momentum was adopted (in Chapter 3) to derive the energy ratio for representing the fraction of initial kinetic energy (delivered by a moving object) that would be transferred to the target. The calculation involved using the principle of equal momentum for determining the velocity of the targeted object (structure), which has been set into motion immediately following the collision.

The equation of angular momentum (Equation (4.1)) is employed for analysing the overturning behaviour of a rectangular object, as shown in Figure 4.1, by determining the angular velocity of the overturning motion ($\dot{\theta}$).

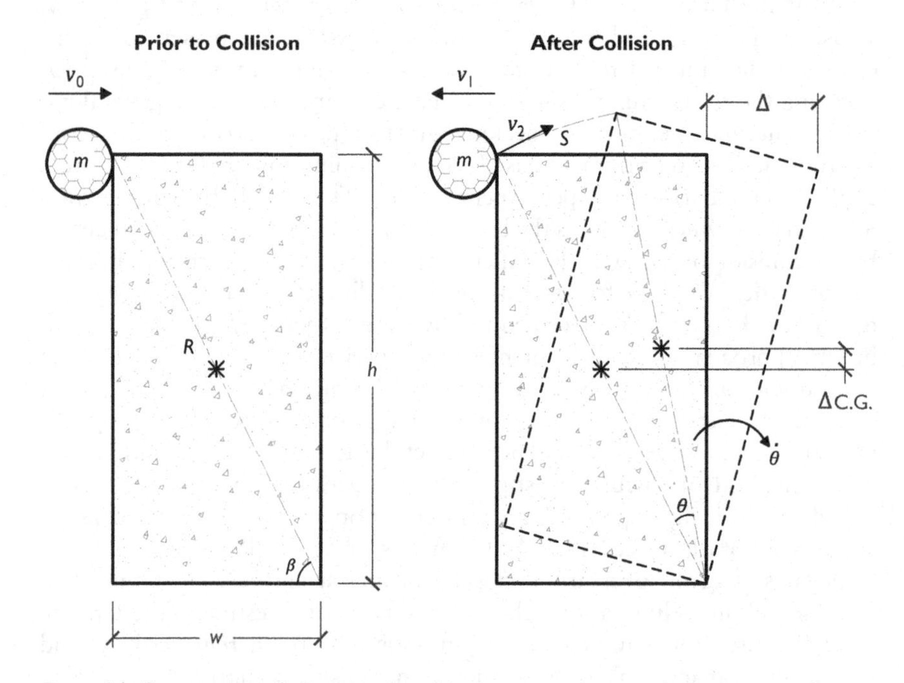

Figure 4.1 Overturning of a rectangular object.

The following illustration is applicable to Type (1) scenarios (as illustrated in Chapter 3). With this type of collision scenario, the mass of the impactor is well exceeded by the mass of the target, which is usually the case of a fallen boulder striking a reinforced concrete (RC) rockfall barrier wall. The impactor in such a scenario is expected to rebound from the barrier following the first contact. There are other Type (2) scenarios where the impactor is modelled to become attached to the target following the collision. A more precise description of the criterion for determining if the scenario is Type (1) or (2) is given in the latter part of the chapter.

$$m(v_0 + v_1)h = I_\theta \dot{\theta} \tag{4.1}$$

where m is mass of the impactor; v_0 is incident velocity of impactor; v_1 is the velocity of the impactor on rebound; h is height of the rectangular section; and I_θ is rotational inertia of the barrier taken about the pivot point of overturning. The value of I_θ can be found using Equation (4.2), which is derived in Section 4.3.2.

$$I_\theta = M\left(\frac{h^2 + w^2}{3}\right) \tag{4.2}$$

where M and w is the mass and width of the rectangular section, respectively.

The arc length, s, representing the motion of the top-left corner of the rectangular section is defined in Equation (4.3a). Rearranging and differentiating Equation (4.3a) with respect to time gives Equation (4.3b) for determining the value of the angular velocity, $\dot{\theta}$.

$$s = v_2 t = R\theta \tag{4.3a}$$

$$\dot{\theta} = \frac{d\theta}{dt} = \frac{v_2}{R} \tag{4.3b}$$

where v_2 is velocity of the barrier undergoing rotation at the point of contact; t is the time when maximum angle of rotation θ is reached; and R is the distance between the point of contact and the pivot point. For a rectangular cross section, $R = \sqrt{h^2 + w^2}$.

Equation (4.4) is obtained by combining Equation (4.1) with Equation (4.3b):

$$v_0 + v_1 = \frac{I_\theta}{mhR} v_2 \tag{4.4}$$

Equation (4.4) can be re-written as Equation (4.5):

$$v_0 + v_1 = \kappa v_2 \tag{4.5}$$

where κ is the dimensionless factor as defined by Equation (4.6):

$$\kappa = \frac{I_\theta}{mhR} \tag{4.6}$$

Adding v_2 to both sides of Equation (4.5) and then dividing every term by v_0 results in Equation (4.7), which provides predictions for the velocity ratio.

$$\frac{v_2}{v_0} = \frac{1 + \text{COR}}{1 + \kappa} \tag{4.7}$$

where COR is the coefficient of restitution as defined by Equation (4.8):

$$\text{COR} = \frac{v_1 + v_2}{v_0} \tag{4.8}$$

The reader is directed to Chapter 3 for a discussion on the coefficient of restitution, which includes broad/general recommendations for the value, for different collision scenarios.

The amount of kinetic energy (KE_2) imposed by the collision action on the targeted object is defined by Equation (4.9a). Substituting Equations (4.3b) and (4.6) into Equation (4.9a) gives Equation (4.9b).

$$\text{KE}_2 = \frac{1}{2} I_\theta \dot{\theta}^2 \tag{4.9a}$$

$$\text{KE}_2 = \frac{1}{2}\left(\frac{\kappa h}{R}\right) m v_2^2 \tag{4.9b}$$

The value of the energy ratio can be found using Equation (4.10):

$$\frac{\text{KE}_2}{\text{KE}_0} = \frac{\kappa h}{R}\left(\frac{1 + \text{COR}}{1 + \kappa}\right)^2 \tag{4.10}$$

where KE_0 is initial kinetic energy and the amount is given by Equation (4.11):

$$\text{KE}_0 = \frac{1}{2} m v_0^2 \tag{4.11}$$

Equation (4.10) is in analogy with Equation (3.3c), which was introduced in Chapter 3 for predicting the energy ratio characterising energy transfer associated with the collision of two colliding lumped masses. Equation (3.3c) has been re-written as Equation (4.12):

$$\frac{KE_2}{KE_0} = \lambda\left(\frac{1 + COR}{1 + \lambda}\right)^2 \tag{4.12}$$

In the case of a flying impactor colliding with a free-standing rectangular object, the transferred kinetic energy, KE_2, is totally transformed into potential energy (PE) as the centre of gravity of the object becomes lifted by the overturning motion. Thus, Equation (4.10) can be re-written into Equation (4.13):

$$\frac{PE}{KE_0} = \frac{\kappa h}{R}\left(\frac{1 + COR}{1 + \kappa}\right)^2 \tag{4.13}$$

The value of PE is simply the product of the weight of the object (Mg) and the vertical displacement of the centre of gravity ($\Delta_{C.G.}$), as shown by Equation (4.14):

$$PE = Mg\Delta_{C.G.} \tag{4.14}$$

Substituting Equations (4.11) and (4.14) into Equation (4.13) and rearranging the terms gives Equation (4.15a), which provides predictions for the amount of uplift at the centre of gravity of the targeted object ($\Delta_{C.G.}$). The derivation of Equation (4.15a), as presented in the foregoing, is based on the assumption of a Type (1) scenario in which the impactor and barrier move in opposite directions following the collision. Equation (4.15b) is to be used instead for a Type (2) scenario, where the impactor is modelled to remain connected to the barrier following the collision. The Type (1) scenario featuring rebound of the impactor is expected where $\kappa \geq 1$, or else a Type (2) scenario is predicted.

$$\Delta_{C.G.} = \frac{mv_0^2}{2Mg}\frac{\kappa h}{R}\left(\frac{1 + COR}{1 + \kappa}\right)^2 \quad \text{·Scenario Type (1)} \tag{4.15a}$$

$$\Delta_{C.G.} = \frac{mv_0^2}{2Mg}\left(1 + \frac{\kappa h}{R}\right)\left(\frac{1}{1 + \kappa}\right)^2 \quad \text{Scenario Type (2)} \tag{4.15b}$$

Equations (4.15a) and (4.15b) were derived based on the assumption that the impactor collides at the top of the barrier. Should the collision occur at a different height, h,Equations (4.15a) and (4.15b) need to be replaced by the collision height measured from the base of the barrier.

The two scenario types as covered by Equations (4.15a) and (4.15b) are in analogy to the two types of collisions between two moving objects (in free space), as depicted in Figure 3.5 in Chapter 3. For Scenario Type (2) collisions where the impactor remains "connected" to the target, the term $\frac{\kappa h}{R}\left(\frac{1+COR}{1+\kappa}\right)^2$ in Equations (4.10) and (4.13) can be replaced by $\left(1+\frac{\kappa h}{R}\right)\left(\frac{1}{1+\kappa}\right)^2$, or simply $\frac{1}{1+\kappa}$ for a slender barrier of rectangular cross-section in which $h \approx R$. The derivation, once modified in the manner as described, results in Equation (4.15b), which can be used for Scenario Type (2) collisions.

The amount of rotation of the object (θ) can be found by the use of Equation (4.16a) to (4.16c) once the value of $\Delta_{C.G.}$ and the gross dimensions of the object are known (refer Figure 4.1). Parameter r is the distance between the centre of gravity of the object and point of pivot, as defined by Equations (4.16b), and β is the angle between diagonal and base of the object as defined by Equation (4.16c).

$$\theta = \sin^{-1}\left(\frac{\frac{h}{2} + \Delta_{C.G.}}{r}\right) - \beta \tag{4.16a}$$

$$r = 0.5\sqrt{h^2 + w^2} \tag{4.16b}$$

$$\beta = \tan^{-1}\left(\frac{h}{w}\right) \tag{4.16c}$$

Given the rotation of the object (θ), the horizontal displacement (Δ) at its upper edge is accordingly found using Equation (4.17):

$$\Delta = h \sin\theta \tag{4.17}$$

In assessing the stability of a free-standing installation (which is idealised as a rectangular object in the presented analyses), the value of $\Delta_{C.G.}$ calculated using Equation (4.15a) can be compared with the limiting (or critical) value for overturning $\Delta_{C.G.(crit)}$, as defined by Equation (4.18):

$$\Delta_{C.G.(crit)} = r - \frac{h}{2} \tag{4.18}$$

It is evident in Figure 4.2 that overturning becomes eminent when the centre of gravity of the object is displaced to a position that is directly above the point of pivot (refer Figure 4.2). The limiting (or critical) angle of rotation (θ_{crit}) can be found using Equation (4.19):

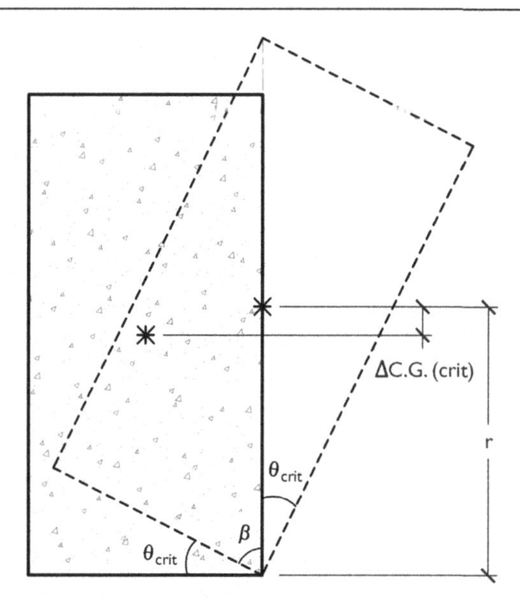

Figure 4.2 Critical angle of rotation.

$$\theta_{\text{crit}} = \frac{\pi}{2} - \beta \tag{4.19}$$

The factor of safety (FOS) against overturning is the ratio of $\Delta_{C.G.(crit)}$ as calculated using Equation (4.18) and $\Delta_{C.G.}$ as calculated using Equation (4.15a). A design chart showing values of FOS as a function of the gross dimensions of the rectangular cross section (h and w) is shown in Figure 4.3. Note that FOS = 1.0 represents critical overturning condition.

In summary, the analytical procedure for predicting overturning is in three steps:

1. Determine the energy ratio using Equation (4.15a), which enables the amount of uplift ($\Delta_{C.G.}$) at the centre of gravity of the targeted object to be found. The value of $\Delta_{C.G.(crit)}/w$ can then be calculated.
2. Determine the aspect ratio h/w.
3. Determine the FOS using the calculated values of $\Delta_{C.G.(crit)}/w$ and h/w from steps (1) and (2), respectively, alongside the use of Figure 4.3.

4.3 BARRIERS OF DIFFERENT TYPES OF CROSS SECTIONS

4.3.1 General

The analytical procedure for predicting overturning was illustrated in Section 4.2 for a free-standing object with a rectangular cross section.

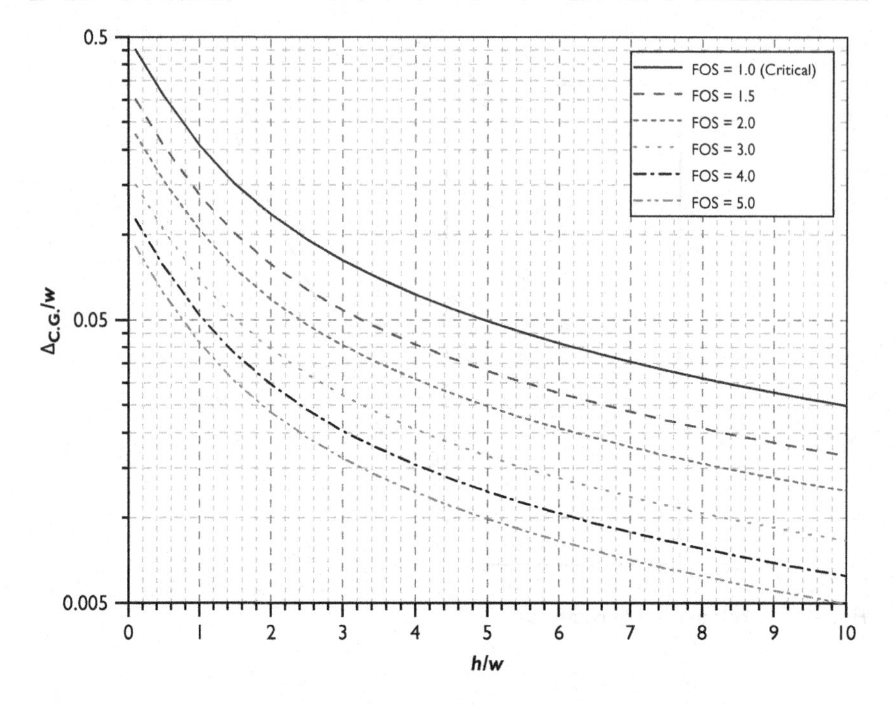

Figure 4.3 Factor of safety against overturning of rectangular block.

Much of the presented modelling approach is generic in nature and is generally applicable to other cross sections, albeit with some minor modifications. Equations (4.15a) or (4.15b) remain part of the procedure for estimating the value of $\Delta_{C.G.}$, and the definition of notation R remains as the distance between the point of contact and point of pivot at the base. However, notations h and w are redefined as $2\bar{x}$ and $2\bar{y}$, respectively, where $\bar{x}$ and $\bar{y}$ are the horizontal and vertical distances, respectively, measured from the centre of gravity of the barrier to the point of pivot at the base of the barrier (e.g. refer Figure 4.4 for an L-shaped barrier). The redefined expressions for finding the value of θ, r and β are listed in Equations (4.20a) to (4.20c). To achieve a safe design that safeguards against overturning, the value of $\Delta_{C.G.}$ as calculated using Equations (4.15a) or (4.15b) must not exceed the limiting value of $\Delta_{C.G.(crit)}$, which is calculated using Equation (4.21).

$$\theta = \sin^{-1}\left(\frac{\bar{y} + \Delta_{C.G.}}{r}\right) - \beta \qquad (4.20a)$$

$$r = \sqrt{\bar{x}^2 + \bar{y}^2} \qquad (4.20b)$$

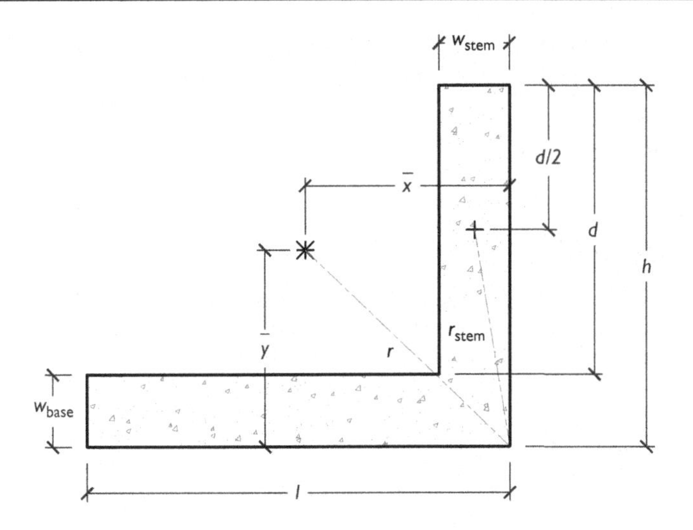

Figure 4.4 Barrier of L-shaped cross section.

$$\beta = \tan^{-1}\left(\frac{\bar{y}}{\bar{x}}\right) \tag{4.20c}$$

$$\Delta_{C.G.(crit)} = r - \bar{y} \tag{4.21}$$

In addition, the expression for determining the rotational inertia (I_θ) about the point of pivot, at the base of the barrier, would also need to be modified as per the geometry of the different cross section. A rundown of a list of expressions for calculating the value of I_θ (and their derivation) for a range of cross sections is shown in the following sub-sections.

4.3.2 Rectangular Cross Section

Derivation of Equation (4.2) for determining the value of I_θ for a rectangular cross section (Figure 4.1) is presented in this section.

First, the rotational inertia of a rectangular cross section about its centroid is defined by Equation (4.22):

$$I_c = M\left(\frac{h^2 + w^2}{12}\right) \tag{4.22}$$

By the parallel axis theorem, the rotational inertia of the rectangular section about the point of pivot at the base is solved using Equation (4.23):

$$I_\theta = I_c + Mr^2 \tag{4.23}$$

Substituting Equations (4.16b) and (4.22) into Equation (4.23) gives Equation (4.24):

$$I_\theta = M\left(\frac{h^2 + w^2}{3}\right) \tag{4.24}$$

4.3.3 L-shaped Cross Section

Consider an L-shaped cross section that has height h, length l, stem wall thickness w_{stem} and base slab thickness w_{base} (Figure 4.4).

The rotational inertia of the stem wall about its centroidal axis is defined by Equation (4.25):

$$I_{c\,(stem)} = M_{stem}\left(\frac{d^2 + w_{stem}^2}{12}\right) \tag{4.25}$$

where M_{stem} and d are the mass and height, respectively, of the stem wall.

By the parallel axis theorem, Equation (4.26) can be used for determining the rotational inertia of the stem wall about the point of pivot at the base. r_{stem} is defined by Equation (4.27), as indicated in Figure 4.4.

$$I_{\theta\,(stem)} = I_{c\,(stem)} + M_{stem}r_{stem}^2 \tag{4.26}$$

$$r_{stem}^2 = \left(\frac{d}{2} + w_{base}\right)^2 + \left(\frac{w_{stem}}{2}\right)^2 \tag{4.27}$$

Substituting Equations (4.25) and (4.27) into Equation (4.26) results in Equation (4.28), which can be used for determining the rotational inertia of the stem wall about the point of pivot at the base.

$$I_{\theta\,(stem)} = M_{stem}\left(\frac{d^2 + w_{stem}^2}{12}\right) + M_{stem}\left[\left(\frac{d^2}{4} + dw_{base} + w_{base}^2\right) + \left(\frac{w_{stem}}{2}\right)^2\right] \tag{4.28}$$

In practice, the value of w_{stem} and w_{base} are typically of a lower order than the value of d. Thus, the contributions of the higher-order terms can be neglected. Thus, Equation (4.28) can be rewritten as Equation (4.29) for obtaining a reasonably accurate estimate:

$$I_{\theta\,(stem)} \approx M_{stem}\left(\frac{d^2}{12}\right) + M_{stem}\left(\frac{d^2}{4} + dw_{base}\right) \tag{4.29}$$

Substituting $d = h - w_{base}$ from Figure 4.4 into Equation (4.29) gives Equation (4.30a). Expanding Equation (4.30a) and neglecting the higher-order terms gives Equation (4.30b), which has fewer terms than Equation (4.29):

$$I_{\theta(stem)} \approx \frac{M_{stem}}{3}(h - w_{base})(h + 2w_{base}) \tag{4.30a}$$

$$I_{\theta(stem)} \approx \frac{M_{stem}}{3}(h^2 + hw_{base}) \tag{4.30b}$$

By adopting the same algebraic procedure as presented in Section 4.3.2, an expression for determining the rotational inertia of the rectangular base slab about the point of pivot at the base is similarly derived as shown by Equation (4.31a). As w_{base} is much smaller than l in practice, its higher-order term is neglected and Equation (4.31a) is re-written as Equation (4.31b):

$$I_{\theta(base)} = M_{base}\left(\frac{w_{base}^2 + l^2}{3}\right) \tag{4.31a}$$

$$I_{\theta(base)} \approx \frac{M_{base}l^2}{3} \tag{4.31b}$$

where M_{base} is the mass of the base slab.

Taking the sum of Equations (4.30b) and (4.31b) gives an estimate for the total rotation inertia of the L-shaped cross section, as shown by Equation (4.32):

$$I_{\theta} \approx \frac{M_{stem}}{3}(h^2 + hw_{base}) + \frac{M_{base}l^2}{3} \tag{4.32}$$

The value of I_{θ} as obtained using Equation (4.32) is slightly lower than the actual value as the higher-order terms have been neglected, and the error is on the safe side.

4.3.4 L-Shaped Cross Section with Reversed Base Section

Consider an L-shaped cross section that is the same as the one presented in Section 4.3.3, but with reversed base section, as shown in Figure 4.5.

The rotational inertia of the stem wall about its centroidal axis remains the same as Equation (4.25).

Equation (4.26) can again be used for determining the rotational inertia of the stem wall about the point of pivot at the base by parallel axis theorem, but r_{stem} needs to be redefined by Equation (4.33) as indicated in Figure 4.5.

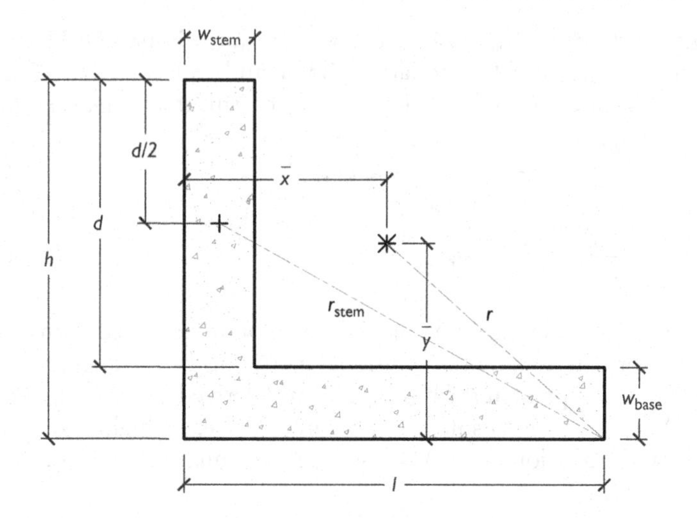

Figure 4.5 Barrier of L-shaped cross section with reversed base section.

$$r_{\text{stem}}^2 = \left(\frac{d}{2} + w_{\text{base}}\right)^2 + \left(l - \frac{w_{\text{stem}}}{2}\right)^2 \tag{4.33}$$

Substituting Equations (4.25) and (4.33) into Equation (4.26) results in Equation (4.34), which can be used for determining the rotational inertia of the stem wall about the point of pivot at the base.

$$I_{\theta(\text{stem})} = M_{\text{stem}}\left(\frac{d^2 + w_{\text{stem}}^2}{12}\right) + M_{\text{stem}}\left[\left(\frac{d^2}{4} + dw_{\text{base}} + w_{\text{base}}^2\right) + \left(l^2 - lw_{\text{stem}} + \frac{w_{\text{stem}}^2}{4}\right)\right] \tag{4.34}$$

By neglecting the contributions of the higher-order terms of w_{stem} and w_{base}, Equation (4.34) can be rewritten as Equation (4.35) for obtaining a reasonably accurate estimate:

$$I_{\theta(\text{stem})} \approx M_{\text{stem}}\left(\frac{d^2}{12}\right) + M_{\text{stem}}\left(\frac{d^2}{4} + dw_{\text{base}} + l^2 - lw_{\text{stem}}\right) \tag{4.35}$$

Substituting $d = h - w_{\text{base}}$ from Figure 4.5 into Equation (4.35) gives Equation (4.36a). Organising Equation (4.36a) and neglecting the higher-order terms gives Equation (4.36b), which has fewer terms than Equation (4.35):

$$I_{\theta(\text{stem})} \approx \frac{M_{\text{stem}}}{3}\left(h^2 - 2hw_{\text{base}} + w_{\text{base}}^2 + 3hw_{\text{base}} - 3w_{\text{base}}^2 + 3l^2 - 3lw_{\text{stem}}\right) \tag{4.36a}$$

$$I_{\theta(\text{stem})} \approx \frac{M_{\text{stem}}}{3}\left(h^2 + hw_{\text{base}} + 3l^2 - 3lw_{\text{stem}}\right) \tag{4.36b}$$

Rotational inertia of the rectangular base slab remains the same as that in Section 4.3.3, as defined by Equation (4.31b).

Taking the sum of Equations (4.31b) and (4.36b) gives an estimate for the total rotation inertia of the L-shaped cross section with reversed base section, as shown by Equation (4.37):

$$I_{\theta} \approx \frac{M_{\text{stem}}}{3}\left(h^2 + hw_{\text{base}} + 3l^2 - 3lw_{\text{stem}}\right) + \frac{M_{\text{base}} l^2}{3} \tag{4.37}$$

As for Equation (4.32), the value of I_{θ} as obtained using Equation (4.37) is slightly lower than the actual value because of the neglect of the higher order terms.

4.3.5 L-Shaped Cross Section with Rectangular Side Walls

Consider an L-shaped cross section of height h, length l, stem wall thickness w_{stem}, base slab thickness w_{base} and with n_s number of side walls, as shown in Figure 4.6.

The rotational inertia of each rectangular side wall as shown in Figure 4.6 about its centroidal axis can be found by the use of Equation (4.38):

$$I_{c(\text{side})} = M_{\text{side}}\left(\frac{c^2 + d^2}{12}\right) \tag{4.38}$$

where M_{side} is the mass of one side wall; $c = l - w_{\text{stem}}$; and $d = h - w_{\text{base}}$.

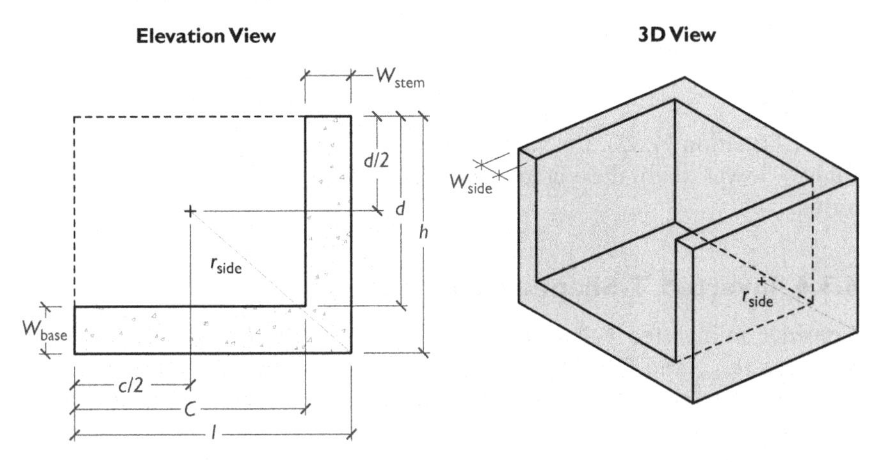

Figure 4.6 Barrier of L-shaped cross section with rectangular side walls.

By the parallel axis theorem, the rotational inertia of the same side wall about the point of pivot at the base is as defined by Equation (4.39). r_{side} is defined by Equation (4.40) as indicated in Figure 4.6.

$$I_{\theta(side)} = I_{c(side)} + M_{side} r_{side}^2 \tag{4.39}$$

$$r_{side}^2 = \left(\frac{c}{2} + w_{stem}\right)^2 + \left(\frac{d}{2} + w_{base}\right)^2 \tag{4.40}$$

where the meaning of r_{side} is as indicated in Figure 4.6.

Substituting Equations (4.38) and (4.40) into Equation (4.39) results in Equation (4.41a) for determining the contribution of one side wall to the rotational inertia. Equation (4.41b) gives the contribution of n_s number of side walls:

$$I_{\theta(side)} = M_{side}\left[\left(\frac{c^2 + d^2}{12}\right) + \left(\frac{c}{2} + w_{stem}\right)^2 + \left(\frac{d}{2} + w_{base}\right)^2\right] \tag{4.41a}$$

$$I_{\theta(side)} = n_s M_{side}\left[\left(\frac{c^2 + d^2}{12}\right) + \left(\frac{c}{2} + w_{stem}\right)^2 + \left(\frac{d}{2} + w_{base}\right)^2\right] \tag{4.41b}$$

The total rotational inertia of the barrier assemblage about the point of pivot at the base is obtained by taking the sum of Equations (4.32) and (4.41b), as shown by Equation (4.42):

$$I_\theta \approx \frac{M_{stem}}{3}\left(h^2 + h w_{base}\right) + \frac{M_{base} l^2}{3} + n_s M_{side}\left[\left(\frac{c^2 + d^2}{12}\right) + \left(\frac{c}{2} + w_{stem}\right)^2 + \left(\frac{d}{2} + w_{base}\right)^2\right] \tag{4.42}$$

As for Equation (4.32), the value of I_θ as obtained using Equation (4.42) is slightly lower than the actual value due to neglecting the higher-order terms.

4.3.6 Inverted T-Shaped Cross Section

Consider an inverted T-shaped cross section of height h, length l, stem wall thickness w_{stem} and base slab thickness w_{base}, as shown in Figure 4.7.

The rotational inertia of the stem wall about its centroidal axis is defined by Equation (4.43):

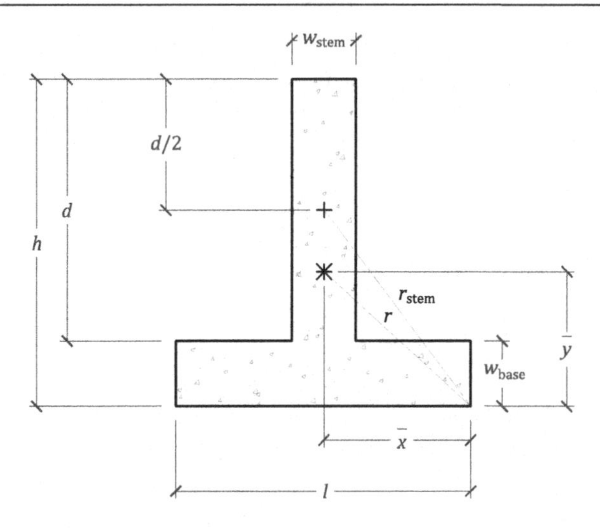

Figure 4.7 Barrier of inverted T-shaped cross section.

$$I_{c\,(\text{stem})} = M_{\text{stem}}\left(\frac{d^2 + w_{\text{stem}}^2}{12}\right) \tag{4.43}$$

By parallel axis theorem, Equation (4.44) can be used for determining the rotational inertia of the stem wall about the point of pivot at the base. r_{stem} is defined by Equation (4.45), as indicated in Figure 4.7.

$$I_{\theta\,(\text{stem})} = I_{c\,(\text{stem})} + M_{\text{stem}} r_{\text{stem}}^2 \tag{4.44}$$

$$r_{\text{stem}}^2 = \left(\frac{d}{2} + w_{\text{base}}\right)^2 + \left(\frac{l}{2}\right)^2 \tag{4.45}$$

Substituting Equations (4.43) and (4.45) into Equation (4.44) results in Equation (4.46), which can be used for determining the rotational inertia of the stem wall about the point of pivot at the base.

$$I_{\theta\,(\text{stem})} = M_{\text{stem}}\left(\frac{d^2 + w_{\text{stem}}^2}{12}\right) + M_{\text{stem}}\left[\left(\frac{d^2}{4} + d w_{\text{base}} + w_{\text{base}}^2\right) + \left(\frac{l}{2}\right)^2\right] \tag{4.46}$$

By neglecting the contributions of the higher-order terms of w_{stem} and w_{base}, Equation (4.46) can be rewritten as Equation (4.47) for obtaining a reasonably accurate estimate:

$$I_{\theta\,(\text{stem})} \approx M_{\text{stem}}\left(\frac{d^2}{12}\right) + M_{\text{stem}}\left(\frac{d^2}{4} + d w_{\text{base}} + \frac{l^2}{4}\right) \tag{4.47}$$

Substituting $d = h - w_{\text{base}}$ from Figure 4.7 into Equation (4.47) gives Equation (4.48a). Organising Equation (4.48a) and neglecting the higher-order terms gives Equation (4.48b), which has fewer terms than Equation (4.47):

$$I_{\theta(\text{stem})} \approx \frac{M_{\text{stem}}}{3}\left(h^2 - 2hw_{\text{base}} + w_{\text{base}}^2 + 3hw_{\text{base}} - 3w_{\text{base}}^2 + \frac{3}{4}l^2\right) \tag{4.48a}$$

$$I_{\theta(\text{stem})} \approx \frac{M_{\text{stem}}}{3}\left(h^2 + hw_{\text{base}} + \frac{3}{4}l^2\right) \tag{4.48b}$$

Rotational inertia of the rectangular base slab can be derived in a similar manner as that in Section 4.3.3, as shown by Equation (4.49):

$$I_{\theta(\text{base})} \approx \frac{M_{\text{base}}l^2}{3} \tag{4.49}$$

Taking the sum of Equations (4.48b) and (4.49) gives an estimate for the total rotation inertia of the inverted T-shaped cross section, as shown by Equation (4.50):

$$I_{\theta} \approx \frac{M_{\text{stem}}}{3}\left(h^2 + hw_{\text{base}} + \frac{3}{4}l^2\right) + \frac{M_{\text{base}}l^2}{3} \tag{4.50}$$

As for Equation (4.32), the value of I_θ as obtained using Equation (4.50) is slightly lower than the actual value because of the neglect of the higher-order terms.

4.3.7 Inverted T-Shaped Cross Section with Side Walls

Consider an inverted T-shaped cross section of height h, length l, stem wall thickness w_{stem}, base slab thickness w_{base} and with n_s number of pairs of side walls, as shown in Figure 4.8.

The rotational inertia of each triangular side wall as shown in Figure 4.8 about its centroidal axis can be found by the use of Equation (4.51):

$$I_{c(\text{side})} = M_{\text{side}}\left(\frac{c^2 + d^2}{36}\right) \tag{4.51}$$

where M_{side} is the mass of one side wall; $c = (l - w_{\text{stem}})/2$; and $d = h - w_{\text{base}}$.

By the parallel axis theorem, the rotational inertia of the same side wall about the point of pivot at the base is as defined by Equation (4.52). r_{side1}

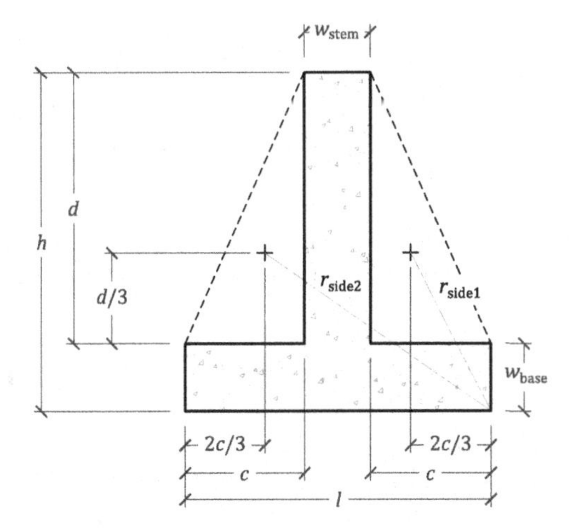

Figure 4.8 Barrier of inverted T-shaped cross section with rectangular side walls.

and r_{side2} are defined by Equations (4.53a) and (4.53b), respectively, as indicated in Figure 4.8.

$$I_{\theta(side)} = 2I_{c(side)} + M_{side}\left(r_{side1}^2 + r_{side2}^2\right) \tag{4.52}$$

$$r_{side1}^2 = \left(\frac{2c}{3}\right)^2 + \left(\frac{d}{3} + w_{base}\right)^2 \tag{4.53a}$$

$$r_{side2}^2 = \left(l - \frac{2c}{3}\right)^2 + \left(\frac{d}{3} + w_{base}\right)^2 \tag{4.53b}$$

Substituting Equations (4.51), (4.53a) and (4.53b) into Equation (4.52) results in Equation (4.54a) for determining the contribution of one side wall to the rotational inertia. Equation (4.54b) gives the contribution of n_s, number pairs of side walls:

$$I_{\theta(side)} = M_{side}\left[\frac{c^2}{2} + \frac{d^2}{18} + \left(l - \frac{2c}{3}\right)^2 + 2\left(\frac{d}{3} + w_{base}\right)^2\right] \tag{4.54a}$$

$$I_{\theta(side)} = n_s M_{side}\left[\frac{c^2}{2} + \frac{d^2}{18} + \left(l - \frac{2c}{3}\right)^2 + 2\left(\frac{d}{3} + w_{base}\right)^2\right] \tag{4.54b}$$

The total rotational inertia of the barrier assemblage about the point of pivot at the base is obtained by taking the sum of Equations (4.50) and (4.54b), as shown by Equation (4.55):

$$I_\theta \approx \frac{M_{\text{stem}}}{3}\left(h^2 + hw_{\text{base}} + \frac{3}{4}l^2\right) + \frac{M_{\text{base}}l^2}{3}$$
$$+ n_s M_{\text{side}}\left[\frac{c^2}{2} + \frac{d^2}{18} + \left(l - \frac{2c}{3}\right)^2 + 2\left(\frac{d}{3} + w_{\text{base}}\right)^2\right] \tag{4.55}$$

As for Equation (4.50), the value of I_θ as obtained using Equation (4.55) is slightly lower than the actual value due to neglecting the higher-order terms.

4.4 SLIDING GENERATED BY SINGLE COLLISION

The kinetic energy ratio was introduced in Chapter 3 for characterising energy transfer to the target in a collision scenario. The transferred energy was then equated to the strain energy of absorption in the structure for predicting the displacement demand generated by the collision. This calculation technique has been extended for predicting the overturning of a free-standing element in the early part of this chapter. The prediction involved equating the transferred energy to the gain in potential energy of a free-standing object experiencing rotation and uplift at its centre of mass. In this section, the methodology is extended further for predicting the sliding of a free-standing element on the soil surface.

To predict the amount of sliding displacement, the transferred energy, KE_2, is equated to the energy of dissipation (W_f) on the slab-soil sliding interface, as shown by Equation (4.56):

$$KE_2 = W_f \tag{4.56}$$

W_f is a product of the frictional force (F_f) and the amount of sliding displacement (Δ). The frictional force is in turn estimated as the product of the normal force applied to the sliding surface (i.e. self-weight of the barrier λmg) and the coefficient of friction (μ). The amount of energy dissipation is therefore given by Equation (4.57):

$$W_f = F_f \Delta = \mu \lambda m g \Delta \tag{4.57}$$

Refer to the annotation in Figure 4.9 of the free-standing object showing the expression for estimating the frictional resistance (F_f).

Equation (4.12), which gives an estimate of the energy ratio for characterising the transfer of energy in a collision scenario, is equally applicable for predicting sliding. The kinetic energy delivered by the impact (KE_0) as estimated using Equation (4.11) may be substituted into Equation (4.12) to give the energy transferred to the target, which is given by Equation (4.58):

$$KE_2 = \lambda\left(\frac{1 + COR}{1 + \lambda}\right)^2\left(\frac{1}{2}mv_0^2\right) \tag{4.58}$$

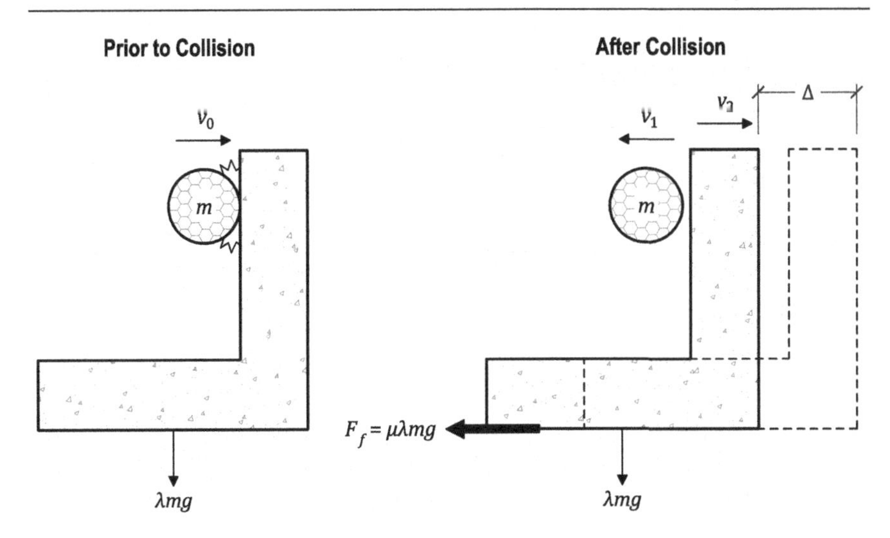

Figure 4.9 Sliding movement generated by impact action.

Substituting Equations (4.57) and (4.58) into Equation (4.56) and rearranging the terms results in Equation (4.59a), which can be used for predicting the amount of sliding displacement in a projected Type (1) collision scenario where the impactor and barrier moves in opposite directions following the collision. Equation (4.59b) is to be used instead for a Type (2) scenario where the impactor is modelled to remain connected to the barrier following the collision. The Type (1) scenario featuring rebound of the impactor is expected where $\lambda \geq 1$, or else a Type (2) scenario is predicted.

$$\Delta = \frac{v_0^2}{2\mu g}\left(\frac{1 + \mathrm{COR}}{1 + \lambda}\right)^2 \quad \text{Scenario Type (1)} \tag{4.59a}$$

$$\Delta = \frac{v_0^2}{2\mu g}\left(\frac{1}{1 + \lambda}\right)^2 \quad \text{Scenario Type(2)} \tag{4.59b}$$

Similar to before, the two types of scenarios are well illustrated in Figure 3.5 in Chapter 3. Equation (4.59a) can be modified to apply to Scenario Type (2) collisions where the impactor "connects" to the target by changing the right-hand portion of the equation. This results in Equation (4.59b) can be used for Scenario Type (2) collisions.

The value of μ representing conditions at the slab-soil interface is best derived from results recorded from dynamic testing in the field. In the absence of such representative information, the value of μ (which is equal to tan (δ_s) where δ_s is the angle of skin friction) can be inferred from the effective angle of shear resistance (ϕ') of the soil as obtained from triaxial testing of representative soil samples. The relationship between μ and ϕ' is

Table 4.1 Values of c_s from GEO [4.8]

Barrier Material	*Note*	c_s
Smooth concrete	Cast with steel formwork	0.8–0.9
Rough concrete	Cast with timber formwork	0.9–1.0
Smooth masonry	Dressed granitic or volcanic blocks	0.5–0.7
Rough masonry	Irregular granitic or volcanic blocks	0.9–1.0
Smooth steel	Polished	0.5–0.6
Rough steel	–	0.7–0.8
Geotexile	–	0.5–0.9

shown by Equation (4.60), where factor c_s may be taken from existing design guidelines. For example, values of c_s between a barrier and granular soil as suggested in GEO [4.8] are listed in Table 4.1.

$$\mu = \tan(c_s\phi') \tag{4.60}$$

In situations where the soils are saturated with water, resulting in significant pore water pressure (u), Equation (4.59a) would need to be modified into Equation (4.61a) to take into account reduction in the normal force at the sliding interface compromising its frictional resistance. Refer to the annotation on the diagram of the free-standing barrier (of Figure 4.10) showing the expression for estimating the frictional resistance (f), taking into account the reduction in resistance as described. Similarly, Equation (4.59b) is also modified into Equation (4.61b):

$$\Delta = \frac{mv_0^2}{2\mu(\lambda mg - uA)} \times \lambda\left(\frac{1 + \text{COR}}{1 + \lambda}\right)^2 \quad \text{Scenario Type (1)} \tag{4.61a}$$

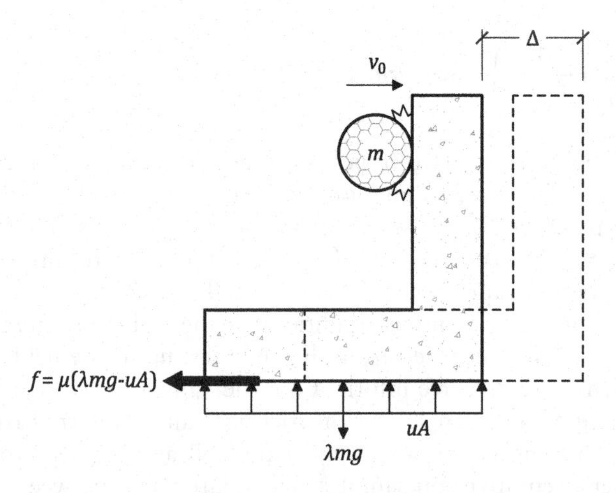

Figure 4.10 Effect of pore water pressure on frictional resistance.

$$\Delta = \frac{mv_0^2}{2\mu((1+\lambda)mg - uA)} \times \left(\frac{1}{1+\lambda}\right) \quad \text{Scenario Type (2)} \qquad (4.61b)$$

where A is the surface area of the underside of the barrier.

4.5 SLIDING GENERATED BY MULTIPLE COLLISIONS

Expressions presented in Section 4.4 are for predicting the amount of sliding caused by a one-off (single) strike on a free-standing barrier. In a rockfall collision scenario, a barrier is likely to be subject to multiple strikes (of Type (1) scenarios). Following each strike by a fallen boulder, the barrier becomes more resistant to further sliding because of two reasons:

1. increase in the mass of the barrier caused by the fallen boulders left on the upper surface of the base slab; and
2. increase in frictional resistance caused by the increase in the normal force applied to the sliding interface. This compounding effect is illustrated in Figures 4.11(a) to 4.11(d) for the first, second, third and the *i*-th impact.

Considering the first occurrence of impact, as shown in Figure 4.11, Equation (4.59a) may be used for estimating the sliding displacement of the barrier. For the second occurrence, assuming that the masses of both impactors are the same, the first impactor will have deposited on the barrier, adding to the total target mass and hence its inertial resistance. The target mass is increased to $m + \lambda m$, and the mass ratio, which is the target mass divided by the impactor mass, is equal to $1 + \lambda$. In a similar manner, the mass ratio for the third occurrence is equal to $2 + \lambda$. Thus, the mass ratio for the *i*-th occurrence may be expressed in the general form of $i - 1 + \lambda$. Replacing the λ term in Equation (4.59a) by the new term $i - 1 + \lambda$ gives Equation (4.62), which can be used for estimating displacement caused by the *i*-th occurrence of impact in isolation. The cumulative displacement of a barrier when subjected to N_{impact} numbers of impact can be estimated using Equation (4.63):

$$\Delta_i = \frac{v_o^2}{2\mu g}\left(\frac{1+COR}{i+\lambda}\right)^2 \qquad (4.62)$$

$$\Delta_c = \frac{v_o^2(1+COR)^2}{2\mu g}\sum_{i=1}^{N_{impact}}\left(\frac{1}{i+\lambda}\right)^2 \qquad (4.63)$$

The decrease in the amount of incremental sliding displacement to the cumulative total sliding displacement with an increasing value of N_{impact} means that the culminated displacement would start to converge as the

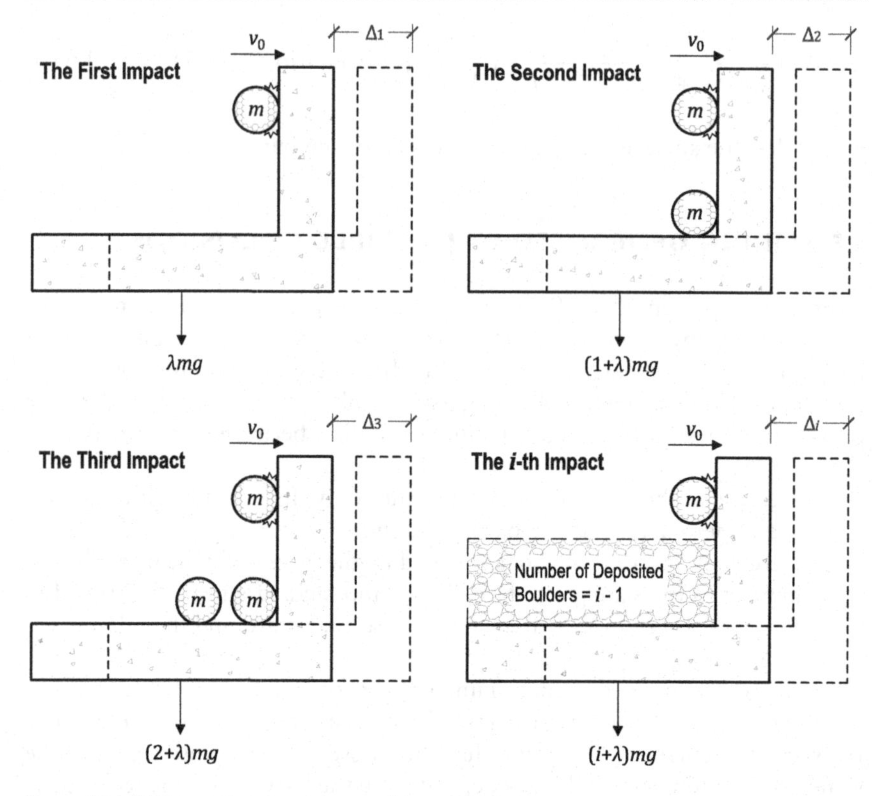

Figure 4.11 Sliding of barrier generated by multiple impact.

value of N_{impact} becomes very large. As the value of N_{impact} tends to infinity, predictions by Equation (4.63) converge. By curve fitting these converged results to a wide range of values for λ (i.e. from 1 to 1,000), it was found that Equation (4.63) can be approximated by Equation (4.64). For a λ value greater than 10, which is typically the case for rockfall barriers, Equation (4.64) can be approximated to Equation (4.65):

$$\Delta_c = \frac{v_o^2(1 + \text{COR})^2}{2\mu g}\left(\frac{1}{0.56 + \lambda}\right) \tag{4.64}$$

$$\Delta_c = \frac{v_o^2(1 + \text{COR})^2}{2\lambda\mu g} \tag{4.65}$$

Equations (4.63) – (4.65) are based on the assumption that the barrier is initially stationary whenever struck by a fallen boulder, which implies that at the time of strike all sliding motions caused by previous boulder strikes on the same barrier have been completed. In cases where the fallen boulders reach the barrier in a swarm, the use of the expressions presented in this sub-section might not be valid, and might understate the total amount of sliding movement. A more conservative prediction based on substituting the total mass of all the striking boulders as m into equations (4.61a) or (4.61b) is warranted. Factor λ in the equations shall be calculated as ratio of the mass of the barrier to the total mass of all the striking boulders.

4.6 WORKED EXAMPLES

4.6.1 Overturning Assessment of a Rectangular Barrier

Consider a free-standing RC barrier of rectangular cross section with the following dimensions: w = 1 m; h = 5 m; and b = 8 m. The RC barrier is presented in Figure 4.12. A boulder with a mass of 2,000 kg moving at a velocity of 10 m/s collides on the barrier at its upper edge. A COR value of 0.2 is assumed. Assess if the barrier is safe from overturning. If so, what is the FOS against overturning?

Solution

$$M = 2400(1)(5)(8) = 96000 \text{ kg}$$

Employing Equation (4.2):

$$I_\theta = M\left(\frac{h^2 + w^2}{3}\right) = 96000\left(\frac{5^2 + 1^2}{3}\right) = 832000 \text{ kgm}^2$$

$$R = \sqrt{h^2 + w^2} = \sqrt{5^2 + 1^2} = 5.1 \text{ m}$$

Using Equation (4.6):

$$\kappa = \frac{I_\theta}{mhR} = \frac{832000}{2000(5)(5.1)} = 16.3$$

The calculated parameters are then substituted into Equation (4.15a):

$$\Delta_{\text{C.G.}} = \frac{mv_0^2}{2Mg}\frac{\kappa h}{R}\left(\frac{1 + \text{COR}}{1 + \kappa}\right)^2 = \frac{2000(10)^2}{2(96000)(9.81)}\left(\frac{16.3(5)}{5.1}\right)\left(\frac{1 + 0.2}{1 + 16.3}\right)^2 = 0.008 \text{ m}$$

The critical value of $\Delta_{\text{C.G.}}$ to cause overturning can be found using Equation (4.18):

$$\Delta_{\text{C.G.(crit)}} = r - \frac{h}{2} = 0.5\sqrt{5^2 + 1^2} - \frac{5}{2} = 0.05 \text{ m}$$

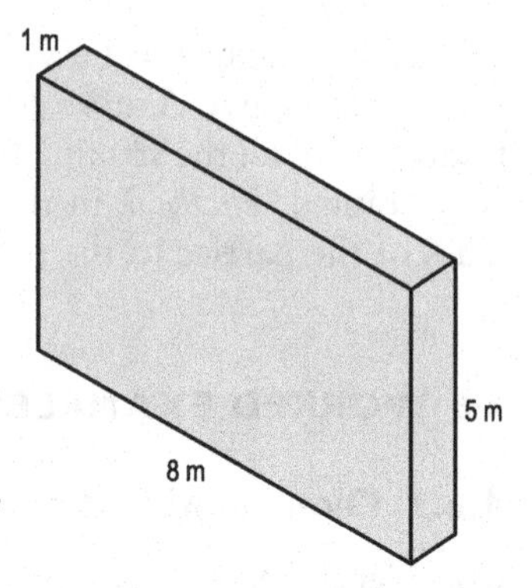

Figure 4.12 Worked example featuring barrier of rectangular cross section.

Thus, the FOS against overturning is equal to 0.05/0.008 = 6.25.

4.6.2 Overturning Assessment of an L-Shaped Barrier with Side Walls

Consider a free-standing barrier assemblage, as shown in Figure 4.13. The barrier has dimensions: h = 5 m; l = 5 m; b = 8 m; w_{stem} = 1 m; w_{base} = 1 m; and w_{side} = 0.5 m. A boulder with a mass of 5,000 kg moving at a velocity of 20 m/s strikes the barrier at its upper edge. A COR value of 0.2 is assumed. Assess if the barrier is safe from overturning. If so, what is the FOS against overturning?

Solution

$$M = M_{stem} + M_{base} + n_s M_{side} = 2400 \times (4 \times 8 \times 1 + 5 \times 8 \times 1 + 2 \times 4 \times 4 \times 0.5) = 76800 + 96000 + 38400 = 211200 \text{ kg}$$

Employing Equation (4.42):

$$I_\theta = \frac{M_{stem}}{3}(h^2 + hw_{base}) + \frac{M_{base}l^2}{3} + n_s M_{side}\left[\left(\frac{c^2 + d^2}{12}\right) + \left(\frac{c}{2} + w_{stem}\right)^2 + \left(\frac{d}{2} + w_{base}\right)^2\right]$$

$$= \frac{76800}{3}(5^2 + 5 \times 1) + \frac{96000}{3}8^2 + 2 \times 38400\left[\left(\frac{4^2 + 4^2}{12}\right) + \left(\frac{4}{2} + 1\right)^2 + \left(\frac{4}{2} + 1\right)^2\right]$$

$$= 768000 + 2048000 + 1586688 = 4402688 \text{ kgm}^2$$

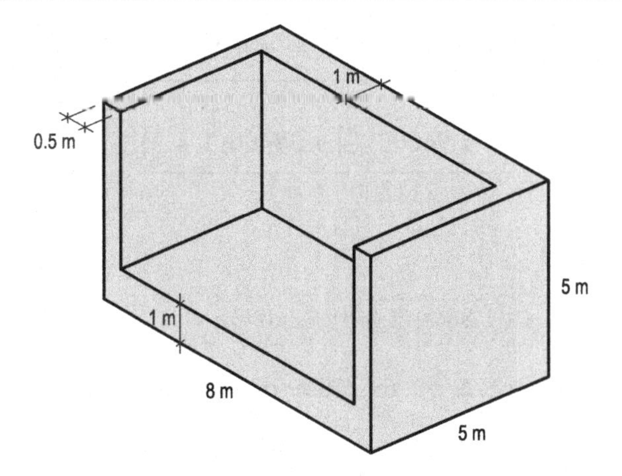

Figure 4.13 Worked example featuring barrier of L-shaped cross section with rectangular side walls.

$$R = \sqrt{h^2 + w_{\text{stem}}^2} = \sqrt{5^2 + 1^2} = 5.1 \text{ m}$$

Using Equation (4.6):

$$\kappa = \frac{I_\theta}{mhR} = \frac{4402688}{5000(5)(5.1)} = 34.53$$

The calculated parameters are then substituted into Equation (4.15a):

$$\Delta_{\text{C.G.}} = \frac{mv_0^2}{2Mg}\frac{\kappa h}{R}\left(\frac{1 + \text{COR}}{1 + \kappa}\right)^2$$

$$= \frac{5000(20)^2}{2(211200)(9.81)}\left(\frac{34.53 \times 5}{5.1}\right)\left(\frac{1 + 0.2}{1 + 34.53}\right)^2 = 0.02 \text{ m}$$

Centre of gravity is determined as follows:

$$c = l - w_{\text{stem}} = 5 - 1 = 4 \text{ m}$$

$$d = h - w_{\text{base}} = 5 - 1 = 4 \text{ m}$$

$$\bar{x} = \frac{M_{\text{stem}}\left(\frac{w_{\text{stem}}}{2}\right) + M_{\text{base}}\left(\frac{l}{2}\right) + n_s M_{\text{side}}\left(w_{\text{stem}} + \frac{c}{2}\right)}{M}$$

$$= \frac{76800\left(\frac{1}{2}\right) + 96000\left(\frac{5}{2}\right) + 38400\left(1 + \frac{4}{2}\right)}{211200} = 1.86 \text{ m}$$

$$\bar{y} = \frac{M_{\text{stem}}\left(w_{\text{base}} + \frac{d}{2}\right) + M_{\text{base}}\left(\frac{w_{\text{base}}}{2}\right) + n_s M_{\text{side}}\left(w_{\text{base}} + \frac{d}{2}\right)}{M}$$

$$= \frac{76800\left(1 + \frac{4}{2}\right) + 96000\left(\frac{1}{2}\right) + 38400\left(1 + \frac{4}{2}\right)}{211200} = 1.86 \text{ m}$$

From Equation (4.20b):

$$r = \sqrt{\bar{x}^2 + \bar{y}^2} = \sqrt{1.86^2 + 1.86^2} = 2.63 \text{ m}$$

The critical value of $\Delta_{\text{C.G.}}$ to cause overturning can be found using Equation (4.21):

$$\Delta_{\text{C.G.(crit)}} = r - \bar{y} = 2.63 - 1.86 = 0.77 \text{ m}$$

Thus, the FOS against overturning is equal to 0.77/0.02 = 38.5.

4.6.3 Prediction of Sliding Movement Generated by Boulder Collision during a Rockfall

A free-standing concrete barrier weighing 100 tonnes is struck by a fallen boulder weighing 2 tonnes moving at a velocity of 10 m/s. Estimate the amount of sliding movement of the barrier from the strike based on the following assumptions:

- Barrier only slides with no overturning motion
- Coefficient of Restitution (COR) = 0.2
- Effective Angle of Shear Resistance $\phi' = 30°$
- The barrier was cast with steel formwork
- Zero pore water pressure
- Dimensions of the barrier are not shown as they are not relevant to the calculation

Solution

$$\lambda = \frac{100}{2} = 50$$

Employing Equation (4.60) and reading the value of c_s from Table 4.1:

$$\mu = \tan(c_s\phi') = \tan(0.8(30)) = 0.45$$

Using Equation (4.59a):

$$\Delta = \frac{v_0^2}{2\mu g}\left(\frac{1 + COR}{1 + \lambda}\right)^2 = \frac{10^2}{2 \times 0.45 \times 9.81}\left(\frac{1 + 0.2}{1 + 50}\right)^2 = 0.006 \text{ m or } 6 \text{ mm}$$

4.6.4 Prediction of Sliding Movement Generated by Multiple Boulder Collisions during a Rockfall

The same barrier from Section 4.6.3 is subject to repetitive strikes occurring four times. Estimate the cumulative sliding movement, assuming that the fallen boulder rests on the upper surface of the base slab following every strike.

Solution
Calculations employing Equation (4.63) are shown:

$$\Delta_c = \frac{v_o^2(1 + COR)^2}{2\mu g}\sum_{i=1}^{N_{impact}}\left(\frac{1}{i + \lambda}\right)^2 = \frac{10^2(1 + 0.2)^2}{2(0.45)(9.81)}\sum_{i=1}^{4}\left(\frac{1}{i + 50}\right)^2 = 16.3 \times 0.00145 = 0.024 \text{ m or } 24 \text{ mm}$$

4.6.5 Prediction of Sliding Movement Generated by Infinite Number of Boulder Collisions during a Rockfall

In a massive rockfall affecting the same barrier from Section 4.6.3 it would be difficult to estimate the number of occurrences of impact the barrier would be subjected to. It is required to estimate the maximum possible cumulative sliding displacement using the same impact parameters as for Section 4.6.3.

Solution
As $\lambda = 50 \gg 10$, Equation (4.65) may be employed as shown:

$$\Delta_c = \frac{v_o^2(1 + COR)^2}{2\lambda\mu g} = \frac{16.3}{50} = 0.326 \text{ m or } 32.6 \text{ mm}$$

4.7 CLOSING REMARKS

This chapter is concerned with the prediction of overturning and sliding a free-standing installation (a barrier, for example) that can be of rectangular or L-shaped cross sections, including L-shaped barriers with sidewalls. Expressions based on applying the equal momentum and equal energy principles are introduced for determining the amount of uplift at the centre of gravity of the barrier for a projected collision scenario.

Given that the critical value of the uplift to cause overturning is known, the factor of safety against overturning can be calculated accordingly. The amount of overturning rotation, and horizontal displacement, at the upper edge of the barrier can also be predicted. Predictive expressions have also been introduced for determining the amount of sliding displacement of a free-standing barrier (or the like) caused by the strike of a fallen object. Further expressions for calculating the amount of cumulative sliding displacement following multiple occurrences of collisions have also been introduced. It has been shown that the culminated sliding displacement would converge as the number of collision occurrences becomes very large (since the prior collisions add additional dead load to the footing of the barrier). An expression that predicts the convergence can be used to estimate the maximum possible culminative sliding displacement. This expression can be used for assessing the performance of a barrier in a massive rockfall scenario.

REFERENCES

4.1 Kwan, J.S.H., *Supplementary technical guidance on design of rigid debris-resisting barriers (GEO Report No. 270)*. 2012: Geotechnical Engineering Office, the Government of the Hong Kong Special Administrative Region.

4.2 Standards Australia, 2017, *AS 5100.2 Bridge Design Part 2: Design Loads*, Standard Australia Limited, New South Wales, Australia.

4.3 American Association of State Highway and Transportation Officials, 2012, *AASHTO LRFD Bridge Design Specifications, 6th edition*, American Association of State Highway and Transportation Officials, Washington DC, U.S.

4.4 Japan Road Association, 2000, "Manual for anti-impact structures against falling rock (in Japanese)". Japan Road Association, Tokyo, Japan.

4.5 ASTRA, 2008, "Einwirkungen infolge Steinschlags auf Schutzgalerien (in German)", *Richtlinie, Bundesamt für Strassen, Baudirektion SBB, Eidgenössische Drucksachen-und Materialzentrale*, Bern, Switzerland.

4.6 Austroads, 2013, *Standardised Bridge Barrier Design*, Austroads Ltd, Sydney.

4.7 Kim, K.-M., Briaud, J.-L., Bligh, R., and Abu-Odeh, A., 2012, "Design guidelines and full-scale verification for MSE walls with traffic barriers", *Journal of Geotechnical and Geoenvironmental Engineering*, Vol. 138(6), pp. 690–699, doi:10.1061/(ASCE)GT.1943-5606.0000642.

4.8 GEO, *Guide to Retaining Wall Design (Geoguide 1) (Continuously Updated E-Version released on 29 August 2017)*. 2017: Geotechnical Engineering Office, Civil Engineering and Development Department, the Government of the Hong Kong Special Administrative Region. p. 245.

Chapter 5

Collision with a Fallen Object

5.1 INTRODUCTION

In a collision scenario where a fallen object lands on a structural element (a horizontal member forming part of a floor structure, for example) and then rebounds and bounces off to the side of the element, the collision is one-off and there is no second occurrence of contact. This is the case when the impactor lands on the target at an angle, and then rebounds from it with a high coefficient of restitution, causing it to land outside the affected part of the structure. Rebound usually occurs when the weight of the impactor is much less than the weight of the target element (i.e. $\lambda > 1$, which is a Scenario Type (1) collision, as presented in Chapter 3). It is shown in Section 5.2.2.1 that the deflection caused to the target in this situation is no different to that of a horizontal collision scenario of the same impactor and incident velocity of impact. Expressions presented in Chapter 3 may be used to estimate the outcome of a vertical collision of this nature.

In a more common collision scenario, the fallen object settles onto the target (e.g. a floor beam) following multiple occurrences of impact and rebound. The exact solution for such a collision scenario is complex. A simplified modelling approach for such a scenario is to consider the fallen object to be embedded into the target, adding to its self-weight after the initial contact/collision occurs. Thus, the participating mass of the affected structural element and the embedded impactor mass is considered as one combined mass. This condition can occur irrespective of the value of λ (i.e. the ratio of the impactor mass to the target mass). This modelling approach has been proven to give reasonable predictions of the deflection based on recordings from impact experiments conducted at different scales [5.1, 5.2]. In collision scenarios where the impactor becomes embedded within the target, or disintegrated on the surface of the target, without experiencing any rebound, the same modelling approach may be adopted. As the structural element deflects vertically, the loss of potential energy in the system (being the product of the combined mass and vertical deflection) would add to the kinetic energy delivered to the target, thereby aggravating the intensity of the collision. It is precisely this aggravating factor (associated with the loss of

DOI: 10.1201/9781003133032-5

potential energy as described) that distinguishes vertical collisions from horizontal collisions.

Collision testings of reinforced concrete (RC) beams and slabs by a dropped weight have been reported in the literature [5.3–5.11]. However, the cited references do not give design recommendations in the form of closed-form expressions for predicting the deflection of the targeted structural element. Closed-form expressions, which are presented in this chapter for predicting the outcome of a collision, are preferred by structural designers because of simplicity in the predictive process and the ability to track the trends and understand how different inputs/variables affect the outcome of the collision.

An analytical model based on considering the idealised collision scenario of a member of negligible self-weight is first introduced (Section 5.2.1). The model is then modified to deal with a structural element, which has a significant mass of its own, where the loss of potential energy that results from vertical displacement of the combined mass is taken into account (Section 5.2.2). The structural element is assumed to respond within the limit of linear elastic behaviour for both types of scenarios. In the latter part of the chapter, the analytical model is developed further to deal with the structural element surpassing the limit of yield (Section 5.3). Applications of the derived expressions are illustrated by worked examples featuring a timber beam, RC beam and steel beam (Section 5.4).

The analysis methodology presented in this chapter is similar to the methodology introduced in Chapter 3, where the deflection demand from the collision action is calculated by idealising the structural element into a spring connected lumped mass system, which is characterised by parameters m^* and k^* (which were first introduced in Chapter 2). The approach presented in Chapter 3 allows for a beam to be idealised for scenarios where the collision occurs at either mid-span or other locations along the span of the beam. Once the deflection is known, internal forces including bending moments and shear forces on a beam, for example, can be found by employing the displacement-based approach as introduced in Section 3.7.

5.2 IMPACT ON A MEMBER RESPONDING WITH ELASTIC BEHAVIOUR

5.2.1 Impact on a Member of Negligible Mass

Consider an experiment where a lumped mass is placed on a simply supported beam at mid-span. The lumped mass was initially suspended from a piece of wire which is severed in order that its weight is transferred to the beam in an abrupt manner/instantaneous (refer Figure 5.1). Importantly, the lumped mass was initially levelled with the upper surface of the beam in order that the drop height was equal to zero. Thus, no collision actually occurs (i.e. the beam was essentially loaded with the lumped mass in an abrupt/instantaneous manner). Equation (5.1a) is derived by simply

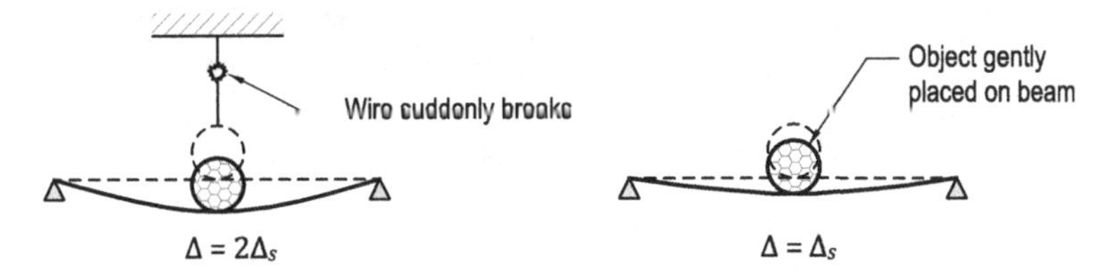

Figure 5.1 Suddenly released load on a beam.

equating the loss in potential energy (i.e. $mg\Delta$) with the elastic energy of absorption (i.e. $0.5k\Delta^2$). It is shown by rearranging the terms in Equation (5.1a) that maximum dynamic deflection (Δ) caused to the beam is double the static deflection, which is $\Delta_s = mg/k$, as shown by Equation (5.1b):

$$mg\Delta = \frac{1}{2}k\Delta^2 \tag{5.1a}$$

$$\Delta = 2\left(\frac{mg}{k}\right) = 2\Delta_s \tag{5.1b}$$

where m is the lumped mass; k is the vertical stiffness of the beam when loaded at mid-span; g is acceleration due to gravity (i.e. 9.81 m/s^2); Δ is the maximum deflection experienced by the beam responding to the abruptly applied load; and Δ_s is the static deflection of the beam under the weight of the lumped mass.

It can be shown by dynamic analysis that the beam would undergo oscillation after reaching maximum deflection and eventually settle at half of the maximum deflection (or the static deflection). Figure 5.2 presents the time-history displacement response of the idealised system of Figure 5.1 that has a span length of 2 m, generalised mass of 12 kg and stiffness of 194 kN/m. The lumped mass is 110 kg. The phenomenon revealed by this

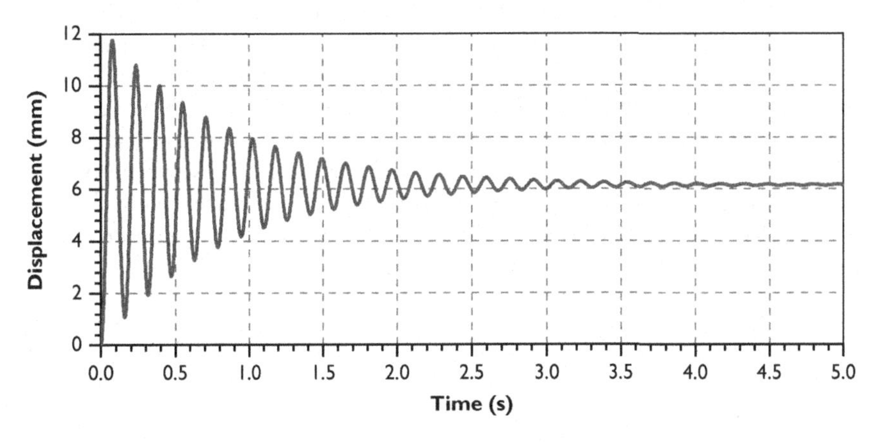

Figure 5.2 Response displacement time-history.

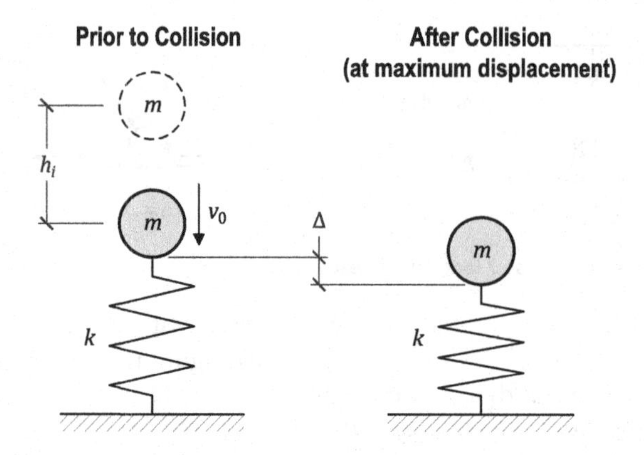

Figure 5.3 Impact of a fallen object on a "massless" target.

simple experiment is an illustration of the aggravating actions of gravity in a vertical collision scenario.

Consider that the experiment is now modified such that the lumped mass positioned above the beam is released from a drop height (h_i), as shown in Figure 5.3. The simply supported beam can be idealised as a spring of stiffness (k), which is equal to the beam stiffness. The idealisation of beam elements into an equivalent spring system is presented in Chapter 3, which includes scenarios where the collision occurs at mid-span or other locations along the span of the beam. The underlying assumption that the impactor becomes attached to the targeted lumped mass following the collision is in analogy to the assumption of a Type (2) scenario as introduced in Chapter 3.

Equation (5.1a) is modified to Equation (5.2a) for incorporating the effects of the drop height (i.e. instead of the potential energy being calculated across the dynamic deflection only; it is calculated based on both the drop height and the dynamic deflection). Through algebraic manipulation of Equation (5.2a), noting the static deflection is equal to $\Delta_s = mg/k$, an expression for solving the maximum deflection (Δ) of the collision scenario is shown in Equation (5.2b). The right-hand side of Equation (5.2b) can be rearranged into the form of Equation (5.2c), which is a function of the static deflection (Δ_s) and the deflection demand of a horizontal collision scenario (δ_h). The value of δ_h may in turn be solved using Equation (5.2d), which was first introduced in Chapter 3 (refer Section 3.5, Equation (3.7b), wherein λ is set to zero since the mass of the beam is considered negligible in this instance). Note that the incident velocity of impact (v_0) can be found from a given value of h_i by the use of the well-known relationship $v_0 = \sqrt{2gh_i}$.

$$mg(h_i + \Delta) = \frac{1}{2}k\Delta^2 \tag{5.2a}$$

$$\Delta = \Delta_s\left(1 + \sqrt{1 + \frac{2h_i}{\Delta_s}}\right) \tag{5.2b}$$

$$\Delta = \Delta_s + \sqrt{\Delta_s^2 + \delta_h^2} \tag{5.2c}$$

$$\delta_h = \frac{mv_0}{\sqrt{mk}} = \sqrt{2\Delta_s h_i} \tag{5.2d}$$

It can be shown that Equation (5.2c) gives the same result as Equation (5.1b) if $h_i = 0$.

5.2.2 Impact on a Member of Significant Self-Weight

5.2.2.1 One-Off Strike by a Fallen Object

Consider the impact scenario of Figure 5.4 where a fallen object initially lands on a member, which has been idealised into a single-degree-of-freedom (SDOF) system, and then bounces off from it, and eventually settles outside the member without making a second contact.

The amount of kinetic energy transferred to the target (KE_2) is taken from Equation (3.3b) of Chapter 3, and is re-written here as Equation (5.3). Recall that the participation (generalised) mass of the targeted member is λm, i.e. $m^* = \lambda m$.

$$KE_2 = \frac{1}{2} m v_0^2 \lambda \left(\frac{1 + COR}{1 + \lambda} \right)^2 \tag{5.3}$$

where COR is the coefficient of restitution as defined by Equation (3.1c).

The loss of potential energy (PE) resulted from the displacement of the participating mass of the member alone (without including the mass of the fallen object since it has bounced off) is defined by Equation (5.4):

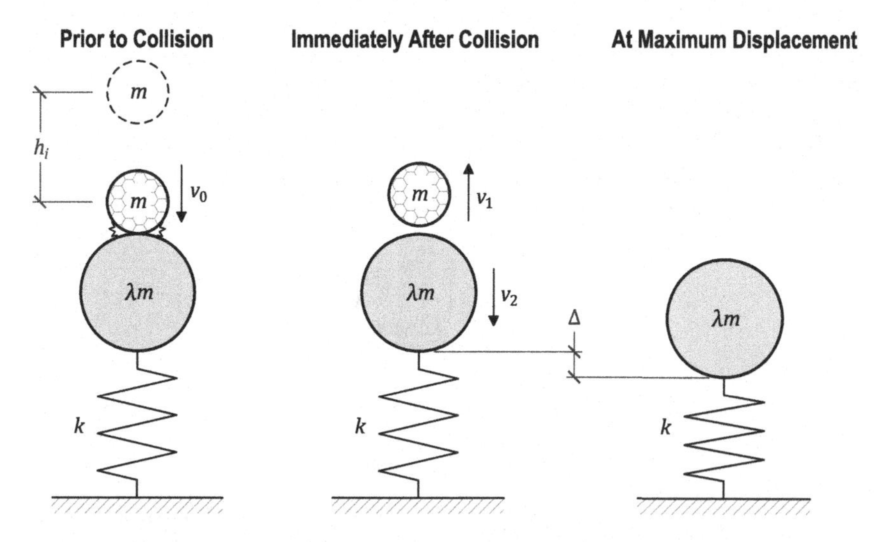

Figure 5.4 One-off impact of a fallen object on a target of significant mass.

$$PE = \lambda mg\Delta \tag{5.4}$$

Summation of KE_2 and PE represents the energy demand imposed by the collision action, which is to be absorbed by the structural element in the form of strain energy (SE), as shown by Equation (5.5):

$$KE_2 + PE = SE \tag{5.5}$$

If linear elastic behaviour is assumed, the amount of absorbed energy is represented by the area of a triangle (refer the schematic diagram of Figure 5.5). Given that the element is prestressed by its own weight, the amount of energy absorbed is represented by the shaded area of the trapezium as shown in the schematic diagram. This area is defined by Equation (5.6):

$$SE = \lambda mg\Delta + \frac{1}{2}k\Delta^2 \tag{5.6}$$

Substituting Equations (5.3), (5.4) and (5.6) into Equation (5.5) gives Equation (5.7a). Rearranging the terms in Equation (5.7a) gives Equation (5.7b), which can be used for finding the deflection generated by the collision:

$$\frac{1}{2}mv_0^2\lambda\left(\frac{1+COR}{1+\lambda}\right)^2 + \lambda mg\Delta = \lambda mg\Delta + \frac{1}{2}k\Delta^2 \tag{5.7a}$$

$$\Delta = \frac{mv_0}{\sqrt{mk}}\sqrt{\lambda\left(\frac{1+COR}{1+\lambda}\right)^2} \tag{5.7b}$$

The velocity of impact (v_o) is equal to $\sqrt{2gh_i}$ for scenarios where the impactor is a fallen object accelerated due to gravity only.

Note that Equation (5.7b) is identical to Equation (3.7a) of Chapter 3 for modelling Type (1) scenarios. Thus, in a single-strike scenario as described, the deflection demand of a collision in the horizontal and vertical direction on an elastically responding element is essentially the same.

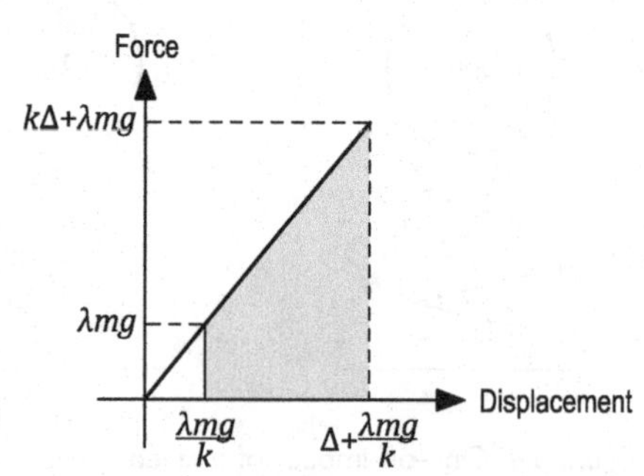

Figure 5.5 Linear elastic system prestressed by its own weight.

5.2.2.2 *Fallen Object Embedded into the Target*

Consider another impact scenario in which the fallen object becomes embedded in the target, forming a combined mass after initial contact is made (Figure 5.6).

Equation (3.4b) was first introduced in Chapter 3 for estimating the amount of kinetic energy (KE_2) developed in the combined mass immediately following a collision. This same expression is re-written here as Equation (5.8):

$$KE_2 = \frac{1}{2}mv_0^2\left(\frac{1}{1+\lambda}\right) \tag{5.8}$$

The potential energy loss of the combined mass in a vertical collision scenario is given by Equation (5.9):

$$PE = (1+\lambda)mg\Delta \tag{5.9}$$

The strain energy of absorption as shown by the shaded area in Figure 5.5 is given by Equation (5.6) as first introduced in Section 5.2.2.1. According to the energy balance defined by Equation (5.5), the energy demand, i.e. KE_2 + PE (as calculated using Equations (5.8) and (5.9)), is to be equated to the energy of absorption, i.e. SE, which can be calculated using Equation (5.6). The result is represented by Equation (5.10a), which can be rearranged into a quadratic expression, as shown in Equation (5.10b):

$$\frac{1}{2}mv_0^2\left(\frac{1}{1+\lambda}\right) + (1+\lambda)mg\Delta = \lambda mg\Delta + \frac{1}{2}k\Delta^2 \tag{5.10a}$$

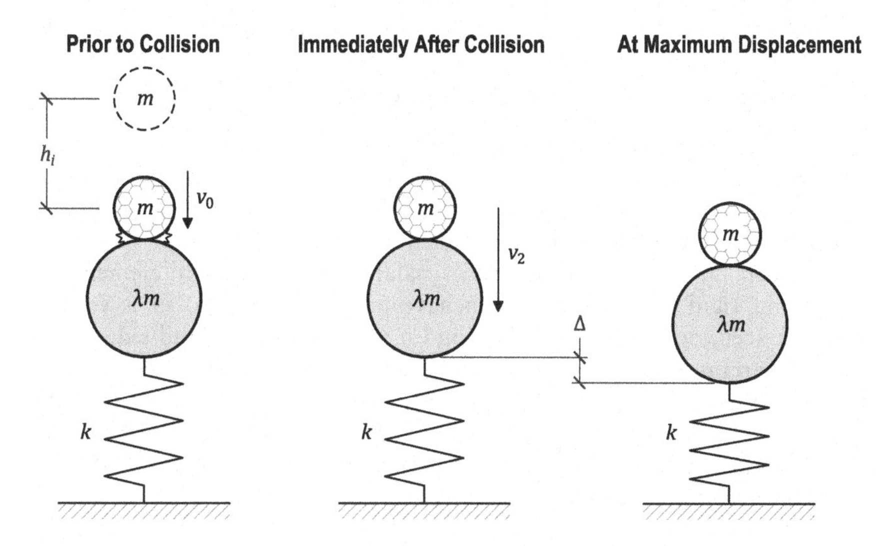

Figure 5.6 Impact of a fallen object embedded into a target of significant mass.

$$\Delta^2 - \frac{2mg}{k}\Delta - \frac{mv_0^2}{k(1+\lambda)} = 0 \tag{5.10b}$$

The solution for the positive root of Equation (5.10b) is given by Equation (5.11):

$$\Delta = \frac{mg}{k} + \sqrt{\left(\frac{mg}{k}\right)^2 + \frac{mv_0^2}{k(1+\lambda)}} \tag{5.11}$$

Equation (5.11) can be re-written as Equation (5.12a), which is function of $\Delta_s (= mg/k)$ and δ_h. The value of δ_h may in turn be solved using Equation (5.12b), which was first introduced in Chapter 3 (refer to Section 3.5).

$$\Delta = \Delta_s + \sqrt{\Delta_s^2 + \delta_h^2} \tag{5.12a}$$

$$\delta_h = \frac{mv_0}{\sqrt{mk}}\sqrt{\frac{1}{1+\lambda}} \tag{5.12b}$$

Note again that the velocity of impact (v_o) is equal to $\sqrt{2gh_i}$ for scenarios where the impactor is a fallen object accelerated due to gravity only.

The similarity between Equations (5.2c) and (5.12a) is noted. The δ term as defined by Equation (5.12b) is identical to Equation (3.7b), which represents the deflection demand of a collision in the horizontal direction. For a member of negligible mass ($\lambda \approx 0$), Equation (5.12b) reverts to Equation (5.2d).

5.3 IMPACT ON A MEMBER RESPONDING WITH RIGID PERFECTLY PLASTIC OR ELASTO-PLASTIC BEHAVIOUR

5.3.1 Impact on a Member of Negligible Mass

Two force-displacement relationships are considered in this section: (1) a rigid perfectly plastic behaviour model, refer Figure 5.7(a); and (2) an elasto-plastic behaviour model, refer Figure 5.7(b). F_y and Δ_y in Figure 5.7 are the yield force and yield displacement, respectively.

Equation (5.13a) shows the energy balance of a structural element experiencing rigid perfectly plastic behaviour (Figure 5.7(a)) in a vertical collision scenario. The first and second terms on the left-hand side of the equal sign represent kinetic energy and loss of potential energy, respectively, whereas the term on the right-hand side of the equal sign represents the energy of absorption (as indicated by the area under the graph of Figure 5.7(a)). Equation (5.13a) can be rearranged into Equation (5.13b) for determining the vertical displacement of the collision:

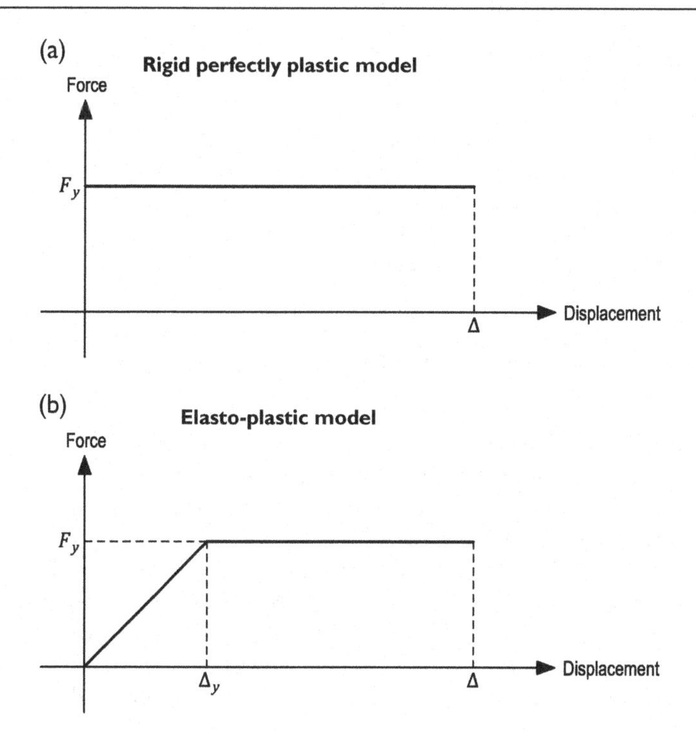

Figure 5.7 Inelastic response of a lumped mass system with negligible mass.

$$\frac{1}{2}mv_o^2 + mg\Delta = F_y\Delta \tag{5.13a}$$

$$\Delta = \frac{mv_o^2}{2(F_y - mg)} \tag{5.13b}$$

Equations (5.14a) and (5.14b) can be derived in the same manner for dealing with the vertical collision on a structural element experiencing elasto-plastic behaviour (Figure 5.7(b)).

$$\frac{1}{2}mv_o^2 + mg\Delta = \frac{1}{2}F_y\Delta_y + F_y(\Delta - \Delta_y) \tag{5.14a}$$

$$\Delta = \frac{mv_o^2}{2(F_y - mg)} + \frac{F_y}{F_y - mg}\frac{\Delta_y}{2} \tag{5.14b}$$

5.3.2 Impact on a Member of Significant Self-Weight

For a vertical collisions scenario where the structural element resisting the collision (i.e. the target) has a significant self-weight, deriving expressions

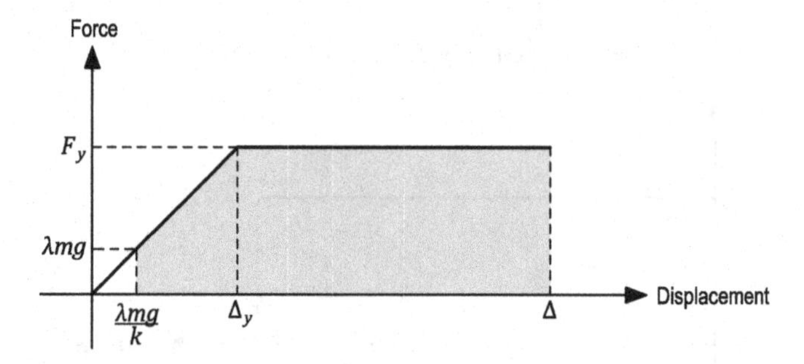

Figure 5.8 Inelastic response of a lumped mass system prestressed by its own weight.

for predicting the displacement demand needs to take into account energy losses as illustrated in Section 5.2.2.2. The prestressing of the element by its own weight also needs to be taken into account when calculating the energy of absorption (refer Figure 5.8).

The amount of energy demand in the collision scenario can be taken as the summation of KE_2 and PE, as defined by Equations (5.8) and (5.15a), respectively. Meanwhile, the amount of energy absorption of a structural element undergoing elasto-plastic behaviour is represented by Equation (5.15b), which is indicated by the shaded area under the graph of Figure 5.8.

$$PE = (1 + \lambda)mg\left(\Delta - \frac{\lambda mg}{k}\right) \tag{5.15a}$$

$$SE = \frac{(\lambda mg + F_y)}{2}\left(\Delta_y - \frac{\lambda mg}{k}\right) + F_y\left(\Delta - \Delta_y\right) \tag{5.15b}$$

According to the energy balance defined by Equation (5.5), the value of KE_2 as calculated using Equation (5.8) and the value of PE as calculated using Equation (5.15a), are to be equated to the value of SE, as calculated using Equation (5.15b). The result is shown by Equation (5.16a). Equation (5.16a) can be rearranged into Equation (5.16b) for determining the vertical displacement of the collision.

$$\frac{1}{2}mv_0^2\left(\frac{1}{1+\lambda}\right) + (1+\lambda)mg\left(\Delta - \frac{\lambda mg}{k}\right) = \frac{(\lambda mg + F_y)}{2}\left(\Delta_y - \frac{\lambda mg}{k}\right) + F_y(\Delta - \Delta_y) \tag{5.16a}$$

$$\Delta = \frac{mv_o^2}{2(F_y - (1+\lambda)mg)}\left(\frac{1}{1+\lambda}\right) + \frac{F_y}{F_y - (1+\lambda)mg}\frac{\Delta_y}{2} - \frac{\lambda(2+\lambda)(mg)^2}{2k(F_y - (1+\lambda)mg)} \tag{5.16b}$$

The third term on the right-hand side of the equal sign in Equation (5.16b) has been found to be insignificant (an order of magnitude lower than the other two terms in the expression). Thus, Equation (5.16b) may be simplified into Equation (5.17). Omission of the third term would result in a more conservative deflection estimation

$$\Delta = \frac{mv_o^2}{2\left(F_y - (1+\lambda)mg\right)}\left(\frac{1}{1+\lambda}\right) + \frac{F_y}{F_y - (1+\lambda)mg}\frac{\Delta_y}{2} \tag{5.17}$$

5.4 WORKED EXAMPLES

5.4.1 Vertical Collision on a Timber Beam

Consider a timber beam with span length of 1.2 m, width of 100 mm and depth of 200 mm. The beam has density ρ = 1,000 kg/m^3, Young's modulus E = 9 GPa and yield strength f_y = 55 MPa. The beam is subject to the collision of an object of mass m = 30 kg released from a drop height of 1 m.

Solution

The generalised mass of a simply supported beam may be taken as 50% of total beam mass, as illustrated in Chapter 2.

$$\lambda m = 0.5\rho BDL = 0.5(1000 \times 0.1 \times 0.2 \times 1.2) = 12 \text{ kg}$$

$$\lambda = \frac{\lambda m}{m} = \frac{12}{30} = 0.4$$

The expression for estimating the generalised stiffness of a beam can also be found in Chapter 2.

$$I = \frac{BD^3}{12} = \frac{(0.1)(0.2)^3}{12} = 66.7 \times 10^{-6} \text{ m}^4$$

$$k = \frac{48EI}{L^3} = \frac{48(9 \times 10^6)(66.7 \times 10^{-6})}{1.2^3} = 16675 \text{ kN/m}$$

The velocity of impact can be calculated by employing the principles of conservation of energy as shown in the following:

$$v_0 = \sqrt{2gh_i} = \sqrt{2(9.81)(1)} = 4.43 \text{ m/s}$$

As the value of λ is less than 1, the impactor is expected not to rebound from the beam as contact is made. Thus, Equations (5.10a) and (5.10b) may be applied, assuming elastic response behaviour of the beam.

$$\Delta_s = \frac{mg}{k} = \frac{30(9.81)}{16675} = 0.02 \text{ mm}$$

$$\delta_h = \frac{mv_0}{\sqrt{mk(1+\lambda)}} = \frac{30(4.43)}{\sqrt{50(16675 \times 10^3)(1+0.24)}} \times 10^3 = 5 \text{ mm}$$

$$\Delta = \Delta_s + \sqrt{\Delta_s^2 + \delta_h^2} = 0.02 + \sqrt{0.02^2 + 5^2} = 5 \text{ mm}$$

The equivalent design static force and bending moment of the beam may be estimated as follows:

$$F^* = k\Delta = 16675(5 \times 10^{-3}) = 84 \text{ kN}$$

$$M^* = \frac{F^*L}{4} = \frac{84(1.2)}{4} = 25.2 \text{ kNm}$$

The bending moment at the limit of yield moment is determined as follows:

$$M_y = f_y Z = 55\left(\frac{0.1 \times 0.2^2}{6}\right) = 36.7 \text{ kNm} > M^*$$

Given that the bending moment at yield is not exceeded by the applied bending moment, the assumption of elastic response behaviour is valid.

5.4.2 Vertical Collision on a RC Beam

Consider a RC beam specimen (used for laboratory testing) with a span length of 1.4 m, width of 150 mm and depth of 250 mm. The beam has density $\rho = 2{,}400 \text{ kg/m}^3$, an effective stiffness at mid-span of $k = 33{,}330$ kN/m, yield displacement of $\Delta_y = 3$ mm and yield strength of $F_y = 100$ kN. The modelling methodology for determining the effective stiffness of a cracked reinforced concrete member is the subject matter of Chapter 8. The beam is subject to the impact of a dropped object of mass $m = 400$ kg, striking the beam with velocity of $v_0 = 4.85$ m/s at mid-span.

Solution

The generalised mass of the beam (λm) is calculated as follows:

$$\lambda m = 0.5\rho BDL = 0.5(2400 \times 0.15 \times 0.25 \times 1.4) = 63 \text{ kg}$$

$$\lambda = \frac{\lambda m}{m} = \frac{63}{400} = 0.16$$

Equation (5.14) is used, as the beam is expected to have surpassed the limit of yield.

$$F_y - (1 + \lambda)mg = 100 - (1 + 0.16)(400)(9.81) \times 10^{-3} = 95.4 \text{ kN}$$

$$\Delta = \frac{mv_o^2}{2(F_y - (1 + \lambda)mg)} \frac{1}{1 + \lambda} + \frac{F_y}{F_y - (1 + \lambda)mg} \frac{\Delta_y}{2}$$

$$= \frac{400(4.85)^2}{2(95.4)}\left(\frac{1}{1 + 0.16}\right) + \frac{100}{95.4}\left(\frac{3}{2}\right) = 44 \text{ mm}$$

The assumption of inelastic response of the beam is valid, as the deflection experienced by the beam is predicted to have surpassed the limit of yield (i.e. $\Delta > \Delta_y$).

5.4.3 Vertical Collision on a Pair of Steel Beams

Consider a pair of steel beams supporting a RC slab, as shown in Figure 5.9. To simplify calculations, composite actions between the concrete and steel are ignored. The pair of beams and the concrete slab has a combined mass of 1,600 kg. The force-deflection relationship is as shown in Figure 5.10. The structure is subject to vertical collision of a fallen object of mass m = 600 kg, dropping from a height of 1 m and 2 m in two separate scenarios.

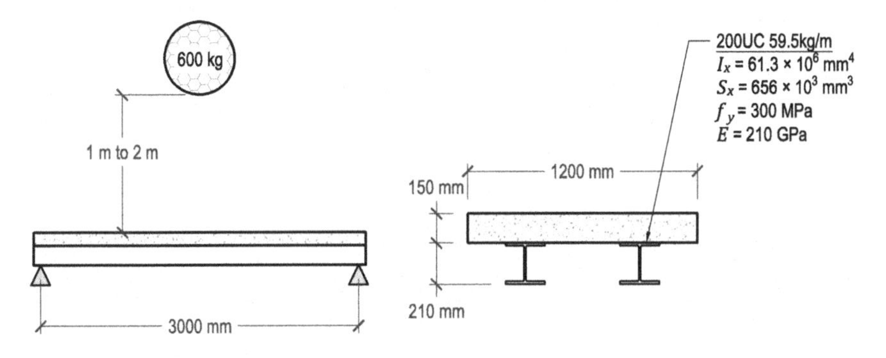

Figure 5.9 Impact on a pair of steel beams.

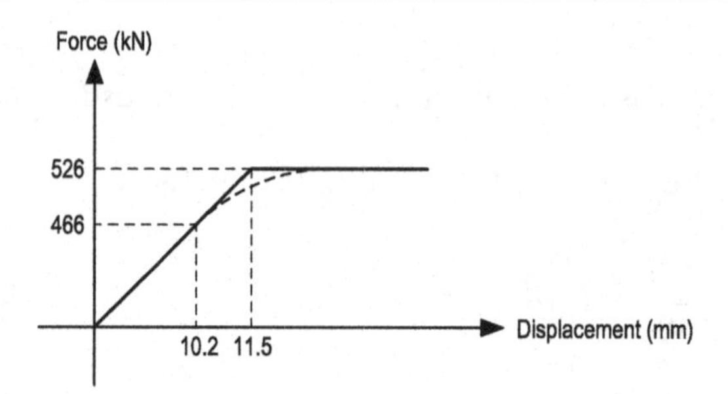

Figure 5.10 Force-displacement relationship of steel beam.

A bi-linear force displacement graph. The point of first yield occurs when force is equal to 466 kN and displacement is equal to 10.2 mm. The point of notional yield occurs when force is equal to 526 kN and displacement is equal to 11.5 mm.

Solution

The generalised mass, λm, may be calculated in the same manner as done for the earlier examples.

$$\lambda m = 0.5(1600) = 800 \text{ kg}$$

$$\lambda = \frac{\lambda m}{m} = \frac{800}{600} = 1.33$$

From Figure 5.10, the structure has stiffness at mid-span of $k = 45{,}700$ kN/m, yield displacement of $\Delta_y = 11.5$ mm and yield strength of $F_y = 526$ kN.

Scenario #1: Drop Height = 1 m

$$v_0 = \sqrt{2gh_i} = \sqrt{2(9.81)(1)} = 4.43 \text{ m/s}$$

Equations (5.10a) and (5.10b) are applied and elastic response behaviour of the structure is assumed.

$$\Delta_s = \frac{mg}{k} = \frac{600(9.81)}{45700} = 0.13 \text{ mm}$$

$$\delta_h = \frac{mv_0}{\sqrt{mk(1+\lambda)}} = \frac{600(4.43)}{\sqrt{600(45700 \times 10^3)(1+1.33)}} \times 10^3 = 10.5 \text{ mm}$$

$$\Delta = \Delta_s + \sqrt{\Delta_s^2 + \delta_h^2} = 0.13 + \sqrt{0.13^2 + 10.5^2} = 10.6 \text{ mm}$$

The calculated vertical deflection of 10.6 mm is close to the notional yield limit and has not exceeded it. Thus, the presented calculation based on the assumption of linear elastic behaviour is valid.

Scenario #2: Drop Height = 2 m

$$v_0 = \sqrt{2gh_i} = \sqrt{2(9.81)(2)} = 6.26 \text{ m/s}$$

Equation (5.14) is used, as inelastic response behaviour of the structure is to be modelled.

$$F_y - (1 + \lambda)mg = 526 - (1 + 1.33)(600)(9.81) \times 10^{-3} = 512 \text{ kN}$$

$$\Delta = \frac{mv_o^2}{2(F_y - (1 + \lambda)mg)} \frac{1}{1 + \lambda} + \frac{F_y}{F_y - (1 + \lambda)mg} \frac{\Delta_y}{2}$$
$$= \frac{600(6.26)^2}{2(512)} \left(\frac{1}{1 + 1.33} \right) + \frac{526}{512} \left(\frac{11.5}{2} \right) = 15.8 \text{ mm}$$

Given that the vertical deflection experienced by the beams has surpassed the limit of yield (of Δ_y = 11.5 mm), the assumption of elasto-plastic behaviour of the beams is valid.

5.5 CLOSING REMARKS

This chapter presents closed-form expressions for predicting maximum deflection of a structural element when subject to the collision of a fallen object. With a scenario where the fallen object is eventually settled on the target following multiple occurrences of collision and rebound, the fallen object and the target is treated as a combined mass in order to circumvent the need of undertaking complex analyses. The loss of potential energy by displacement of the combined mass (due to gravity) is shown to aggravate the outcome (i.e. the displacement demand) of the impact action. The early sections of this chapter deal with elements responding within the limit of linear elastic behaviour. The latter section of this chapter deals with elements responding in their inelastic range. Two types of non-linear force-displacement relationships have been covered: (1) rigid perfectly plastic behaviour, which is suited to modelling ductile elements; and (2) elasto-plastic behaviour, which is suited to modelling elements with limited ductility such as RC elements that have not been designed/detailed for ductility enhancement. Worked examples featuring vertical collisions on a timber beam, an RC beam and a steel beam are presented at the end of the chapter.

REFERENCES

5.1 Ali, M., Sun, J., Lam, N., Zhang, L., and Gad, E., 2014, "Simple hand calculation method for estimating deflection generated by the low velocity impact of a solid object", *Australian Journal of Structural Engineering*, Vol. 15(3), pp. 243–259, doi:10.7158/13287982.2014.11465162

5.2 Yong, A.C.Y., Lam, N.T.K., and Menegon, S.J., 2021, "Closed-form expressions for improved impact resistant design of reinforced concrete beams", *Structures*, Vol. 29, pp. 1828–1836, doi:10.1016/j.istruc.2020.12.041

5.3 Chen, Y. and May, I.M., 2009, "Reinforced concrete members under drop-weight impacts", *Proceedings of the ICE - Structures and Buildings*, Vol. 162(1), pp. 45–56, doi:10.1680/stbu.2009.162.1.45

5.4 Othman, H. and Marzouk, H., 2016, "An experimental investigation on the effect of steel reinforcement on impact response of reinforced concrete plates", *International Journal of Impact Engineering*, Vol. 88, pp. 12–21, doi:10.1016/j.ijimpeng.2015.08.015

5.5 Hummeltenberg, A., Beckmann, B., Weber, T., and Curbach, M., 2011, "Investigation of concrete slabs under impact load", *Applied Mechanics and Materials*, Vol. 82, pp. 398–403, doi:10.4028/www.scientific.net/AMM.82.398

5.6 Zineddin, M. and Krauthammer, T., 2007, "Dynamic response and behavior of reinforced concrete slabs under impact loading", *International Journal of Impact Engineering*, Vol. 34(9), pp. 1517–1534, doi:10.1016/j.ijimpeng.2006.10.012

5.7 Fujikake, K., Li, B., and Soeun, S., 2009, "Impact response of reinforced concrete beam and its analytical evaluation", *Journal of Structural Engineering*, Vol. 135(8), pp. 938–950, doi:10.1061/(ASCE)ST.1943-541X.0000039

5.8 Kishi, N. and Bhatti, A.Q., 2010, "An equivalent fracture energy concept for nonlinear dynamic response analysis of prototype RC girders subjected to falling-weight impact loading", *International Journal of Impact Engineering*, Vol. 37(1), pp. 103–113, doi:10.1016/j.ijimpeng.2009.07.007

5.9 Saatci, S. and Vecchio, F.J., 2009, "Effects of shear mechanisms on impact behavior of reinforced concrete beams", *ACI Structural Journal*, Vol. 106(1), pp. 78–86.

5.10 Adhikary, S.D., Li, B., and Fujikake, K., 2015, "Low velocity impact response of reinforced concrete beams: experimental and numerical investigation", *International Journal of Protective Structures*, Vol. 6(1), pp. 81–111, doi:10.1260%2F2041-4196.6.1.81

5.11 Bhatti, A.Q., Kishi, N., Mikami, H., and Ando, T., 2009, "Elasto-plastic impact response analysis of shear-failure-type RC beams with shear rebars", *Materials & Design*, Vol. 30(3), pp. 502–510, doi:10.1016/j.matdes.2008.05.068

Chapter 6

Contact Force of Impact

6.1 INTRODUCTION

The localised action of a collision can cause damage to the target in the form of indentation, spalling or punching depending on the type of material that is impacted upon. The extent of damage is principally controlled by the magnitude of the force experienced at the point of contact between the impactor and the surface of the target. The type of force is abbreviated as the "contact force," which typically lasts for up to tens of milliseconds, except in the scenario of a cushioned impact (that is dealt with in Chapter 7).

The contact force is not to be confused with a "design collision force," such as an equivalent static force that has been calculated (e.g. using the method presented in Chapter 3) to result in the deflection of a structural element matching the maximum collision generated displacement. An equivalent static force representing a collision action is commonly stipulated by contemporary codes of practice for designing a barrier, or for checking the capacity of a structural element to withstand a projected collision scenario. The magnitude of the contact force is normally much higher than the (code stipulated) equivalent static force for a projected collision scenario. Certain failure mechanisms resulted from a collision scenario, such as spalling, or punching, cannot be emulated by static testing nor simulations. Localised damage to concrete is treated in detail in Chapter 8 and damage to cladding panels in Chapter 12. This chapter is concerned mainly with the estimation of the peak value of the contact force in a projected collision scenario.

Experimental measurements can be used for validating the accuracy of predictions by analytical modelling or numerical simulations. The first section in this chapter is devoted to introducing a viable method for measuring contact force (Section 6.2). A number of contact force models that have been reported in the literature are then reviewed (Section 6.3). Detailed descriptions of two classes of contact force models and their implementation for dealing with dynamic conditions are then presented in the subsequent sections (Sections 6.4 to 6.6). This chapter concludes with worked examples illustrating the use of the presented contact force models (Section 6.7).

DOI: 10.1201/9781003133032-6

6.2 MEASUREMENT OF CONTACT FORCE

Drop hammer tests are commonly employed for dynamic testing and typically make use of a load cell to capture transient forces generated by the impact action. Given the need to avoid damage to the load cell (or force sensor), the measurement device is normally placed underneath the test specimen in order that the impactor would not come in direct contact with the device. Inertial resistance developed within the specimen would result in a reduction in transient forces transmitted to the device. The forcing function so recorded by the device would therefore not represent the actual contact force experienced by the specimen at the point of contact. The same is said of an experimental setup in which the measurement device is placed behind, or within, the drop hammer, in which case transient forces transmitted to the device would be distorted by inertial resistances developed within the hammer.

Contact force can be captured by an accelerometer, which needs to be secured firmly to the impactor during testing. Whilst the principle of calculating impact force, as a product of the acceleration and mass of the impactor, is straightforward there are challenges with this measurement approach. Given that the duration of contact in a collision scenario can be as short-lived as a few milliseconds, or tens of milliseconds, the rate of data acquisition would need to be at least 10 kHz. Securing the accelerometer firmly onto the impactor can also be an issue. For example, it is clearly not feasible to attach an accelerometer on a piece of windborne debris nor on a hailstone. Even if that was possible, the motion behaviour of the impactor would have been modified by the presence of the accelerometer. As a result of the aforementioned challenges in measuring the contact force generated by a collision, empirical data/measurements of contact force are not abundant.

A piece of equipment for measuring contact forces generated by the impact of windborne debris (or hail) has been custom designed, and built, by the authors and their co-workers (Figure 6.1(a)). The measurement device features a dummy mass (m_t) that is fitted with a helical spring. The nozzle of a gas gun for propelling the impactor is placed close to the dummy mass. The time-history of shortening of the helical spring, $x_t(t)$, is monitored by the use of a high-speed camera (HSC) with a frame rate of up to 25,000 frames per second. The dummy mass can be fitted with an accelerometer for monitoring the acceleration time-history, $a_t(t)$. The product of $x_t(t)$ and spring constant (k) gives the spring force whereas the product of $a_t(t)$ and m_t gives the inertial force. The sum of the spring force and inertial force gives the contact force $F_c(t)$, as illustrated in Figure 6.1(b). The HSC can also be used to capture images during the millisecond/s of contact. An example image capture of the impact scenario of a granite specimen propelled by the gas gun against the dummy mass is presented in Figure 6.2. The duration of the loading phase can be as short-lived as a millisecond.

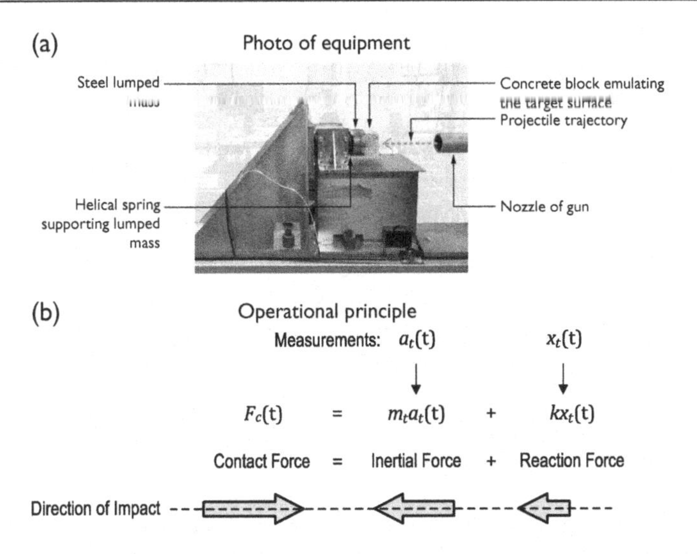

Figure 6.1 Equipment for measuring contact force generated by a flying object.

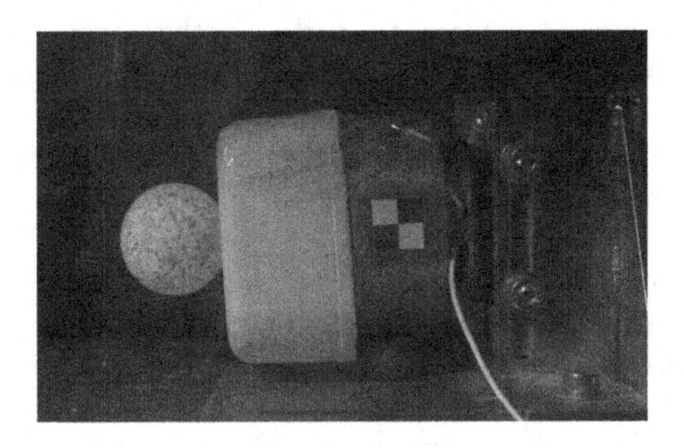

Figure 6.2 Impact by a granite specimen on the dummy mass.

There are other successful attempts at measuring the forcing function at contact adopting similar principles by other research teams [6.1]. Measurements of contact force in a collision scenario have also been accomplished by use of the Hopkinson Bar [6.2].

6.3 CONTACT FORCE MODELS

6.3.1 General

Four contact force models are described briefly in this section: (1) linear elastic model, (2) *Kelvin-Voight* model, (3) non-linear elastic model and (4) non-linear visco-elastic model. The contact force (F_c) of impact on the surface of the target and the amount of indentation into its surface (δ) can be modelled using a spring-connected lumped mass, as depicted in Figure 6.3. Dashpot shown in

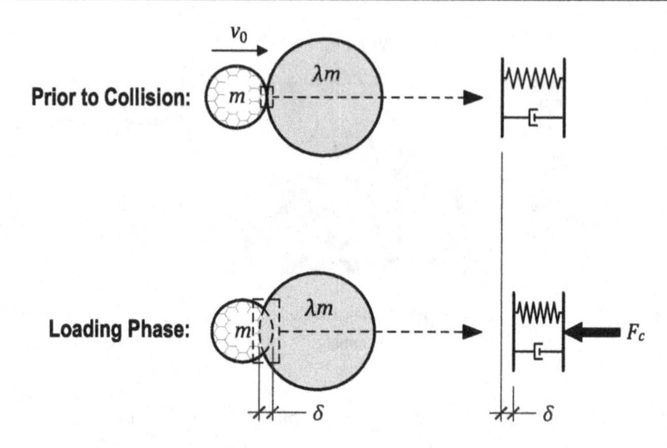

Figure 6.3 Contact force and indentation.

the figure represents damping and is only present in the *Kelvin-Voight* model and non-linear visco-elastic model out of the four models to be introduced in the following sub-sections. The modelling may be based on the assumption that the impactor is rigid, whereas the surface of the target is deformable. The same modelling principle may be applied if only the impactor is deformable, or if both impactor and target surface are deformable.

Each contact model attempts to describe the relationship between F_c and δ, as shown in Figure 6.3, across two phases:

- The loading phase or compression phase. This commences immediately after the impactor comes in contact with the target and ends when δ reaches its maximum value (δ_{max}), i.e. when the relative velocity between the two objects is equal to zero.
- The unloading phase or restitution phase. This commences immediately after the end of the loading phase and ends when $F_c = 0$.

6.3.2 Linear Elastic Model

The linear elastic model for predicting contact force for a given amount of indentation has the merit of simplicity. However, the linear behaviour represented by the model contradicts with the well-known phenomenon of a rising rate of increase in the magnitude of the contact force with increasing indentation; this phenomenon is also known as hardening. The omission of the consideration of loading rate dependence is another shortcoming of the model. The linear elastic model is illustrated in Figure 6.4.

6.3.3 Kelvin-Voight Model

The *Kelvin-Voight* model expresses contact force as the sum of a linear elastic term and a rate dependent term. The two terms are similar to that

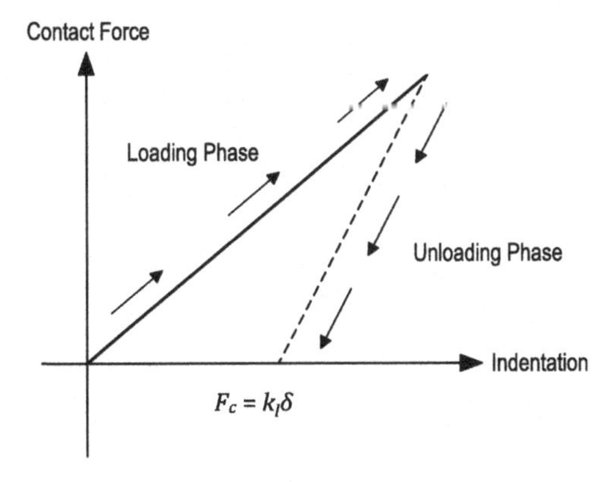

Figure 6.4 Linear elastic model.

representing elastic and damping forces in a structural system. The incorporation of the rate-dependent term in the model represents an improvement over the linear elastic model. However, the hardening behaviour described previously is not accounted for. Another shortcoming with this model is the prediction of negative contact force on unloading, which contradicts the real behaviour of the collision. The *Kelvin-Voight* model is illustrated in Figure 6.5.

6.3.4 Non-Linear Elastic Model

The non-linear elastic contact force model features a rising rate of increase in the contact force with increasing indentation. This hardening behaviour is controlled by the exponent p in the expression for defining the F_c and δ relationship. The *Hertzian* contact theory is a non-linear elastic contact force model, which takes $p = 1.5$ [6.3]. The non-linear elastic model is illustrated in Figure 6.6.

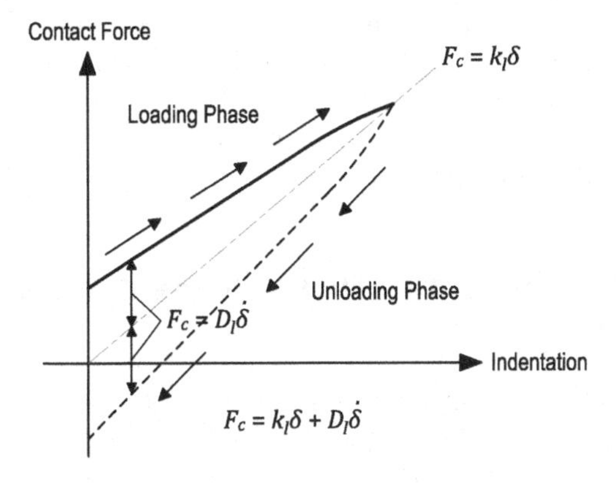

Figure 6.5 Kelvin-Voight model.

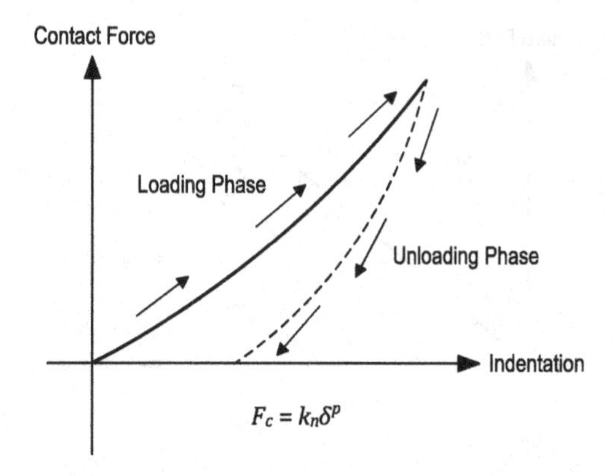

Figure 6.6 Non-linear elastic model.

Hertzian contact theory has received a great deal of research attention and is widely known and reported in the literature. The adaptation of this model and implementation in practice for predicting the contact force generated by a collision is presented in Section 6.4. There is however a lack of comparison of model predictions with measurements from impact experiments. Adapting this class of models for making predictions of the contact force demand in a dynamic event is purely intuitive and is without proper experimental validation.

6.3.5 Non-Linear Visco-Elastic Model

The non-linear visco-elastic model, also known as the *Hunt and Crossley* (H&C) model [6.4], expresses contact force as the sum of a *Hertzian* term consistent in form to the non-linear elastic model, and has a rate-dependent term. The implementation of the non-linear visco-elastic model for making predictions of the contact force generated by a collision is well supported by comparison of predictions with experimental measurements, as outlined in the literature review presented in Chapter 1. The non-linear visco-elastic model is illustrated in Figure 6.7.

A hybrid model based on modifying the (more user-friendly) non-linear elastic contact model to give the same predictions as the non-linear visco-elastic model is presented in Section 6.6.

6.4 APPLICATION OF THE NON-LINEAR ELASTIC CONTACT FORCE MODEL

The non-linear force-displacement (F_c vs δ) relationship as defined by Equation (6.1) can be used for determining the maximum indentation (δ_{max}) and maximum contact force ($F_{c,\,max}$).

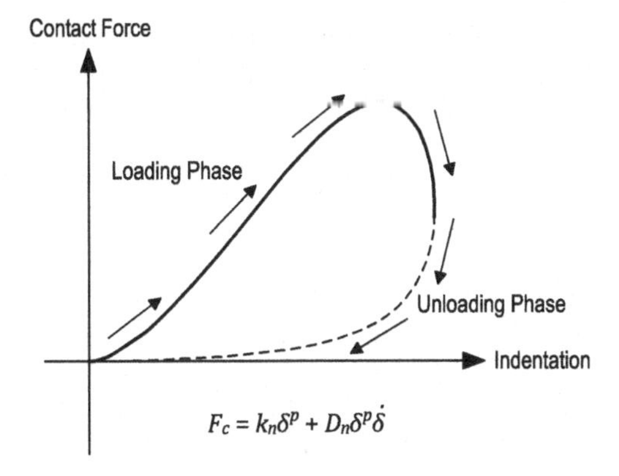

Figure 6.7 Non-linear visco-elastic model.

$$F_c = k_n \delta^p \tag{6.1}$$

where F_c is the contact force; k_n is the contact stiffness; δ is the indentation; and p is an exponent that characterises the non-linear behaviour of the material/components.

Equating the kinetic energy delivered by the collision to the elastic energy of absorption gives Equation (6.2a). Substituting Equation (6.1) into Equation (6.2a) and integrating the right-hand side of the equal sign of Equation (6.2a) gives Equation (6.2b):

$$\frac{1}{2}mv_0^2 = \int_0^{\delta_{\max}} F_c \, d\delta \tag{6.2a}$$

$$\frac{1}{2}mv_0^2 = \frac{k_n}{p+1}\delta_{\max}^{p+1} \tag{6.2b}$$

where m and v_0 are the mass and incident velocity of the impactor, respectively.

Rearranging Equation (6.2b) gives Equation (6.3a), which can be used for determining the value of $\delta_{\max}$. Since $F_{c,\max}$ and $\delta_{\max}$ occur at the same instance, Equation (6.3a) can be substituted into Equation (6.1) for determining $F_{c,\max}$, as shown by Equation (6.3b):

$$\delta_{\max} = \left[\left(\frac{p+1}{2k_n}\right)mv_0^2\right]^{\frac{1}{p+1}} \tag{6.3a}$$

$$F_{c,\max} = k_n\left[\left(\frac{p+1}{2k_n}\right)mv_0^2\right]^{\frac{p}{p+1}} \tag{6.3b}$$

Equations (6.4a) and (6.4b) may be written in accordance with the *Hertzian* contact theory, which can be used to equate the radius of a spherical impactor, the Young's modulus of the surface of the target and the contact stiffness.

$$k_n = \frac{4}{3}E\sqrt{R_c} \tag{6.4a}$$

$$p = 1.5 \text{ (for the } \textit{Hertzian} \text{ contact theory, refer to Section 6.3.4)} \tag{6.4b}$$

where E is the Young's modulus of the surface of the target, and R_c is the radius of curvature of the spherical object which makes contact with the target.

Substituting both Equations (6.4a) and (6.4b) into Equation (6.3a) and Equation (6.3b) gives Equation (6.5a) and Equation (6.5b), respectively.

$$\delta_{\max} \approx \left(\frac{mv_0^2}{E\sqrt{R_c}}\right)^{\frac{2}{5}} \tag{6.5a}$$

$$F_{c,\max} \approx \frac{4}{3}E\sqrt{R_c}\left(\frac{mv_0^2}{E\sqrt{R_c}}\right)^{\frac{3}{5}} \tag{6.5b}$$

The expressions presented are based on the assumption that the impactor is non-deformable in order that deformation mainly takes place as indentation into the surface of the target. Conversely, the same expressions may also be used if only the impactor is deformable, in which case E would instead be the Young's modulus of the impactor. In a general situation where the impactor and the target are both deformable, the same expressions may be used for giving predictions provided that the value of E in Equations (6.5a) and (6.5b) are replaced by the transformed Young's modulus (E_T), as defined by Equation (6.6a). The term R_c represents the radius of curvature of the impactor only if the surface of the target is flat, or else Equation (6.6b) should be used for determining its value when applying the contact force model.

$$E_T = \frac{1}{\frac{1-\nu_i^2}{E_i} + \frac{1-\nu_t^2}{E_t}} \tag{6.6a}$$

$$R_c = \frac{1}{\frac{1}{R_i} + \frac{1}{R_t}} \tag{6.6b}$$

where ν is Poisson's ratio; and subscripts "i" and "t" denote the impactor and the target, respectively.

The adaptation of the *Hertzian* contact model for making predictions of a collision-generated indentation and contact force is intuitive and has not been validated by comparison with experimental measurements.

6.5 ENERGY ABSORBED BY MATERIAL DEFORMATION AT CONTACT

Expressions presented in Section 6.4 may only be applicable to collision scenarios where the target is of a size that can be treated as an infinitely large mass, in which case all the kinetic energy delivered by the collision may be taken to be totally absorbed by the deformation of materials. In a situation where the target is of a finite size, only a fraction of the kinetic energy is absorbed by the deformation of materials. The rest of the energy would remain in the system as kinetic energy following the collision, given that the target has also been excited into motion with a velocity (v_2), as shown in Figure 6.8. The phenomenon as described is illustrated in the figure that depicts the collision of two spring-connected lumped masses (m and λm).

In the course of a collision, the initial kinetic energy KE_0 (as defined by Equation (6.7a)) is partly absorbed by the deformation of materials, which is denoted as ΔE, whereas the remaining energy KE_2 (as defined by Equation (6.7b)) is expanded as kinetic energy of the target (combined with the impactor) when excited into motion.

$$KE_0 = \frac{1}{2} m v_0^2 \tag{6.7a}$$

$$KE_2 = \frac{1}{2} (m + \lambda m) v_2^2 \tag{6.7b}$$

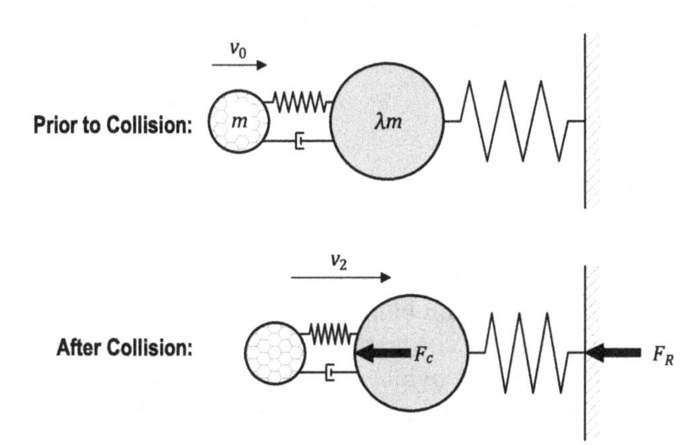

Figure 6.8 Basis of the derivation of the combinational mass.

The energy balance as described is represented by Equation (6.8a). Substituting Equations (6.7a) and (6.7b) into Equation (6.8a) results in Equation (6.8b).

$$\mathrm{KE}_0 = \Delta E + \mathrm{KE}_2 \tag{6.8a}$$

$$\frac{1}{2}mv_0^2 = \Delta E + \frac{1}{2}(m + \lambda m)v_2^2 \tag{6.8b}$$

It is noted that ΔE in Equations (6.8a) and (6.8b) is referring to the amount of energy loss from the instance when the impactor first makes contact with the target to the instance when the centre of the impactor comes closest to the target. As this happens, there is no relative motion between the two colliding objects that travel at a common velocity (v_2).

The transfer of momentum occurring on collision is represented by Equation (6.9a), which can be rearranged into Equation (6.9b) for determining v_2 as described.

$$mv_0 = (m + \lambda m)v_2 \tag{6.9a}$$

$$v_2 = \left(\frac{1}{1 + \lambda}\right)v_0 \tag{6.9b}$$

Combining Equation (6.8b) with Equation (6.9b) gives Equations (6.10a) and (6.10b) to deal with impact scenarios where the coefficient of restitution (COR) is taken as zero. Equation (6.10a) can be modified into Equation (6.10c) to deal with impact scenarios where COR > 0. The derivation of Equation (6.10c) is presented in Section 12.7 of Chapter 12.

$$\Delta E = \frac{1}{2}M_c v_0^2 \tag{6.10a}$$

$$M_c = \left(\frac{\lambda}{1 + \lambda}\right)m \tag{6.10b}$$

$$\Delta E = \frac{1}{2}M_c v_0^2(1 - \mathrm{COR}^2) \tag{6.10c}$$

where ΔE is the amount of energy to be absorbed in the form of material deformation, and M_c is the combinational mass.

The combinational mass concept conveniently presents the energy of absorption (ΔE) as a function of mass and velocity of the impactor, in order that Equation (6.10a) is of the same form as an expression for quantifying the kinetic energy delivered by the impactor, i.e. Equation (6.7a). It is noted that in situations where $m \ll \lambda m$ (or $\lambda \gg 1$), $M_c \approx m$, which means that the target comes close to behaving like an infinitely large mass. In this condition, estimates as derived from the use of Equation (6.10a) approaches that from Equation (6.7a).

Note, ΔE, which can also be found in Chapter 12, is used in that chapter to represent the condition at the end of the restitution phase of the impact.

The illustration presented in this section is based on treating the target as a lumped mass. In a situation where a slab or a plate like structure is impacted upon, a fraction of the target mass surrounding the point of contact (referred herein as participating mass) may be treated as the lumped mass. The determination of the participating mass on an aluminium plate is treated in Chapter 12. As a general rule, over-stating the participating mass would result in a conservative estimate of the contact force.

6.6 APPLICATION OF THE NON-LINEAR VISCO-ELASTIC CONTACT FORCE MODEL

6.6.1 Analytical Solution

The non-linear force-displacement (F_c vs δ) relationship of the H&C contact force model [6.4] is defined by Equation (6.11a) and illustrated in Figure 6.9. Equation (6.11b) for solving D_n was developed by Sun et al. [6.5], which is different from its original form presented by Hunt and Crossley [6.4]. Equations (6.11a) and (6.11b) may be used to simulate contact forcing functions for any given collision scenarios:

$$F_c = k_n\delta^p + D_n\delta^p\dot{\delta} \tag{6.11a}$$

$$D_n = (0.2p + 1.3)\left(\frac{1 - \text{COR}}{\text{COR}}\right)\left(\frac{k_n}{\dot{\delta}_0}\right) \tag{6.11b}$$

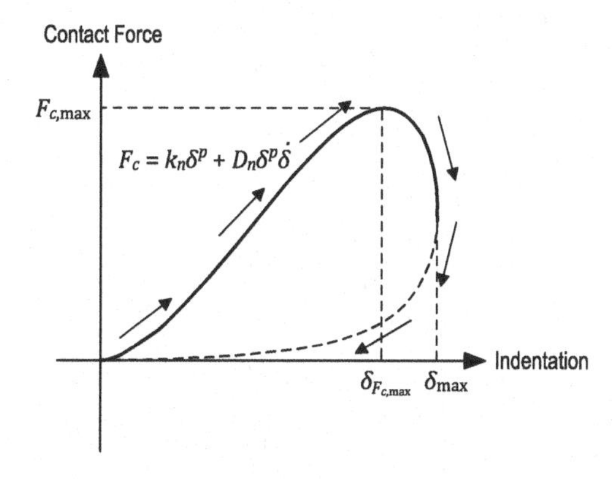

Figure 6.9 Maximum contact force and maximum indentation in Hunt and Crossley model.

where $\dot{\delta}$ is the rate of indentation; COR is the coefficient of restitution; $\dot{\delta}_0$ is the initial rate of indentation that is also the incident velocity of impact v_0; and D_n is the damping coefficient.

This section is focused on developing closed-form expressions for estimating the maximum contact force, $F_{c,\max}$, without having to carry out numerical simulations. Details of the application of the H&C model for numerical simulations in a Microsoft Excel spreadsheet are presented in Chapter 13. The derivation may appear to be very mathematically involved to some readers. The model that is developed at the end is easy to apply as is illustrated with a few worked examples. In the first step of the derivation, the contact forcing function, as defined by Equation (6.11a), is differentiated with respect to δ, and equated to zero, as shown by Equation (6.12). Note that Equation (6.11b) has been substituted into Equation (6.11a) prior to applying the differentiation.

$$\frac{\partial F_c}{\partial \delta} = pk_n\delta^{p-1} + k_n(0.2p + 1.3)\left(\frac{1 - \text{COR}}{\text{COR}}\right)\left(\frac{1}{\dot{\delta}_o}\right)\left(p\delta^{p-1}\dot{\delta} + \delta^p\frac{\partial \dot{\delta}}{\partial \delta}\right) = 0 \tag{6.12}$$

Define x and y by Equations (6.13a) and (6.13b), respectively.

$$x = \frac{\delta}{\delta_{\max}} \tag{6.13a}$$

$$y = \frac{\dot{\delta}}{\dot{\delta}_0} \tag{6.13b}$$

Differentiating $\dot{\delta}$ with respect to δ, and making use of the relationships defined by Equations (6.13a) and (6.13b), results in Equation (6.14):

$$\frac{\partial \dot{\delta}}{\partial \delta} = \frac{\partial y}{\partial x}\left(\frac{\dot{\delta}_0}{\delta_{\max}}\right) \tag{6.14}$$

Substituting Equations (6.13a), (6.13b) and (6.14) into Equation (6.12) and rearranging gives Equation (6.15):

$$p + (0.2p + 1.3)\left(\frac{1 - \text{COR}}{\text{COR}}\right)\left(py + x\frac{\partial y}{\partial x}\right) = 0 \tag{6.15}$$

The elliptical relationship between indentation, δ, and its rate, $\dot{\delta}$, as expressed by Equation (6.16a), as illustrated in Figure 6.10(a), is adopted. Substituting Equations (6.13a) and (6.13b) into Equation (6.16a) gives Equation (6.16b), as illustrated in Figure 6.10(b).

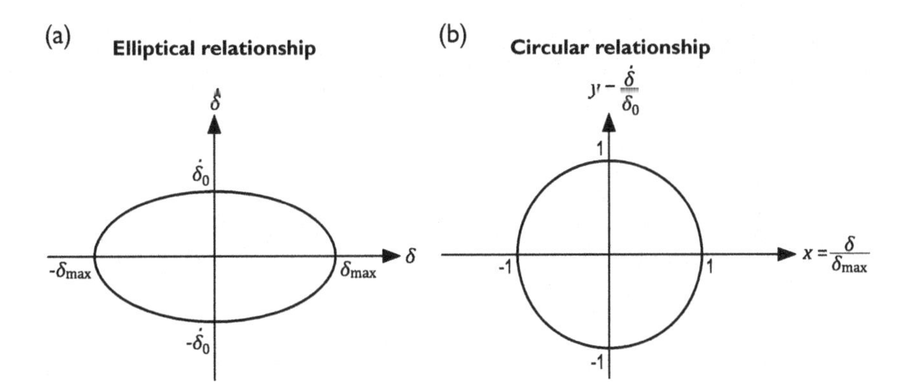

Figure 6.10 Adopted relationship of indentation with rate of indentation.

$$\left(\frac{\delta}{\delta_{max}}\right)^2 + \left(\frac{\dot{\delta}}{\dot{\delta}_0}\right)^2 = 1 \tag{6.16a}$$

$$x^2 + y^2 = 1 \tag{6.16b}$$

Differentiating Equation (6.16b) with respect to x results in Equation (6.17):

$$\frac{\partial y}{\partial x} = -\frac{x}{\sqrt{1 - x^2}} \tag{6.17}$$

Substituting Equation (6.17) into Equation (6.15) and making use of the stated x-y relationship gives Equation (6.18):

$$py + (0.2p + 1.3)\left(\frac{1 - \text{COR}}{\text{COR}}\right)(py^2 - 1 + y^2) = 0 \tag{6.18}$$

Equation (6.18) can be algebraically manipulated into Equation (6.19):

$$y^2 + \frac{p}{(p + 1)(0.2p + 1.3)}\left(\frac{\text{COR}}{1 - \text{COR}}\right)y - \frac{1}{p + 1} = 0 \tag{6.19}$$

The positive root of Equation (6.19), $y*$, can be expressed as Equations (6.20a) to (6.20c):

$$y* = \frac{-b + \sqrt{b^2 + 4c}}{2} \tag{6.20a}$$

$$b = \frac{p\text{COR}}{(p + 1)(0.2p + 1.3)(1 - \text{COR})} \tag{6.20b}$$

$$c = \frac{1}{p + 1} \tag{6.20c}$$

Substituting Equation (6.13a) into Equation (6.16b), and rearranging the terms, gives Equation (6.21):

$$\delta = \delta_{\max}\sqrt{1 - y^2} \tag{6.21}$$

Substituting Equations (6.11b), (6.13b) and (6.21) into Equation (6.11a), and rearranging the terms gives Equation (6.22). Note that Equation (6.22) is based on the case when $y = y*$ and $F_c = F_{c,\max}$.

$$F_{c,\max} = k_n \delta_{\max}^p (1 - y*^2)^{\frac{p}{2}} \left[1 + (0.2p + 1.3)\left(\frac{1 - \text{COR}}{\text{COR}}\right) y* \right] \tag{6.22}$$

where $y*$ is given by Equation (6.20a), and the value of $\delta_{\max}$ can be found using the expression as derived in the following.

By equal energy principles, ΔE from Equation (6.10a), is equated to the energy of absorption up to the point of maximum indentation (when there is no relative velocity between the two objects), as expressed by Equation (6.23). The right-hand side of the equal sign of Equation (6.23) is the integral of F_c, as defined by Equation (6.11a):

$$\frac{1}{2} M_c v_0^2 = \int_0^{\delta_{\max}} k_n \delta^p d\delta + \int_0^{\delta_{\max}} D_n \delta^p \dot{\delta} d\delta \tag{6.23}$$

Substituting Equation (6.11b) into Equation (6.23) and solving the integrals results in Equation (6.24):

$$\frac{1}{2} M_c v_0^2 = \frac{k_n \delta_{\max}^{p+1}}{p + 1} + k_n (0.2p + 1.3)\left(\frac{1 - \text{COR}}{\text{COR}}\right) \int_0^{\delta_{\max}} \frac{\delta^p \dot{\delta}}{\dot{\delta}_0} d\delta \tag{6.24}$$

Substituting Equation (6.13b) into Equation (6.16b), and rearranging the terms, gives Equation (6.25):

$$\dot{\delta} = \dot{\delta}_0 \sqrt{1 - x^2} \tag{6.25}$$

Substituting Equations (6.13a), (6.25) and $d\delta = \delta_{\max} dx$ from Equation (6.13a) into Equation (6.24), and rearranging the terms, gives Equation (6.26):

$$\frac{1}{2}M_c v_0^2 = \frac{k_n \delta_{max}^{p+1}}{p+1}\left[1 + (0.2p + 1.3)\left(\frac{1 - COR}{COR}\right)(p+1)\int_0^1 x^p\sqrt{1 - x^2}\,dx\right] \tag{6.26}$$

Solving for the remaining integral term gives the *beta* function, as shown by Equation (6.27):

$$\int_0^1 x^p\sqrt{1 - x^2}\,dx = \frac{1}{2}B\left(\frac{3}{2}, \frac{p+1}{2}\right) \tag{6.27}$$

Substituting Equation (6.27) into Equation (6.26) gives Equation (6.28):

$$\frac{1}{2}M_c v_0^2 = \frac{k_n \delta_{max}^{p+1}}{p+1}\left[1 + (0.2p + 1.3)\left(\frac{1 - COR}{COR}\right)\frac{(p+1)B\left(\frac{3}{2}, \frac{p+1}{2}\right)}{2}\right] \tag{6.28}$$

It is found that the last term of Equation (6.28), which involves the *beta* function, can be simplified into Equation (6.29):

$$\frac{(p+1)B\left(\frac{3}{2}, \frac{p+1}{2}\right)}{2} = \frac{1}{0.2p + 1.3} \tag{6.29}$$

Substituting Equation (6.29) into Equation (6.28) gives Equation (6.30):

$$\frac{1}{2}M_c v_0^2 = \frac{k_n \delta_{max}^{p+1}}{p+1}\left[1 + \left(\frac{1 - COR}{COR}\right)\right] \tag{6.30}$$

Equation (6.30) is then rearranged into Equation (6.31), which can be used for determining the amount of maximum indentation.

$$\delta_{max} = \left[\left(\frac{p+1}{2k_n}\right)M_c v_0^2\,COR\right]^{\frac{1}{p+1}} \tag{6.31}$$

In summary, the value of $F_{c,\max}$ may be calculated by combining the use of Equations (6.20a) to (6.20c), (6.22) and (6.31).

The accuracy of predictions obtained from the presented expressions have been validated by comparison with experimental measurements from collision testing of impactors made of ice, granite, concrete, masonry and timber [6.5, 6.6]. None of the models that have been proposed fully represents real behaviour as their derivations always involved approximations. Preference by the authors to adopt the H&C model over other models is mainly based on evidence of good agreement between the model predictions and experimental measurements (refer to Sections 14.6 and 14.7 for details). The use of the H&C model to approximate the forcing function generated by the collision of a vehicle onto the crash barrier of a bridge is presented in Chapter 10.

6.6.2 Hybrid Model for Prediction of the Maximum Contact Force

The non-linear elastic contact force model presented in Section 6.4 has the merit of simplicity, whereas the non-linear visco-elastic (H&C) model is able to represent real behaviour accurately. The non-linear elastic model has been modified into a hybrid model by incorporating a correction factor in order to give the same result as the H&C model. Derivations of the hybrid model are as follows.

Substituting Equation (6.31) into Equation (6.22), and rearranging the terms, gives Equation (6.32):

$$F_{c,\max} = k_n \left[\left(\frac{p+1}{2k_n} \right) M_c v_0^2 \right]^{\frac{p}{p+1}} (\mathrm{COR})^{\frac{p}{p+1}} \left[1 + (0.2p + 1.3) \left(\frac{1 - \mathrm{COR}}{\mathrm{COR}} \right) y_* \right] (1 - y_*^2)^{\frac{p}{2}} \tag{6.32}$$

Equation (6.32) can then be written in the form of Equation (6.33) by the introduction of the correction factor, $C_{F_{c,\max}}$. Note the resemblance between Equation (6.33) and Equation (6.3b).

$$F_{c,\max} = C_{F_{c,\max}} k_n \left[\left(\frac{p+1}{2k_n} \right) M_c v_0^2 \right]^{\frac{p}{p+1}} \tag{6.33}$$

$C_{F_{c,\max}}$ is defined by Equation (6.34):

$$C_{F_{c,\max}} = (\mathrm{COR})^{\frac{p}{p+1}} \left[1 + (0.2p + 1.3) \left(\frac{1 - \mathrm{COR}}{\mathrm{COR}} \right) y_* \right] (1 - y_*^2)^{\frac{p}{2}} \tag{6.34}$$

where y_* is as defined previously by Equations (6.20a)–(6.20c).

The hybrid model as defined by Equations (6.33) and (6.34) combines the merit of both models. Although a rather lengthy algebraic expression has been presented for the correction factor, it is a function of only two parameters: p and COR.

Given that the correction factor is only parameterised by p and COR, it can be presented in the form of design charts as shown in Figures 6.11(a)

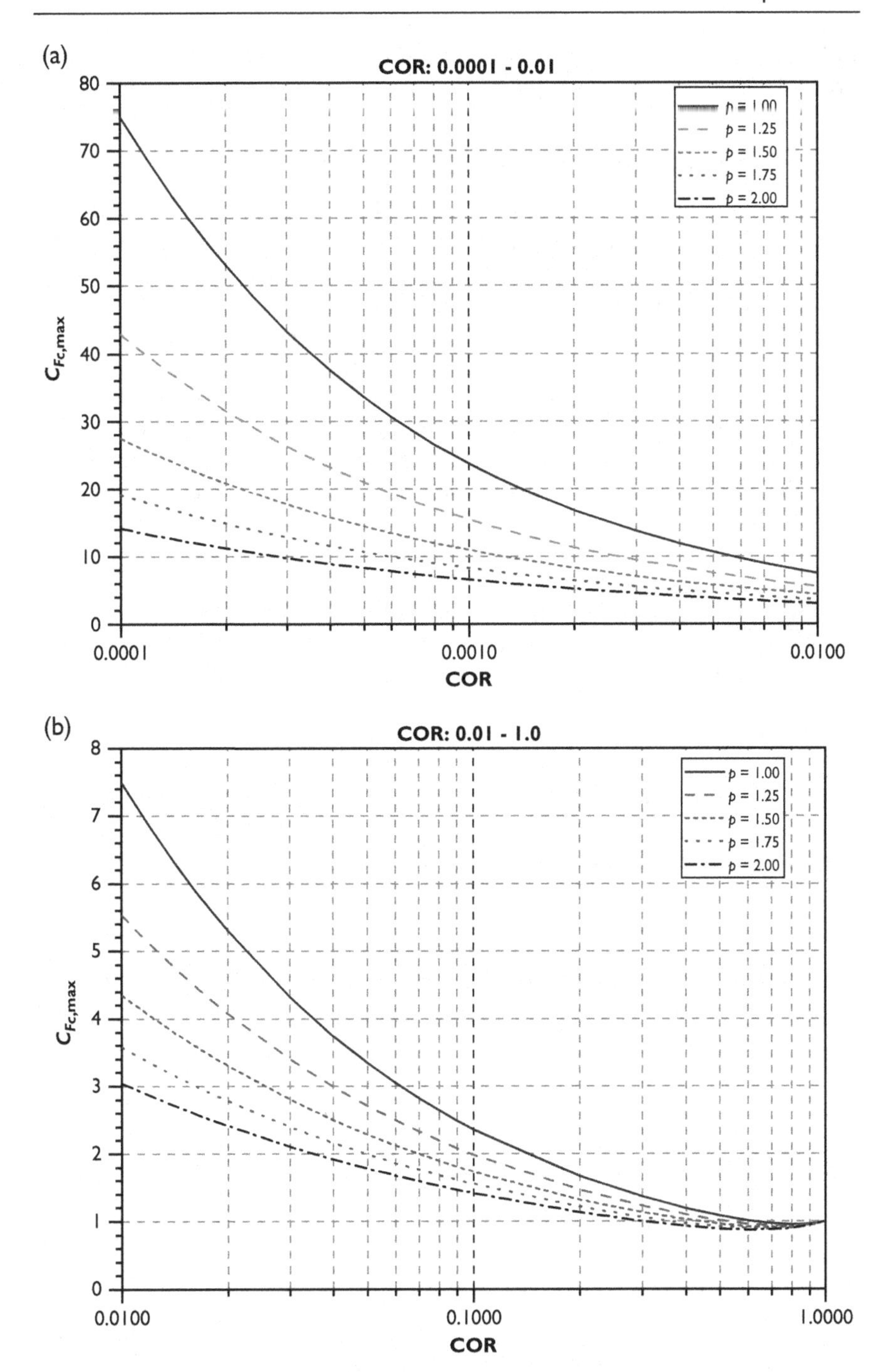

Figure 6.11 Correction factor ($C_{F_{c,\max}}$) in hybrid model.

and 6.11(b) for COR ranging from 0.0001 to 0.01 and 0.01 to 1, respectively. The value of $C_{F_{c,max}}$ is shown to vary between 1 and 75.

6.6.3 Determination of Model Parameters

Input parameters for characterising material properties in the H&C model are k_n, p and COR. Representative values of these parameters can be obtained from dynamic calibration involving the use of a gas gun to accelerate impactor specimens of up to 100 mm in diameter onto the measurement device, as illustrated in Figure 6.1. Impactor specimens that have been tested for calibration included spherical specimens of ice [6.5] and spherical specimens of windborne debris made of concrete, bricks and timber materials [6.6]. It was revealed in parametric studies that the value of k_n and p increased linearly with increasing velocity of the collision, whereas the value of COR representing energy dissipation decreases with increasing velocity of the collision. These correlation relationships are presented in Equations (6.35) to (6.38). Note that v_0 and k_n are in unit of m/s and MN/m^p, respectively, in the presented correlation relationships.

Correlation relationships derived for hailstones (ice) [6.5]:

$$k_n = 0.0022v_0 + 0.17 \tag{6.35a}$$

$$p = 0.01v_0 + 1.263 \tag{6.35b}$$

$$\text{COR} = -0.001v_0 + 0.049 \tag{6.35c}$$

Correlation relationships derived for concrete [6.6]:

$$k_n = 0.237v_0 + 2 \tag{6.36a}$$

$$p = 0.0209v_0 + 0.858 \tag{6.36b}$$

$$\text{COR} = -0.0043v_0 + 0.192 \tag{6.36c}$$

Correlation relationships derived for birch wood [6.6]:

$$k_n = 0.473v_0 + 0.2 \tag{6.37a}$$

$$p = 0.01v_0 + 0.91 \tag{6.37b}$$

$$\text{COR} = -0.0064v_0 + 0.503 \tag{6.37c}$$

Correlation relationships derived for brick [6.6]:

$$k_n = 0.177v_0 + 2 \tag{6.38a}$$

$$p = 0.0136v_0 + 1.116 \qquad (6.38b)$$

$$COR = -0.0012v_0 + 0.094 \qquad (6.38c)$$

When applying the listed correlation relationships, it is important to first consult the respective studies [6.5, 6.6] to acknowledge their limitations in terms of the size of the impactor and the velocity limits that the calibration studies were based upon. It is noted that gas gun testing has limitations, as an impactor of a size exceeding 100 mm in diameter cannot be fitted into the gun barrel. Drop tests or swing pendulum tests also have limitations because the velocity of impact that can be achieved is limited by the available headroom in the test venue. Impact tests conducted in the open can be very costly if the tests are to be sufficiently well instrumented to capture information required for calibration.

In a more recent investigation, undertaken by the authors and their co-workers to develop correlation relationships for the k_n, p and COR parameters for characterising the impact conditions of a rockfall scenario, static load tests were conducted on cylindrical specimens that were cored from a granite boulder [6.7]. To obtain the contact stiffness of a granite boulder impacting on a concrete surface, cylindrical specimens derived from both materials were tested in compression in a quasi-static manner using the configuration shown in Figure 6.12.

Results of the force-displacement relationships so recorded from the cylinder tests results have been subject to corrections to account for differences between conditions of the test rig and that of the collision scenario. The corrections were to allow for significant differences in the rate of loading and also differences in their respective boundary conditions. This newly introduced

Figure 6.12 Quasi-static testing of cylindrical cores.

technique of using quasi-static testings to predict dynamic behaviour was illustrated in Majeed et al. [6.7] based on testing materials cored from a granite boulder. Correlation relationships that have been derived in this cited reference for determining the k_n, p and COR parameters take into account the effects of the size of the impactor, as well as the velocity of impact; refer to Equations (6.39a) to (6.39d). Thus, the listed relationships are able to deal with boulders of any size.

Correlation relationships derived for granite colliding with concrete [6.7]:

$$k_n = k_{n100}\sqrt{\frac{D_i}{100}} \tag{6.39a}$$

$$k_{n100} = 84.273v_0 + 160.86 \tag{6.39b}$$

$$p = 1.21 + 0.0175v_0 + 0.0005D_i \tag{6.39c}$$

$$\text{COR} = 0.068\text{KE}_0^{-0.433} \tag{6.39d}$$

where k_n is the stiffness parameter of specimen with diameter D_i (in mm), k_{n100} (in MN/m^p) is the stiffness parameter of specimen of 100 mm in diameter and KE_0 is the impact energy (in kJ).

6.7 WORKED EXAMPLES

6.7.1 Hailstone Impact

A spherical piece of ice (i.e. hailstone) that is 62.5 mm in diameter, weighting 0.125 kg and moving with a velocity of 24.5 m/s is accelerated to strike a lumped mass, which weighs 1 kg and is connected to a spring with spring constant (k) of 50 kN/m. Estimate the value of the maximum contact force using the hybrid model presented in the chapter.

Solution

$$\lambda = \frac{1}{0.125} = 8$$

By Equation (6.10b):

$$M_c = \left(\frac{\lambda}{1+\lambda}\right)m = \left(\frac{8}{1+8}\right)(0.125) = 0.111 \text{ kg}$$

Equations (6.35a) to (6.35c) are used to determine the modelling parameters for hailstone impact.

$$k_n = 0.0022v_0 + 0.17 = 0.0022(24.5) + 0.17 = 0.224\ \text{MN/m}^p$$

$$p = 0.01v_0 + 1.263 = 0.01(24.5) + 1.263 = 1.508$$

$$\text{COR} = -0.001v_0 + 0.049 = -0.001(24.5) + 0.049 = 0.0245$$

By the use of the calculated values of p and COR, along with the use of Figure 6.11(b), the value of $C_{F_{c,\max}}$ is estimated to be of a value of around 3.

The maximum contact force is estimated with the use of Equation (6.33).

$$F_{c,\max} = C_{F_{c,\max}} k_n \left[\left(\frac{p+1}{2k_n} \right) M_c v_0^2 \right]^{\frac{p}{p+1}}$$

$$= 3(0.224 \times 10^6) \left[\left(\frac{1.508 + 1}{2(0.224 \times 10^6)} \right) (0.111)(24.5)^2 \right]^{\frac{1.508}{1.508+1}}$$

$$= 5.8\,\text{kN}$$

The simulation of the impact scenario using procedures presented in Chapter 13 gives the contact force vs indentation relationship (as shown in Figure 6.13) and the time-history of the contact force development. The calculation method presented in this chapter only gives estimates of the peak contact force and the maximum indentation.

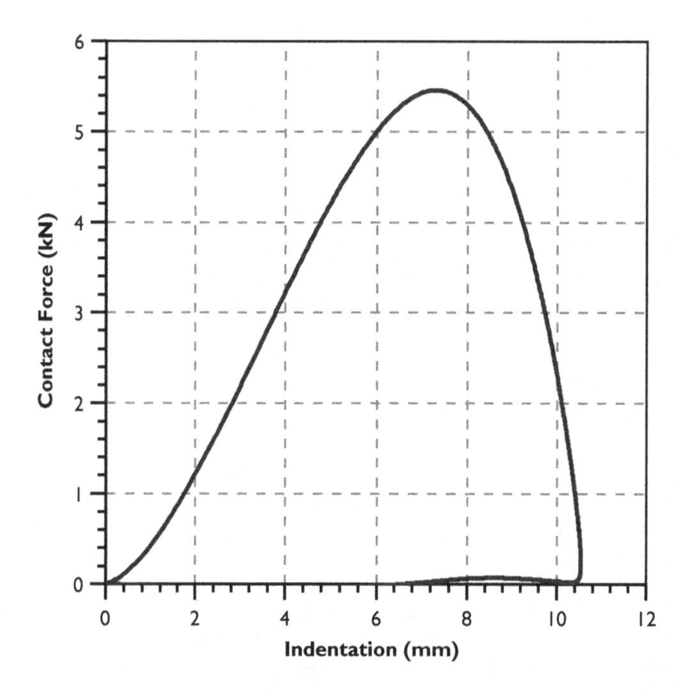

Figure 6.13 Contact force vs indentation of a hailstone impact.

The estimated $F_{c,\max}$ value of 5.8 kN is close to the simulated value of 5.5 kN, as shown in the graph of Figure 6.13. Equation (6.10a) may then be employed for estimating the energy of absorption by the deformation of materials (ΔE).

$$\Delta E = \frac{1}{2} M_c v_0^2 = \frac{1}{2}(0.111)(24.5)^2 = 33.3\,\text{J}$$

The amount of energy absorbed based on the calculation presented is consistent with the alternative approach of estimating the energy of absorption, where it is taken as the area under the graph in Figure 6.13. The latter approach gives an estimate of about 33.6 J, which is very similar to the value above of 33.3 J.

6.7.2 Concrete Debris Impact

A spherical concrete element that is 62.5 mm in diameter and weighing 0.3 kg is accelerated to strike a spring-connected lumped mass weighing 1 kg at an impact velocity of 24.5 m/s. Estimate the maximum contact force using the hybrid model presented in Section 6.6.2. Compare the estimated contact force with the quasi-static force developed in the rear spring, which has a spring constant (k) of 50 kN/m using expressions introduced in Chapter 3.

Solution

Use of non-linear visco-elastic (and hybrid) model for predicting $F_{c,\max}$:

$$\lambda = \frac{1}{0.3} = 3.33$$

By the use of Equation (6.10b):

$$M_c = \left(\frac{\lambda}{1+\lambda}\right)m = \left(\frac{3.33}{1+3.33}\right)(0.3) = 0.231\,\text{kg}$$

Equations (6.36a) to (6.36c) are used to determine the modelling parameters for concrete debris impact.

$$k_n = 0.237v_0 + 2 = 0.237(24.5) + 2 = 7.8\,\text{MN/m}^p$$

$$p = 0.0209v_0 + 0.858 = 0.0209(24.5) + 0.858 = 1.37$$

$$\text{COR} = -0.0043v_0 + 0.192 = -0.0043(24.5) + 0.192 = 0.087$$

By the use of the calculated values of p and COR, alongside the use of Figure 6.11(b), $C_{F_{c,\max}}$ is estimated to be of a value of around 2.

The maximum contact force is estimated using Equation (6.33).

$$F_{c,\,\max} = C_{F_{c,\,\max}} k_n \left[\left(\frac{p+1}{2k_n} \right) M_c v_0^2 \right]^{\frac{p}{p+1}}$$

$$= 2(7.8 \times 10^6) \left[\left(\frac{1.37+1}{2(7.8 \times 10^6)} \right) (0.231)(24.5)^2 \right]^{\frac{1.37}{1.37+1}} = 30.9\,\text{kN}$$

The use of equal momentum and energy principles for predicting the quasi-static force (V_b) for the same impact scenario is shown as follows. By the use of Equation (3.9a):

$$V_b = v_0 \sqrt{mk} \sqrt{\lambda \left(\frac{1+COR}{1+\lambda} \right)^2} = 24.5\sqrt{0.3(50 \times 10^3)} \sqrt{3.33 \left(\frac{1+0.087}{1+3.33} \right)^2}$$

$$= 1.4\,\text{kN}$$

It is shown that V_b, representing the quasi-static force, is an order of magnitude lower in intensity than $F_{c,\,\max}$, representing the maximum contact force.

Simulation of the impact scenario gives the contact force vs indentation relationship shown in Figure 6.14.

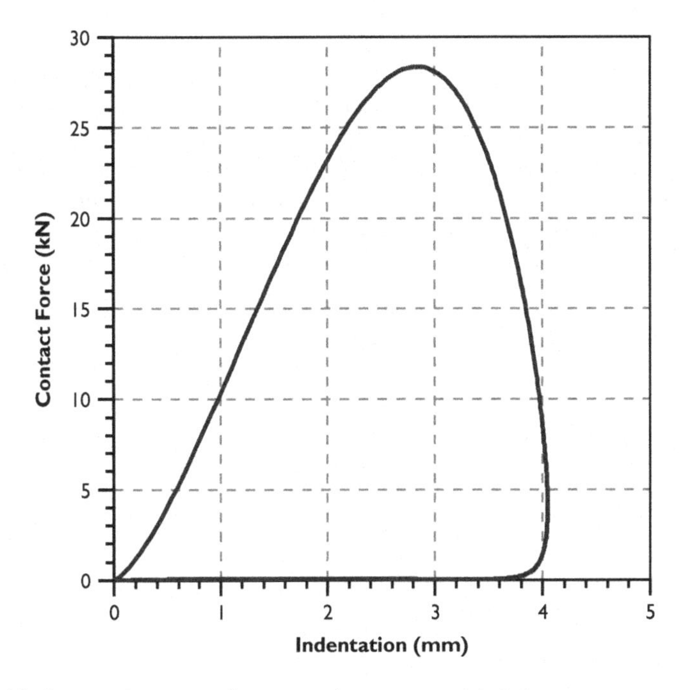

Figure 6.14 Contact force vs indentation of a concrete debris impact.

The estimated $F_{c,\max}$ value of 30.9 kN is close to the simulated value of 28.4 kN (based on employing the computational procedure presented in Chapter 13, as shown in Figure 6.14). Equation (6.10a) may then be used to estimate the amount of energy of absorption in the form of deformation of materials (ΔE).

$$\Delta E = \frac{1}{2}M_c v_0^2 = \frac{1}{2}(0.231)(24.5)^2 = 69.3\,\text{J}$$

The amount of energy absorbed based on the calculation presented is consistent with the alternative approach of estimating the energy of absorption, where it is taken as the area under the graph in Figure 6.14. The latter method gives an estimate of about 69.8 J, which is very similar to the value above of 69.3 J.

6.7.3 Granite Boulder Impact

A spherical granite boulder that is 1 m in diameter and weighing 1.4 tonnes strikes a "strong wall," which can be considered to have an infinitely large mass, at an impact velocity of 10 m/s. Estimate the maximum contact force using the presented hybrid model. Compare this estimate with that obtained using the *Hertzian* contact model, assuming a Young's modulus of 40 GPa for the boulder.

Solution

Use of a non-linear visco-elastic (and hybrid) model for estimating the maximum contact force:

As the target is considered have an infinitely large mass, $M_c \approx m = 1400$ kg.

Equations (6.39a) to (6.39d) are used to determine the modelling parameters of a granite boulder.

$$k_{n100} = 84.273v_0 + 160.86 = 84.273(10) + 160.86 = 1003.7\,\text{MN/m}^p$$

$$k_n = k_{n100}\sqrt{\frac{D_i}{100}} = 1003.7\sqrt{\frac{1000}{100}} = 3174\,\text{MN/m}^p$$

$$p = 1.21 + 0.0175v_0 + 0.0005D_i = 1.21 + 0.0175(10) + 0.0005(1000)$$
$$= 1.885$$

$$\text{KE}_0 = \frac{1}{2}mv_0^2 = \frac{1}{2}(1400)(10)^2 = 70\,k\text{J}$$

$$\text{COR} = 0.068\text{KE}_0^{-0.433} = 0.068(70)^{-0.433} = 0.0108$$

By the use of the calculated values of p and COR, alongside Figure 6.11(a), the value of $C_{F_{c,\max}}$ is estimated to be close to 3. The maximum contact force is estimated using Equation (6.33).

$$F_{c,\max} = C_{F_{c,\max}} k_n \left[\left(\frac{p+1}{2k_n} \right) M_c v_0^2 \right]^{\frac{p}{p+1}}$$

$$= 3(3174 \times 10^6) \left[\left(\frac{1.885+1}{2(3174 \times 10^6)} \right) (1400)(10)^2 \right]^{\frac{1.885}{1.885+1}} = 17.3\text{MN}$$

Use of the *Hertzian* contact model for estimating the maximum contact force is shown:

$$F_{c,\max} = \frac{4}{3} E\sqrt{R_c} \left(\frac{m v_0^2}{E\sqrt{R_c}} \right)^{\frac{3}{5}} = \frac{4}{3}(40 \times 10^9)\sqrt{0.5} \left(\frac{1400(10)^2}{40 \times 10^9 \sqrt{0.5}} \right)^{\frac{3}{5}}$$

$$= 24.7\text{MN}$$

This value is some 43% higher than that predicted by the hybrid model.

Simulation of the impact scenario employing the computational procedure presented in Chapter 13 gives the contact force vs indentation relationship shown in Figure 6.15.

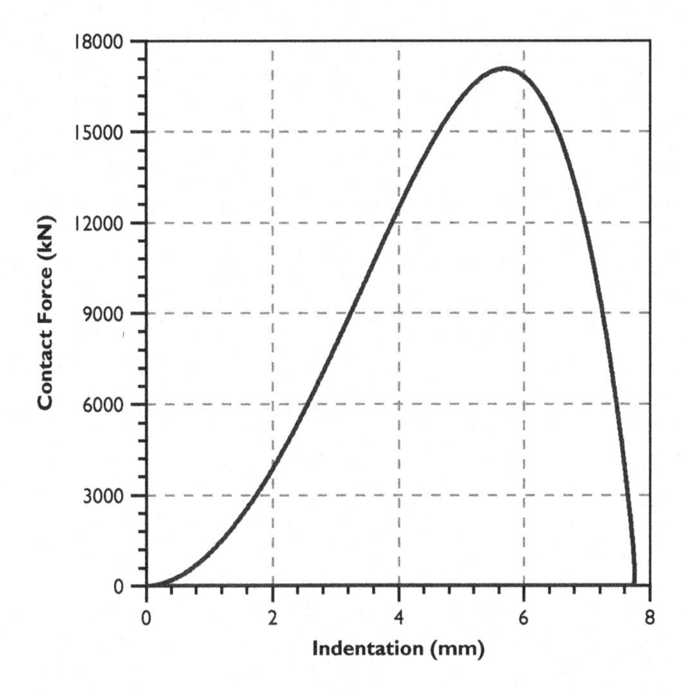

Figure 6.15 Contact force vs indentation of a granite boulder impact.

The estimated $F_{c,\max}$ value of 17.3 MN, as obtained from the hybrid model, is close to the simulated value of 17.1 MN, as shown in Figure 6.15. Equation (6.10a) may then be used to estimate the amount of energy of absorption in the form of deformation of materials (ΔE).

$$\Delta E = \frac{1}{2}M_c v_0^2 = \frac{1}{2}(1400)(10)^2 = 70\,\text{kJ}$$

The amount of energy absorbed based on the calculation presented above is consistent with the alternative approach of estimating the energy of absorption, where it is taken as the area under the graph in Figure 6.15. The two methods give almost identical estimates for the value of ΔE.

6.8 CLOSING REMARKS

This chapter initially defines what the contact force is in a collision scenario and then presents a method of measurement (for measuring the contact force) and the type of localised damage that can be caused by the localised action of a collision. A range of contact force models are then reviewed to discuss their merits and shortcomings. The adaptation of the *Hertzian* contact model for predicting the maximum contact force generated by a collision is introduced. A more elaborate model that can be used to give accurate predictions of the contact force is the visco-elastic non-linear contact force model. The derivation of a closed-form expression for predicting the value of the maximum contact force is presented in detail given that no such details have been reported in the literature. A hybrid model is then introduced to expedite calculations using the presented closed-form expressions. Numerous worked examples employing the presented contact force models are provided in the later part of the chapter to illustrate how the model is applied for different collision scenarios. With concrete debris as impactor, the magnitude of the contact force is shown to be an order of magnitude higher than the quasi-static force used to model the global behaviour of target element for the same collision scenario. With granite as the impactor, the maximum contact force predicted by the non-linear contact force model is shown to be some 43% higher than that predicted by the non-linear visco-elastic contact force model.

REFERENCES

6.1 Carney, K.S., Benson, D.J., DuBois, P., and Lee, R., 2006, "A phenomenological high strain rate model with failure for ice", *International Journal of Solids and Structures*, Vol. 43(25), pp. 7820–7839, doi:10.1016/j.ijsolstr.2006.04.005

6.2 Dousset, S., Girardot, J., Dau, F., and Gakwaya, A., 2018, "Prediction procedure for hail impact", *EPJ Web Conferen*, Vol. 183, p. 01046, doi:10.1051/epjconf/201818301046
6.3 Fischer-Cripps, A.C., 2007, *Introduction to Contact Mechanics*, Springer, U.S.
6.4 Hunt, K.H. and Crossley, F.R.E., 1975, "Coefficient of Restitution Interpreted as Damping in Vibroimpact", *Journal of Applied Mechanics*, Vol. 42(2), p. 440, doi:10.1115/1.3423596
6.5 Sun, J., Lam, N., Zhang, L., Ruan, D., and Gad, E., 2015, "Contact forces generated by hailstone impact", *International Journal of Impact Engineering*, Vol. 84, pp. 145–158, doi:10.1016/j.ijimpeng.2015.05.015
6.6 Perera, S., Lam, N., Pathirana, M., Zhang, L., Ruan, D., and Gad, E., 2016, "Deterministic solutions for contact force generated by impact of windborne debris", *International Journal of Impact Engineering*, Vol. 91, pp. 126–141, doi:10.1016/j.ijimpeng.2016.01.002
6.7 Majeed, Z.Z.A., Lam, N.T.K., Lam, C., Gad, E., and Kwan, J.S.H., 2019, "Contact force generated by impact of boulder on concrete surface", *International Journal of Impact Engineering*, Vol. 132, p. 103324, doi:10.1016/j.ijimpeng.2019.103324

Chapter 7

Cushioning of a Collision

7.1 INTRODUCTION

A layer of cushioning material can be placed in front of a barrier wall, or the external face of a facility exposed to collision hazards, to mitigate damage caused by a collision. Such a layer of cushioning can be a row of gabions, which are made up of steel cages filled with granular materials like crushed rocks or pebbles. Used tyres and bags of recycled glass or shredded rubber may also be used as fill materials for cushioning a collision. It is common practice to place gabions filled with granular materials in front of reinforced concrete (RC) hillside barriers to improve their ability to resist collision scenarios such as landslides and rockfalls (Figure 7.1).

Various mechanisms associated with cushioning are first explained (Section 7.2). An approach for modelling the cushioning of a collision action involves simulating the response behaviour of lumped masses, which are connected by springs, is then presented (Section 7.3). A parametric study employing this modelling methodology was undertaken to quantify the beneficial effects of cushioning for a range of parameters characterising both the impactor and the barrier. The deflection demand of a collision, as calculated by the use of expressions presented in Chapter 3, is modified by a calibrated reduction factor (γ) to represent the beneficial effects of cushioning. The application of this calculation procedure is illustrated with a worked example, which is concerned with a fallen boulder colliding on a hillside barrier protected by gabions (Section 7.4).

7.2 THE MECHANISMS OF CUSHIONING

The mechanisms of cushioning can be resolved into three components, which are presented in the following three points and illustrated in Figure 7.2.

1. **Pressure distribution mechanism.** The contact force is distributed onto a larger surface area in order that the intensity of the contact pressure transmitted to the surface of the target element behind the gabions is

DOI: 10.1201/9781003133032-7

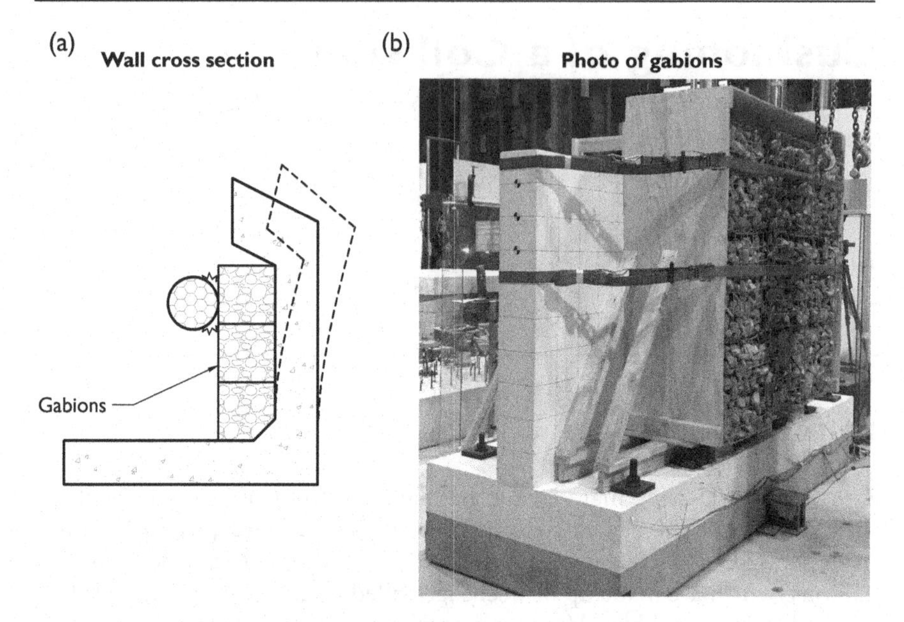

Figure 7.1 Gabion cushion of a rockfall barrier.

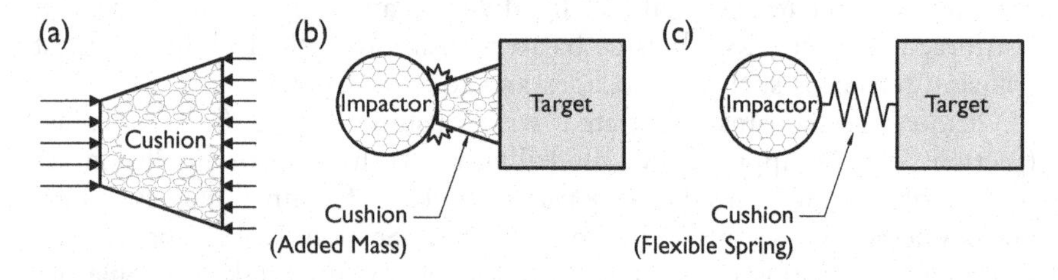

Figure 7.2 Mechanisms of cushioning.

reduced. However, the amount of impulse delivered to the target placed behind the cushion is not reduced. Refer to Figure 7.2(a).

2. **Added mass mechanism.** A conical volume of the cushion material surrounding the point of contact becomes an added mass to the target, thereby enhancing its inertial resistance to the impulse delivered by the collision. Refer to Figure 7.2(b).
3. **Delaying the transfer of momentum mechanism.** The cushion material essentially behaves as a flexible spring to prolong the duration of contact and hence the transfer of momentum. This delaying action serves to reduce the deflection demand of the collision. Refer to Figure 7.2(c).

In summary, both the localised actions and the impulsive actions of the collision are mitigated by cushioning. The amount of reduction in the contact pressure can be assessed easily, whereas quantifying influences by the last two mechanisms is less straightforward and is presented in detail in the following section.

7.3 MODELLING OF CUSHIONING ACTIONS

7.3.1 Two-Degree-of-Freedom Spring-Connected Lumped Mass System

A numerical simulation based on two spring-connected lumped masses, as shown in Figure 7.3, is employed for modelling the mitigating effects of cushioning. The lumped mass at the front represents the impactor, whereas the lumped mass at the rear represents the combined mass of the target (being the added mass of the cushion combined with the generalised mass of the stem wall of the barrier). A method for estimating the added mass (provided by the cushioning material) is presented in Section 7.3.2. The model configuration and derivation of the modelling parameters are presented in Section 7.3.3. The hysteresis behaviour of the frontal spring connecting the two lumped masses is represented by the non-linear visco-elastic model, which was first introduced in Chapter 6 (refer Section 6.6) for estimating the localised contact force. The stiffness parameter of the rear spring (k), characterising the flexural stiffness of the barrier wall, should take into account the effects of crack formation if the barrier wall is of RC construction. A simplified calculation methodology, which was derived by calibrating predictions against results from numerical simulations, is presented in Section 7.3.4. Details of modelling the flexural behaviour of a cracked RC element can be found in Chapter 8.

7.3.2 Added Mass of a Granular Material Filled Gabion Cushion

The added mass that acts as part of the combined target mass (λm), in response to the collision action, is a 20-degree cone of material formed

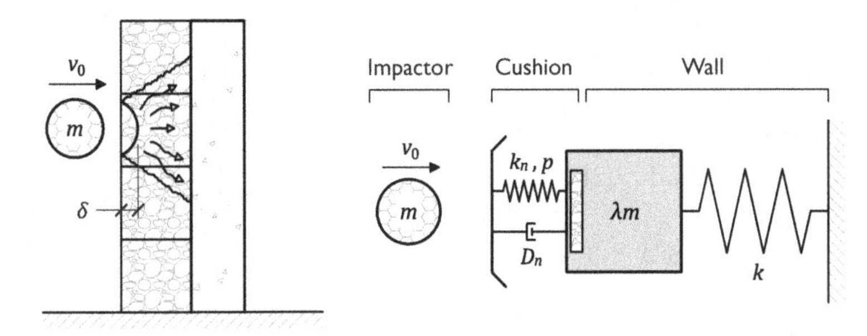

Figure 7.3 Modelling the mechanisms of cushioning.

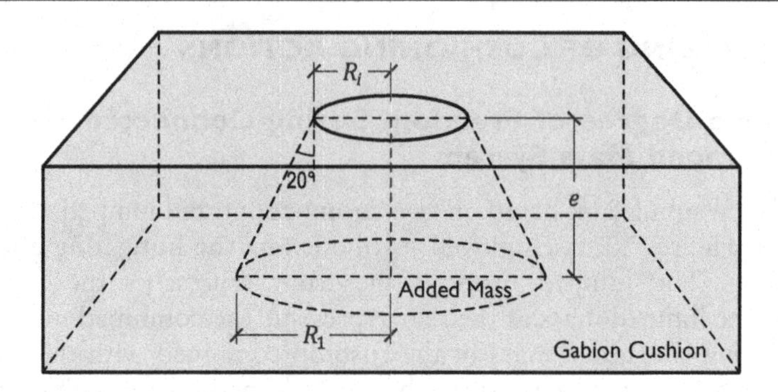

Figure 7.4 Conical model of added mass.

within the granular fill materials, as shown in Figure 7.4. The 20-degree cone was determined from experimental investigations. Thus, λm is the generalised mass of the stem wall combined with the added mass, as represented by Equation (7.1a). The generalised mass of a cantilevered element is approximately equal to 25% of the mass of the cantilever wall (i.e. m_w), as presented in Chapter 2. The added mass is taken as the product of the volume of the cone and the gross density of the granular fill material (ρ_g) as calculated using Equations (7.1b) and (7.1c):

$$\lambda m = 0.25 m_w + m_g \tag{7.1a}$$

$$m_g = \frac{\pi}{3}(R_1^2 + R_1 R_i + R_i^2) e \rho_g \tag{7.1b}$$

$$R_1 = R_i + e \tan 20° \tag{7.1c}$$

where R_i is radius of the equivalent sphere of the impactor; and e and R_1 are as defined in Figure 7.4.

7.3.3 Stiffness Behaviour of a Granular Material Filled Gabion Cushion

The non-linear visco-elastic model of the frontal spring, as defined by Equations (6.11a) and (6.11b) of Chapter 6, are re-written herein as Equations (7.2a) and (7.2b):

$$F_c = k_n \delta^p + D_n \delta^p \dot{\delta} \tag{7.2a}$$

$$D_n = (0.2p + 1.3)\left(\frac{1 - \mathrm{COR}}{\mathrm{COR}}\right)\left(\frac{k_n}{\dot{\delta}_0}\right) \tag{7.2b}$$

Exponent p in the expressions may be taken to be equal to unity in view of the observed linear behaviour from impact experiments [7.1]. The coefficient of restitution (COR) is very close to zero in view of the conditions of no rebound, i.e. impactor becomes embedded into the cushion material, as has been observed with cushioned impact. To avoid singularity, the value of COR = 0.01 (as opposed to COR = 0) is recommended. Numerical simulations based on the use of the two-degree-of-freedom (2DOF) spring-connected lumped mass model of Figure 7.3, and the assumption of $p = 1$ and COR = 0.01, may be adopted for estimating the deflection demand of a collision taking into account the mitigating effects of cushioning. The rest of this section is concerned with manual calculations of the modelling parameters.

The calculation of the maximum contact force, $F_{c,max}$, and maximum indentation, δ_{max}, from Equation (7.2a) was first introduced in Chapter 6. Equations (6.22) and (6.31) can be re-written as Equations (7.3a) and 7.3b, respectively. The combinational mass (M_c), as defined by Equation (6.10b), is re-written as 7.3c. Equations (6.20a) to (6.20c) are also re-written below as Equations (7.3d) to (7.3f):

$$F_{c,max} = k_n \delta_{max}^p (1 - y^{*2})^{\frac{p}{2}} \left[1 + (0.2p + 1.3) \left(\frac{1 - \text{COR}}{\text{COR}} \right) y^* \right] \tag{7.3a}$$

$$\delta_{max} = \left[\left(\frac{p + 1}{2k_n} \right) M_c v_0^2 \text{COR} \right]^{\frac{1}{p+1}} \tag{7.3b}$$

$$M_c = \left(\frac{\lambda}{1 + \lambda} \right) m \tag{7.3c}$$

$$y^* = \frac{-b + \sqrt{b^2 + 4c}}{2} \tag{7.3d}$$

$$b = \frac{p\text{COR}}{(p + 1)(0.2p + 1.3)(1 - \text{COR})} \tag{7.3e}$$

$$c = \frac{1}{p + 1} \tag{7.3f}$$

Equation (7.4) is derived by substituting 7.3b to (7.3f) into Equation (7.3a) and taking $p = 1$ and COR = 0.01. It is shown in the expression of Equation (7.4) that the value of k_n can be found once the amount of contact force generated by a pre-defined collision scenario is known. The scenario is characterised by the kinetic energy demand delivered by the impactor (KE_0).

$$k_n = \frac{F_{c,max}^2}{56 m v_0^2} \left(\frac{1 + \lambda}{\lambda} \right) \quad \text{or} \quad \frac{F_{c,max}^2}{112 \text{KE}_0} \left(\frac{1 + \lambda}{\lambda} \right) \tag{7.4}$$

Correlation between the amount of kinetic energy delivered by a collision and the maximum contact force generated in the granular fill materials, which are placed in a confined condition, is given by Equation (7.5):

$$F_{c,\max} = 1.82e^{-0.5}R_i^{0.7}M_E^{0.4}\tan\phi_k(\mathrm{KE}_0)^{0.6} \quad (7.5)$$

This closed-form expression is a modified version of an expression for predicting the value of $F_{c,\max}$ that was presented originally in the Swiss Code [7.2] as a function of the thickness of the cushion (e), radius of the impactor (R_i), modulus of elasticity (M_E) of the cushion and angle of shear resistance (ϕ_k) of the granular materials filling the gabions and the initial kinetic energy (KE_0). Accuracy of Equation (7.5), which is for dealing with horizontal impact, has been verified experimentally, with good correlations between the calculated values and the experiment results observed [7.1].

The value of ϕ_k is normally taken as 35 degrees for common soil materials, and 40 to 50 degrees for crushed rocks or pebbles. Gabions filled with crushed rocks (refer photo of Figure 7.1(b)) have been found by the authors and co-workers to have the value of M_E averaged at around 3,000 kPa in unconfined conditions and the value of ϕ_k averages at around 40 degrees.

The value of M_E may increase as a result of compaction of the granular fill materials following a previous collision in confined conditions. The amount of incremental increase is represented by a magnification factor, C_n, as defined by Equation (7.6), in which subscript "n" is the number of occurrences of collisions. For instance, C_2 should be taken as the magnification factor (i.e. $n = 2$) if the gabion cushion, which has already been collided upon, is anticipating a second collision.

$$M_{E,n} = C_n M_E \quad (7.6)$$

where $M_{E,n}$ is the value of M_E prior to the n-th occurrence of collision.

The values of C_n have been determined empirically and listed in Table 7.1 for $n = 1$ to 4 from a series of collision experiments that were performed on crushed rock gabions (refer Figure 7.1(b)).

Table 7.1 Values of magnification factor C_n

n	C_n
1	1.0
2	2.0
3	3.5
4	5.0

7.3.4 Simplified Method of Estimating Displacement Demand

Numerical simulations by the 2DOF spring-connected lumped mass system (Figure 7.3) were used in a parametric study to calculate the displacement demand of a cushioned impact for a range of parameters characterising the collision. Results obtained from the study are presented in the form of Equation (7.7), which is consistent in form to Equation (3.7b) of Chapter 3 except that the calibration factor (γ) is only introduced in this chapter to account for the effects of cushioning, i.e. the delay in the transfer of momentum as described in Section 7.2. Equation (7.7) assumes there is no rebound of the impactor and it becomes embedded into the cushioning material.

$$\Delta = \frac{mv_0}{\sqrt{mk}}\sqrt{\frac{1}{1+\lambda}} \times \gamma \tag{7.7}$$

The calibrated values of γ are presented in the form of a design chart (refer to Figure 7.5). If the value of λ is greater than 10, the λ = 10 line may be adopted given that the value of γ is not sensitive to variations in the value of λ when λ > 10. The effects of cushioning have been found to be controlled by the effective natural period of the frontal spring (T_m), as defined by

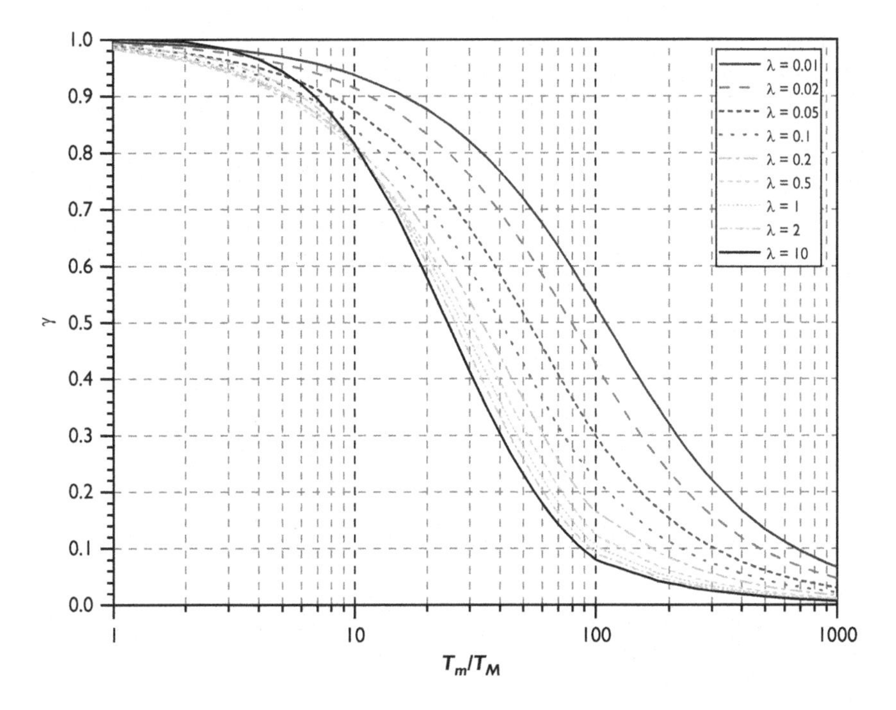

Figure 7.5 Cushion reduction factor, γ.

Equation (7.8a), in comparison with the natural period of vibration of the targeted structural element (T_M), as defined by Equation (7.8b). Thus, the design chart features the ratio of the natural periods ($T_m : T_M$) as the governing parameter alongside the mass ratio (λ). The use of Equation (7.7) alongside Figure 7.5 has been shown to provide accurate estimates of the deflection demand in large-scale collision experiments involving an RC wall, which is cushioned by granular material filled gabions [7.1], such as those presented in Figure 7.1(b).

$$T_m = 2\pi\sqrt{\frac{m}{k_n}} \tag{7.8a}$$

$$T_M = 2\pi\sqrt{\frac{\lambda m}{k}} \tag{7.8b}$$

7.4 WORKED EXAMPLE

A boulder weighing 4,680 kg and 1.5 m in diameter collides into an RC barrier wall at a velocity of 7 m/s, as shown in Figure 7.6. The stem wall, which is 0.8 m thick, 9 m long and 4.5 m tall (measured from the upper surface of the base slab) is protected by a 0.5 m thick gabion cushion. The granular fill (in the gabions) has a gross density of 1,500 kg/m^3, a modulus of elasticity of 3,000 kPa and a shear angle of resistance of 40 degrees. The stiffness behaviour of a cracked RC barrier wall is a topic to be dealt

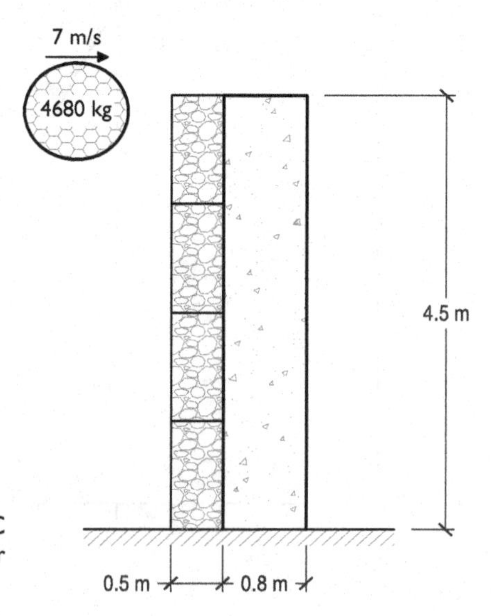

Figure 7.6 Collision of boulder on a RC stem wall cushioned by a layer of gabions.

with in detail in Chapter 8. In this example, the effective flexural rigidity (EI_{eff}) of the barrier wall, taking into account the cracking of RC, is taken as 2,730 MNm^2. Estimate the maximum deflection at the top of the wall resulting from the collision.

Solution

$$R_i = \frac{1.5}{2} = 0.75 \text{ m}$$

Equations (7.1a) to (7.1c) are used to estimate the added mass and the mass ratio.

$$R_1 = R_i + e \tan 20° = 0.75 + 0.5 \tan 20° = 0.93 \text{ m}$$

$$m_g = \frac{\pi}{3}(R_1^2 + R_1 R_i + R_i^2)e\rho_g = \frac{\pi}{3}(0.93^2 + 0.93(0.75) + 0.75^2)(0.5)(1500)$$
$$= 1672 \text{kg}$$

$$m_w = 2400(4.5)(9)(0.8) = 77760 \text{kg}$$

$$\lambda m = 0.25 m_w + m_g = 0.25(77760) + 1672 = 21112 \text{kg}$$

$$\lambda = \frac{\lambda m}{m} = \frac{21112}{4680} = 4.5$$

Equation (7.5) is used to determine the maximum contact force ($F_{c,max}$).

$$\text{KE}_0 = \frac{1}{2} m v_0^2 = \frac{1}{2}(4680)(7)^2 = 114.66 \text{kJ}$$

$$F_{c,max} = 1.82 e^{-0.5} R_i^{0.7} M_E^{0.4} \tan\phi_k (\text{KE}_0)^{0.6}$$
$$= 1.82(0.5)^{-0.5}(0.75)^{0.7}(3000 \times 10^3)^{0.4} \tan 40°(114.66 \times 10^3)^{0.6}$$
$$= 747 \text{kN}$$

The stiffness of the cushion can then be estimated by the use of Equation (7.4).

$$k_n = \frac{F_{c,max}^2}{112 \text{KE}_0}\left(\frac{1+\lambda}{\lambda}\right) = \frac{747^2}{112(114.66)}\left(\frac{1+4.5}{4.5}\right) = 53 \text{kN/m}$$

Equations (7.8a) and (7.8b) are then used for calculating the natural period of the cushion and the target, respectively.

$$T_m = 2\pi\sqrt{\frac{m}{k_n}} = 2\pi\sqrt{\frac{4680}{53 \times 10^3}} = 1.87\,\text{s}$$

The generalised stiffness of the stem wall can be calculated using expressions presented in Table 2.1 of Chapter 2.

$$k = \frac{3EI_{\text{eff}}}{H_w^3} = \frac{3(2730 \times 10^3)}{4.5^3} = 89877\,\text{kN/m}$$

$$T_M = 2\pi\sqrt{\frac{\lambda m}{k}} = 2\pi\sqrt{\frac{21112}{89877 \times 10^3}} = 0.096\,\text{s}$$

$$\frac{T_m}{T_M} = \frac{1.87}{0.096} = 19.5$$

By the use of the calculated values of λ and T_m/T_M, and alongside the design chart of Figure 7.5, a γ value of 0.6 is determined.

The maximum deflection generated by the cushioned collision action is then calculated using Equation (7.7), as shown in the following:

$$\Delta = \frac{mv_0}{\sqrt{mk}}\sqrt{\frac{1}{1+\lambda}} \times \gamma = \frac{4680(7)}{\sqrt{4680(89877 \times 10^3)}}\sqrt{\frac{1}{1+4.5}} \times 0.6$$
$$= 0.0129\,\text{m or } 12.9\,\text{mm}$$

7.5 CLOSING REMARKS

The use of gabions filled with crushed rocks, or pebbles, to provide extra protection to a RC hillside barrier is first introduced in the beginning of this chapter. Three mechanisms of cushioning are described. Two of the mitigation mechanisms, the added mass mechanism and the delaying of transfer of momentum mechanism, both contribute to the mitigation of the impulsive action of a collision. A numerical simulation involving the use of a spring-connected lumped mass system was introduced to represent these two mechanisms. The amount of displacement demand generated by the collision can be predicted accurately using the model, which takes into account the mitigating actions of cushioning. The concept of the natural period of the cushion (T_m) is then introduced as a parameter characterising the mechanism, which prolongs the transfer of momentum to the barrier. The effectiveness of cushioning can be characterised as function of the period ratio, which is T_m divided by the fundamental natural period of vibration of the targeted structural system (T_M). Results are presented in the form of a closed-form expression along with a design chart, which presents the cushioning reduction factor as a function of the period ratio (T_m/T_M) and the mass ratio (λ). The application of the modelling methodology is illustrated by a worked example, which deals with the collision of a fallen boulder on a hillside RC barrier wall.

REFERENCES

7.1 Perera, J.S., Lam, N., Disfani, M.M., and Gad, E., 2021, "Experimental and analytical investigation of a RC wall with a Gabion cushion subjected to boulder impact", *International Journal of Impact Engineering*, Vol. 151, p. 103823, doi:10.1016/j.ijimpeng.2021.103823

7.2 ASTRA, 2008, "Einwirkungen infolge Steinschlags auf Schutzgalerien (in German)", *Richtlinie, Bundesamt für Strassen, Baudirektion SBB, Eidgenössische Drucksachen-und Materialzentrale, Bern.*

Chapter 8

Collision on Concrete

8.1 INTRODUCTION

Analytical models presented in the earlier chapters of the book cover both the global actions (Chapters 3–5 and 7) and localised actions (Chapter 6) of a collision event. The predicted deflection, bending moment and contact force are not material specific, and the material presented in those chapters is applicable to all structural elements, irrespective of their material composition. However, there is also material-specific information for reinforced concrete (RC) that is relevant to its impact-resistant behaviour. Thus, a chapter that is devoted to collision actions on RC specifically is warranted, and more so given that RC is the most common material used in civil engineering construction.

This chapter deals with collision-inflicted damage to RC of different forms, ranging from localised responses such as denting, spalling or punching, to global responses such as yielding and plastic deformation, which are associated with excessive flexural actions. When assessing the likelihood of localised damage from a collision, the parameter characterising the demand of the impact is the maximum contact force ($F_{c,\max}$). Details concerning the determination of $F_{c,\max}$, which has been treated in Chapter 6, will not be repeated in this chapter. This chapter initially presents methodologies to assess denting, spalling and punching behaviour against $F_{c,\max}$. The chapter then presents a methodology to calculate the non-linear force-displacement response of RC barrier walls, including simplified empirical equations that can approximate the response, which can be used in conjunction with the displacement-based assessment procedures presented in Chapter 3. This chapter concludes with a series of worked examples.

8.2 DENTING AND SPALLING

Denting and spalling are two localised responses of RC from collision actions. Denting occurs when the surface of the concrete becomes

DOI: 10.1201/9781003133032-8

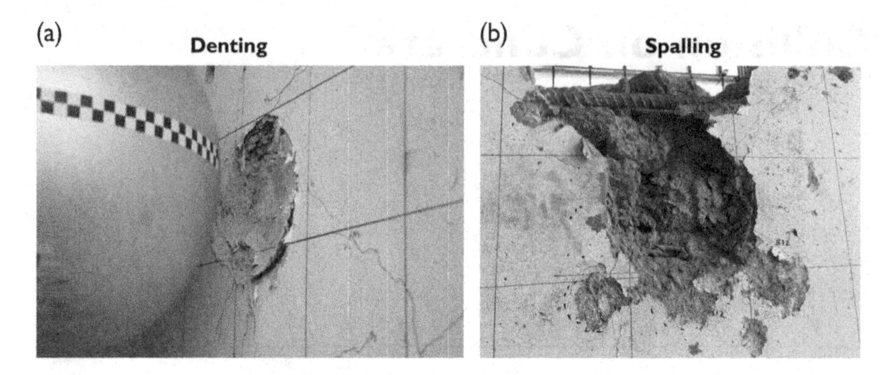

Figure 8.1 Localised behaviour of RC elements subject to collision actions.

permanently indented as a result of the localised action of a collision, yet no other visible signs of damage or concrete crushing can be observed, as shown in Figure 8.1(a). Spalling on the other hand occurs when the surface of the concrete becomes locally crushed, and indented, with visible damage around the point of collision, as shown in Figure 8.1(b). Denting should not result in the creation of any concrete debris, whereas spalling will result in some level of concrete debris and rubble being created.

The parameters that characterise the localised impact behaviour on the concrete are the maximum contact force ($F_{c,\,\max}$), the maximum contact stress (P_m) and the radius of the surface area of contact (r_a). Consider an impact scenario where the concrete surface is flat, and the part of the impactor that collides with the concrete is hemispherical in shape. The contact stress (P_m) experienced at the surface of the concrete (i.e. the target surface) can be determined using Equation (8.1), which is a function of the maximum contact force ($F_{c,\,\max}$). Readers are directed to Chapter 6, where further details on the determination of $F_{c,\,\max}$ can be found.

$$P_m = \frac{F_{c,\,\max}}{\pi r_a^{\,2}} \tag{8.1}$$

Using Hertzian contact theory [8.1,8.2], the value of r_a can be found using Equation (8.2) for an infinitely hard impactor, which makes contact with the surface of the concrete.

$$r_a = \left[\frac{3}{4}\left(\frac{R_i F_{c,\,\max}}{E_c}\right)\right]^{\frac{1}{3}} \tag{8.2}$$

where E_c is the Young's modulus of the concrete (i.e. the target material); and R_i is the radius of curvature of the hemispherical impactor that collides with the concrete surface.

For an impactor that has finite stiffness, the increase in size of the contact surface caused by the deformation of the impactor needs be accounted for. Thus, Equation (8.2) needs to be modified to incorporate deformation of the impactor, as opposed to just the concrete target surface. This is done by replacing parameter E_c with a transformed Young's modulus, E_T, from Equation (6.6a) of Chapter 6, as rewritten here by Equation (8.3):

$$E_T = \frac{1}{\frac{1-\nu_1^2}{E_1} + \frac{1-\nu_2^2}{E_2}} \tag{8.3}$$

where ν_1 and υ_2 are the Poisson's ratio of the impactor and concrete target surface, respectively; and E_1 and E_2 are the Young's modulus of the impactor and concrete target surface, respectively.

Substituting Equation (8.3) into Equation (8.2) and introducing a factor, φ, gives Equation (8.4):

$$r_a = \varphi \left[\frac{3}{4} R_i F_{c,\,\max} \left(\frac{1-\nu_1^2}{E_1} + \frac{1-\nu_2^2}{E_2} \right) \right]^{\frac{1}{3}} \tag{8.4}$$

where φ is an empirical based correction factor, which has its value ranging between 1.7 and 2.5 [8.3]. Large-scale collision testing by Majeed et al. [8.4] on a 230 mm thick RC wall revealed the φ value varying between 1.96 to 2.06 for a granite impactor and 2.34 to 2.48 for a harder structural steel impactor. In view of information that is available, adopting a rounded-off value of 2.0 for the value of φ is reasonable and conservative.

Combining Equations (8.1) and (8.4) provides a prediction of the P_m that is correlated with the deviatoric stress (σ_{vm}), which is also known as the *Von Mises* stress. The deviatoric stress is a significant parameter that controls the likelihood of denting (i.e. permanent deformation) into the surface of the concrete or the amount of spalling around the contact area. The spatial distribution of σ_{vm} (normalised with respect to P_m) is plotted in Figure 8.2 based on the use of expressions that were derived from classical principles of mechanics, as presented in Fischer-Cripps [8.5]. The figure shows how the normalised stress (σ_{vm}/P_m) varies across the depth (thickness) of the concrete element from the contact surface. The vertical axis is depth into the concrete target element (y), which is normalised against the radius of the surface area of contact (r_a). The various lines show the stress distribution at various distances of r (expressed as a fraction of r_a) away from the centre point of the impact.

It is shown in Figure 8.2 that high-stress intensity occurs within a distance of r_a (measured from the centre point of contact) and up to a depth of around 1.0 to 1.5 r_a into the target element. Within the zone of high-stress intensity, the value of σ_{vm} varies between 0.4 P_m and 1.0 P_m. In theory, the 1.0 P_m limit is

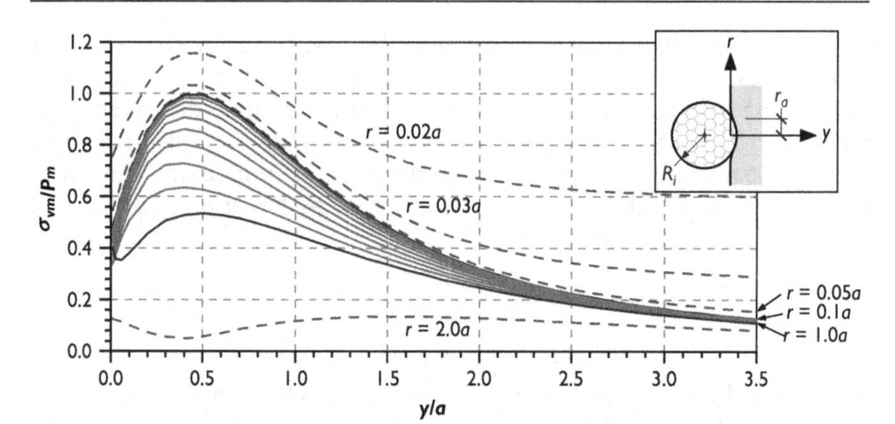

Figure 8.2 Normalised Deviatoric (*Von Mises*) stress in an elastic half space.

exceeded at the centre of the contact area (e.g. refer the dashed line $r = 0.02r_a$ in Figure 8.2). However, the exceedance can be neglected for the purpose of damage predictions because of its very localised nature (i.e. while the stress exceeds 1.0 P_m at $r = 0.02r_a$ it is reduced back to 1.0 P_m at $r = 0.06r_a$, as shown in Figure 8.2). A Poisson's ratio of 0.2 may be adopted for input into the expressions. As the value of ν deviates from 0.2 (and varies within the limits 0.15 to 0.3), the simulated stress intensities differ from what is shown in Figure 8.2 by only up to 4%. Thus, the simulated stress distribution is only weakly dependent on the value of ν. In summary, it is conservative enough to take the maximum value of $\sigma_{vm} = 1.0P_m$ for comparison with damage thresholds.

The occurrence of denting and spalling in concrete is controlled by the peak *Brinell* hardness (H_B) and yield strength (Y). Correlations between H_B and the compressive strength of concrete (f_c) can be found in Szilágyi et al. [8.6], as shown by Equation (8.5):

$$H_B = \frac{f_c - 10.829}{0.0684} \tag{8.5}$$

The actual in-situ concrete compressive strength should be used to determine H_B by use of Equation (8.5). In lieu of specific information being available for the concrete in question, the in-situ compressive strength may be taken as 0.88 times the specified characteristic cylinder strength (f'_c). Further, the "actual" strength can simply be taken as the mean strength, which can be approximated as $1.2f'_c$ for normal strength concrete (with characteristic strength of up to 65 MPa) and $1.1f'_c$ for higher-strength concrete. Longer-term development of in-situ strength should be ignored in view of the uncertainties. In summary, the value of f_c for input into Equation (8.5) may be taken as $1.06\,f'_c$ for normal strength concrete and $0.97\,f'_c$ for high-strength concrete.

Table 8.1 Prediction of denting and spalling

Condition	Results
$\sigma_{vm} < \frac{Y}{\sqrt{3}}$	No damage at the contact region
$\frac{Y}{\sqrt{3}} \leq \sigma_{vm} < \frac{H_B}{\sqrt{3}}$	Denting
$\sigma_{vm} \geq \frac{H_B}{\sqrt{3}}$	Spalling

The quoted conversion factors have been derived from a study by Menegon et al. [8.7] and are based on concrete test data in Australia. These values are likely to differ in other countries around the world and guidance can be found in other relevant documents (e.g. Eurocode 2 [8.8]).

Parameter Y is a ratio of H_B to C, where C is the constrain factor. The value of C may be taken as 1.5 for concrete given that 1.5 is applicable to brittle material [8.5]. Thus, the value of Y can be estimated using Equation (8.6):

$$Y = \frac{H_B}{1.5} \tag{8.6}$$

Expressions for predicting the occurrence of denting and spalling in concrete involving H_B and Y as input parameters are summarised in Table 8.1, which were originally recommended in Majeed et al. [8.4]. The use of the presented expressions for assessing the likelihood of denting and spalling is illustrated with a worked example in Section 8.5.

8.3 PUNCHING

Another form of localised damage to structural concrete that is inflicted by collision actions is punching failure. The maximum contact force ($F_{c,\max}$) is again used to characterise the localised demand of the collision event. To assess the likelihood of punching failure, the calculated value of $F_{c,\max}$ is to be compared with the calculated punching (shear) resistant/capacity, as defined by Equation (8.7):

$$F_{\text{capacity}} = K_t f_{ct} A_v \sin \alpha_v + F_{I,p} \tag{8.7}$$

where K_t is the *dynamic increase factor* [8.9]; f_{ct} is the direct tensile strength of concrete; A_v is the area of the shear failure surface; α_v is the angle of inclination of the shear cracks in the "shear plug" and can be taken as 45 degree for design purposes based on reports of observations from experimental investigations [8.10–8.13]; and $F_{I,p}$ is the inertial resistance of the shear plug, which is product of the mass of the shear plug and its acceleration at the instance when the maximum contact force occurs. K_t is dependent on strain rate ($\dot{\varepsilon}$) and f'_c as defined by Equations (8.8a) and (8.8b). Additional parameters α_t and β_t are defined by Equations (8.8c) and (8.8d), respectively. Typical values of strain rate for vehicular and rockfall collisions scenarios are, respectively, in the range of 10^{-5} to 10^{-3} (adopt 10^{-5} conservatively) and 10^{0} to 10^{2} (adopt 10^{0}, i.e. 1, conservatively) [8.14–8.16].

$$K_t = \left(\frac{\dot{\varepsilon}}{10^{-6}}\right)^{\alpha_t} \text{for } \dot{\varepsilon} \le 1\text{s}^{-1} \tag{8.8a}$$

$$K_t = \beta_t\left(\frac{\dot{\varepsilon}}{10^{-6}}\right)^{\frac{1}{3}} \text{for } \dot{\varepsilon} > 1\text{ s}^{-1} \tag{8.8b}$$

$$\alpha_t = \frac{1}{1 + 0.8f'_c} \tag{8.8c}$$

$$\log \beta_t = 6\alpha_t - 2 \tag{8.8d}$$

where $\dot{\varepsilon}$ is in s^{-1}; and f'_c is in MPa.

For a beam of rectangular cross section where punching failure is likely to mobilise a shear plug that extends across the full width of the beam, as shown in Figure 8.3, the area of the shear failure surface (A_v) can be found using Equation (8.9):

$$A_v = bd_o\sec\alpha_v \tag{8.9}$$

where d_o is the depth to the centroid of the outermost layer of longitudinal reinforcement from the extreme compressive fibre of the cross section; b is the width of the beam; and α_v is the angle of inclination, which is taken as 45 degrees.

For a flat slab or cantilever barrier wall where punching failure will mobilise a shear plug that takes the shape of a truncated cone, as shown in Figure 8.4(a), the area of the shear failure surface can be found using Equation (8.10a). In situations where the impact occurs near the edge of the concrete (e.g. near the top of a cantilever barrier wall), the reduced area of

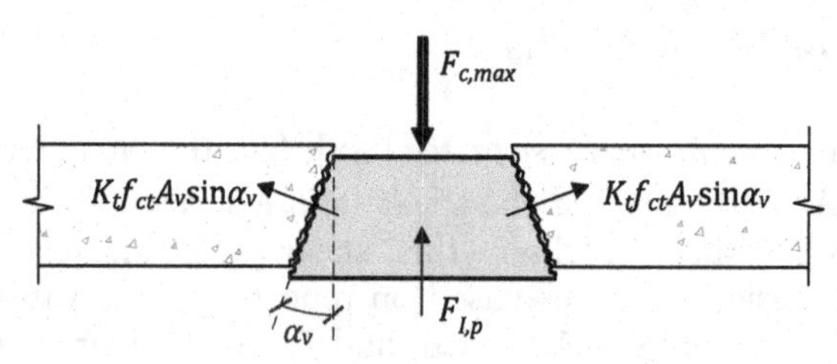

Figure 8.3 Punching of a RC beam of rectangular cross section.

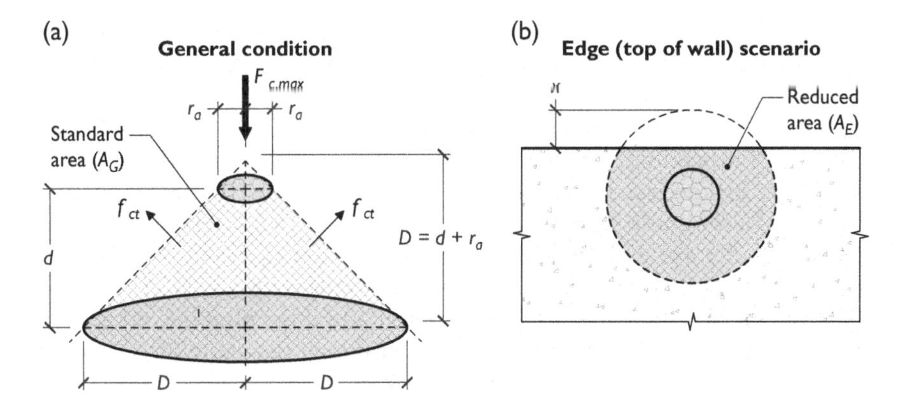

Figure 8.4 Punching of a RC flat slab. (a) General condition. (b) Edge (top of wall) scenario.

the truncated cone needs to be adopted, as shown in Figure 8.4(b), which can be calculated using Equation (8.10b). The area A_x is defined by Equation (8.10c):

$$A_G = \pi\sqrt{2}\,(d_{om}^2 + 2r_a d_{om}) \tag{8.10a}$$

$$A_E = A_G - A_x \tag{8.10b}$$

$$A_x = D^2\left[-2.14\left(\frac{x}{D}\right)^3 + 3.37\left(\frac{x}{D}\right)^2 + \frac{x}{D} - 0.01\right] \text{for } \frac{x}{D} \geq 0.01 \tag{8.10c}$$

where d_{om} is the mean value of depth to the longitudinal reinforcement in each direction; x is the distance from the edge of the concrete to the perimeter of the base of the theoretical cone of the shear plug (refer to Figure 8.4(b)).

The lower characteristic tensile strength of concrete (f_{ct}) stipulated by various codes of practices for normal strength concrete with f'_c of up to 50 MPa are summarised in Table 8.2, i.e. Equations (8.11a) to (8.11c).

Table 8.2 Equations of f_{ct} from codes of practices

References	f_{ct}	f_{ctm}	*Equation No.*
AS 3600 [8.17]	$0.36\sqrt{f'_c}$	$0.50\sqrt{f'_c}$	(8.11a)
NZS 3101.1 [8.18]	$0.38\sqrt{f'_c}$	$0.54\sqrt{f'_c}$	(8.11b)
fib 2010 Model Code [8.19] and Eurocode 2 [8.8]	$0.21(f'_c)^{\frac{2}{3}}$	$0.30(f'_c)^{\frac{2}{3}}$	(8.11c)

The value of f_{ct} as stipulated by these codes of practice are in the range of 2.0–2.1 MPa for grade 30 concrete and 2.3–2.5 MPa for grade 40 concrete. The mean tensile strength (f_{ctm}) can be taken as 1.4 times f_{ct} according to AS 3600 or 1/0.7 = 1.43 times f_{ct} according to NZS 3101.1, fib 2010 Model Code and Eurocode 2. The value of f_{ctm} as stipulated by these codes provisions is in the range of 2.8–2.9 MPa for grade 30 concrete and 3.2–3.5MPa for grade 40 concrete. Stipulations by AS 3600 [8.17] for both concrete grades are slightly more conservative than other code stipulations.

In this chapter, conservative estimates are made without taking into account the inertial resistance developed within the shear plug (i.e. taking $F_{I,p}$ in Equation (8.7) to be equal to zero). Determining the value of $F_{I,p}$ would involve dynamic simulations, which are covered in Chapter 13. In summary, the information presented enables the value of input parameters into Equation (8.7) for determining the value of $F_{capacity}$ to be found.

Additional measures may be adopted from a robustness perspective to ensure that if punching failure does occur, the impactor (e.g. a fallen boulder) would not go through the barrier wall following the occurrence of punching failure. Localised failure of the impacted element (e.g. a cantilevered RC barrier wall) may be acceptable with certain rare scenarios provided that the impactor can be contained. In such a scenario, containment can be achieved by the provision of a continuous matt of reinforcement on the front face (impacting face) of the wall. The reinforcement matt restrains the shear plug by catenary action following the occurrence of punching, as shown in Figure 8.5(a). This concept is likened to the "structural integrity" reinforcement that is recommended in flat slab construction (in RC buildings), which consists of continuous bottom reinforcement running through the column joint. The structural integrity reinforcement prevents catastrophic collapse of the floor in the event of a localised punching shear failure around the column by allowing the slab to hang from the column in catenary action, as shown in Figure 8.5(b) and further discussed in [8.20–8.22]. In the latter scenario, the bottom bars are assumed to "hang" at a 30-degree angle downwards from the column joint

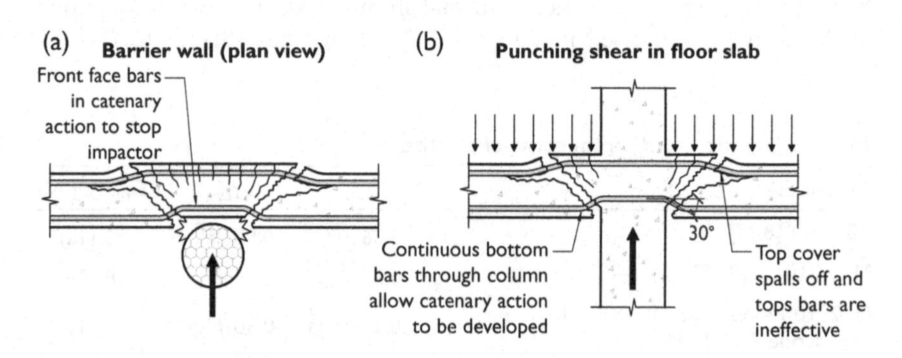

Figure 8.5 Punching of a RC flat slab.

and therefore the area of reinforcement required can be calculated accordingly, i.e. Equation (8.12):

$$A_s = \frac{2N^*}{\phi f_{sy}} \tag{8.12}$$

where A_s is the cumulative area of steel from each face of the column that has continuous bottom bars running through it; N^* is the total floor load that needs to be supported; f_{sy} is the yield stress of the reinforcement; and ϕ is a capacity reduction factor (i.e. safety factor).

In the context of a collision scenario causing punching failure, unlike the floor scenario as depicted in Figure 8.5(b) where the load supported in catenary action is a sustained load from gravity, the collision action is transient in nature, and the outcome of the collision can be sensitive to the (displacement) behaviour of the target element. As such, the initial maximum contact force ($F_{c,\max}$) that would be used to check the wall for punching failure, would not necessarily be the appropriate force to use in Equation (8.12), since the energy required to cause punching failure in the first instance would also absorb a portion of the energy associated with $F_{c,\max}$. Therefore, substituting $F_{c,\max}$ straight into Equation (8.12) might yield a very conservative result. However, in lieu of any research efforts being undertaken in this respect, the front face reinforcement could conservatively be quantified using $F_{c,\max}$ and Equation (8.12).

The behaviour as described was observed in a full-scale collision test performed by the authors [8.23], as shown in Figure 8.6, which is discussed in detail in Chapter 14. The full-scale test consisted of a 230 mm thick cast in-situ RC barrier wall that was 1.5 m tall and 3 m long. The wall was impacted with different size impactors that were dropped from various heights. The initial collision scenarios caused a linear elastic response of the wall. The final collision scenario consisted of a 1 tonne steel impactor released from a height of 1.575 m and resulted in an inelastic flexural

Figure 8.6 Catenary action prevented punching failure in a full-scale impact test.

response of the wall together with punching failure at the point of contact. While punching failure occurred, and a shear plug matching what is shown in Figure 8.4 was mobilised, the shear plug remained attached to the wall and the impactor did not go through the wall, as shown in Figure 8.5 (showing the final state of the wall following the collision event). The satisfactory performance in containment was due to the continuous matt of front face reinforcement in the wall across the point of impact.

8.4 BENDING (FLEXURAL RESPONSE)

The displacement of reinforcement concrete (RC) elements consists of a combination of flexural deformation and shear deformation. For slender elements, which would usually be associated with collision scenarios, the flexural deformation dominates the response and therefore the shear deformation can usually be ignored. However, where collision scenarios are considered for short (or stocky) elements, which would be dominated by shear deformations, the overall displacement response of the element is generally of less concern and localised behaviour (as discussed in Sections 8.2 and 8.3) would be the primary concern. The method presented in this section was originally intended for calculating the flexural displacement behaviour of a barrier wall, which is essentially a cantilevered element. The method can be extended to a pole that is fixed at the base, or to a cantilevered beam.

Reinforced concrete differs from an isotropic material (e.g. steel or plastic) as it typically relies on concrete for its compressive strength and then its steel reinforcement for its tensile strength. The steel reinforcement does not take the entirety of the tensile load until after the concrete cracks in tension. The compressive behaviour (e.g. stress-strain behaviour) is different to its tensile behaviour, of which the latter is further complicated since the concrete can either be cracked or uncracked. As such, RC is a non-linear responding material, even under low magnitudes of loading/displacement. The flexural displacement behaviour is usually simplified into a bilinear response (i.e. elastic and inelastic response) or trilinear response (i.e. uncracked, cracked/elastic and inelastic response).

8.4.1 Flexural Displacement Behaviour of a Barrier Wall

The focus herein will be placed on the displacement behaviour of cantilever barrier walls, i.e. an RC wall being bent about its minor (or weak) axis. The reader is directed to other texts for modelling the in-plane displacement behaviour of RC walls (e.g. Menegon et al. [8.24] and Priestley et al. [8.25]). The response of barrier walls can usually be approximated to a reasonable level of accuracy using a bilinear model. The bilinear response is defined by an elastic branch (i.e. Equations (8.13a) and (8.13b)) and

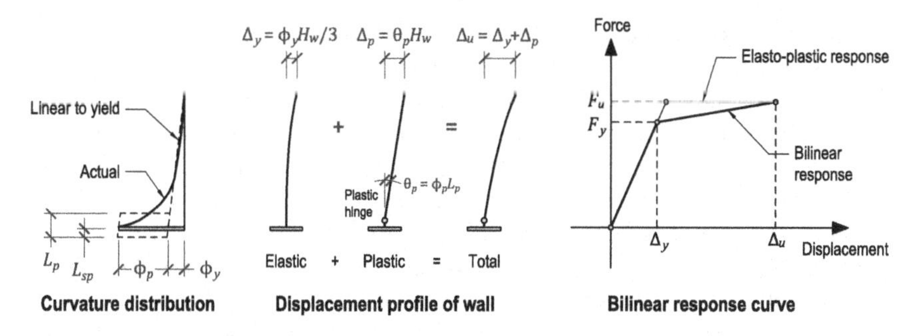

Figure 8.7 Bilinear linear flexural response of a barrier wall.

inelastic branch (i.e. Equations (8.13d) to (8.13f)), as shown in Figure 8.7. The effective stiffness of the barrier wall can be taken as the slope of the elastic branch, i.e. Equation (8.13c). The bilinear response can be further simplified to an elasto-plastic behaviour (as presented and used in Chapter 3, Section 3.9 for modelling inelastic collision scenarios) by extending the elastic branch up to F_u and making the inelastic branch a horizontal line, as shown in Figure 8.7.

$$\Delta_y = \frac{\phi_y H_w^2}{3} \tag{8.13a}$$

$$F_y = \frac{M_y}{H_w} \tag{8.13b}$$

$$k_{\text{eff}} = \frac{F_y}{\Delta_y} \tag{8.13c}$$

$$\Delta_u = \Delta_y + \Delta_p \tag{8.13d}$$

$$\Delta_p = \theta_p H_w = \phi_p L_p H_w \tag{8.13e}$$

$$F_u = \frac{M_u}{H_w} \tag{8.13f}$$

where Δ_y, Δ_u and Δ_p are the yield, ultimate and plastic displacement of the wall, respectively; ϕ_y, ϕ_u and ϕ_p are the yield, ultimate and plastic curvature of the wall, respectively; θ_p is the plastic rotation at the base of the wall; L_p is the plastic hinge length at the base of the wall; M_y and M_u are the yield and ultimate moment capacity of the wall, respectively; F_y and F_u are the yield and ultimate lateral capacity of the wall, respectively; and H_w is the height of the wall.

The yield displacement is calculated based on the assumption of a linear curvature distribution up the height of the wall (as shown in Figure 8.7) and the plastic displacement is calculated by assuming a uniform distribution of inelastic curvature concentrated across a region at the bottom of the wall referred to as the plastic hinge. This approach is broadly in line with the recommendations in Priestley et al. [8.25].

The yield, ultimate and plastic curvatures can be calculated by performing a moment-curvature analysis on the cross-section of the barrier wall, which can simultaneously be used to calculate the yield moment and ultimate moment capacity. Alternatively, these parameters can be approximated using empirical equations. Both the former and the latter will be presented in the subsequent sub-sections, respectively. When calculating the capacity of the barrier wall, the width (or length) of the wall assumed to resist the collision scenario can be taken as two times the height of the wall, where the height of the wall is taken as the point of collision, as per the recommendations presented in Yong et al. [8.23] and Yong [8.26]. Similarly, when calculating the capacity of the supporting cantilevered deck slab adjacent to a barrier, the width (or length) of the slab can be taken as twice the height of the barrier.

The plastic hinge length can be calculated using Equation (8.14a), which was proposed by Priestley et al. [8.25]. The equation was originally developed for RC columns; however, in lieu of a specifically developed equation for barrier walls, Equation (8.14a) should provide an adequate approximation:

$$L_p = kH_w + L_{sp} \geq 2L_{sp} \tag{8.14a}$$

$$k = 0.2\left(\frac{f_{su}}{f_{sy}} - 1\right) \leq 0.08 \tag{8.14b}$$

$$L_{sp} = 0.022 f_{sy} d_b \tag{8.14c}$$

where L_{sp} is the strain penetration in the supporting element (in unit of mm when the input parameters into Equation (8.14c) are expressed in units of MPa and mm); H_w is the wall height; f_{sy} and f_{su} are the yield and ultimate stress of the vertical reinforcement, respectively; and d_b is the bar diameter of the vertical reinforcement.

The approach outlined above assumes two key factors. The first factor is the assumption that the vertical reinforcement is fully developed at the base of the wall (to prevent pull-out failure) and that ductile reinforcement is specified. For example, in the Australian context, ductile reinforcement broadly refers to grade N reinforcement to AS/NZS 4671 [8.27], which has a minimum characteristic strain hardening ratio (f_{su}/f_{sy}) of 1.08 and

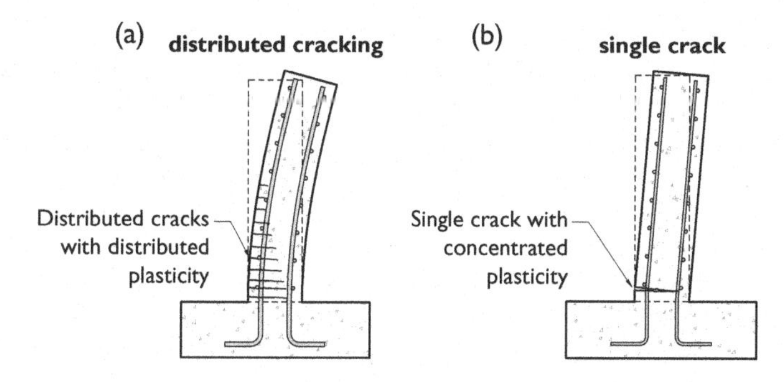

Figure 8.8 Wall cross sections showing distributed and single crack formation.

minimum characteristic ultimate strain (or uniform elongation) of 5%. Grade N reinforcement typically has a mean (or expected) strain hardening ratio and ultimate strain of 1.13 to 1.29 and 6.6% to 12.0%, respectively, depending on the suppliers manufacturing targets [8.7]. The adoption of low ductility reinforcement (e.g. welded mesh or grade L reinforcement in the Australian context) significantly limits the walls ductility and sudden fracturing of the reinforcement would be very likely as soon as the yield displacement of the wall is exceeded. Further, low ductility reinforcement would similarly not be effective at developing the catenary behaviour discussed previously, with respect to punching failure and Figure 8.5(a).

The second factor is the assumption that multiple cracks can form in the wall and a single crack mechanism is not developed, as illustrated in Figures 8.8(a) and 8.8(b). A single crack mechanism develops when the tensile strength of the concrete is greater than the tensile capacity of the reinforcement (i.e. the cracking moment of the section exceeds the yield/ultimate moment of the section) and is an undesirable mechanism as the subsequent plastic tensile strains are concentrated in a single location, which reduces the displacement capacity of the wall. Distributed cracking should be achieved in barrier walls when the yield moment of the wall is at least 1.2 times greater than the cracking moment capacity of the wall, which should be achieved when the reinforcement ratio, i.e. Equation (8.15), exceeds the threshold values listed in Table 8.3. The reinforcement ratios

Table 8.3 Minimum reinforcement content for barrier walls reinforced with 500 MPa reinforcement

Characteristic Strength of Concrete (MPa)	*Minimum Reinforcement Ratio (%)*
32	0.35
40	0.45
50	0.55

are dependent on the concrete grade, since as the concrete compressive strength increases, the tensile strength and cracking moment increase accordingly. Further, they are also dependent on the yield stress of the vertical reinforcement, since as the yield stress increases, a reduced area of reinforcement is required to achieve the same yield moment in the wall. The values in Table 8.3 are for reinforcement with a yield stress of 500 MPa. The values can be modified to suit other grades of reinforcement by multiplying them by the ratio of 500 divided by the yield stress of the different grade.

The concept of distributed cracking is discussed in further detail in Menegon et al. [8.28], with respect to the in-plane behaviour RC walls. The minimum reinforcement ratios in Menegon et al. [8.28] are different than Table 8.3, since they are related to the overall reinforcement ratio (i.e. the reinforcement on both faces of the wall contribute) and are developed with respect to in-plane behaviour.

$$p_{st} = \frac{A_{st}}{bd_o} \tag{8.15}$$

where A_{st} is the total area of vertical reinforcement on the tension face of the wall; b is width of the wall; and d_o is the depth to the centroid of the outermost layer of vertical (tension) reinforcement from the extreme compressive fibre of the wall.

8.4.2 Moment-Curvature Behaviour

Accurate information about the moment-curvature behaviour of the barrier wall is essential for determining the input parameters into Equations (8.13a) to (8.13f). The moment-curvature behaviour of the wall can be determined by undertaking a non-linear fibre-element analysis on the typical cross section of the wall. A fibre-element analysis involves breaking the cross section into a series of discrete concrete slices with the reinforcement superimposed on top. Non-linear stress-strain models are adopted for both the concrete and reinforcement, which are used to calculate the force at each respective concrete slice and reinforcing bar location. The analysis is an iterative process where pairs of moment-curvature response points are calculated by balancing the forces in the cross section using the non-linear stress-strain material models for a given reference strain for each respective pair of points.

There are many software packages available for undertaking moment-curvature analysis on RC sections, including the program WHAM [8.29], which can be downloaded from [8.30]. WHAM was primarily developed as a user-friendly and transparent non-linear analysis program for RC sections. The program uses Microsoft Excel spreadsheets and Visual Basic for Applications (VBA) subroutines to complete the analysis.

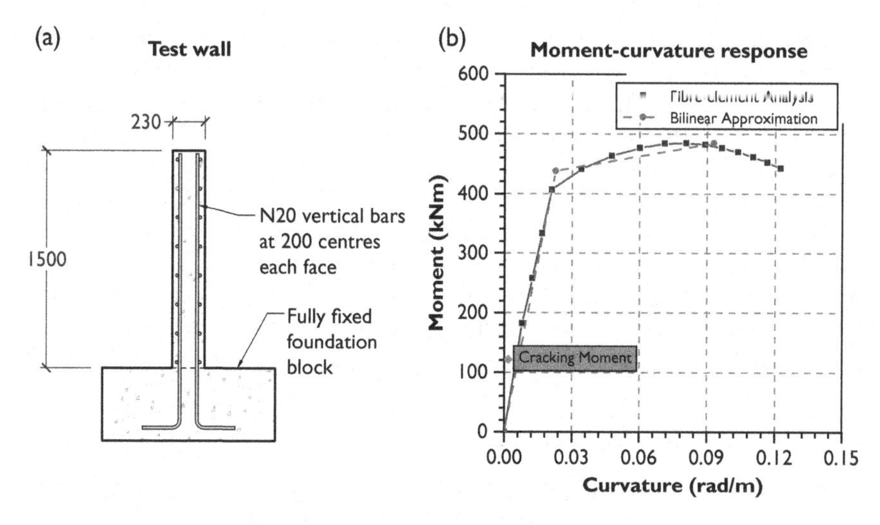

Figure 8.9 Moment-curvature output from the program WHAM for the barrier wall tested in Yong et al. [8.23].

WHAM adopts the very popular and widely cited Mander et al. [8.31] stress-strain model for concrete and a simple bilinear stress-strain model for reinforcement (although it does allow the user to input custom models for concrete and reinforcement if they so choose). The program accounts for tension stiffening using the model developed by Menegon et al. [8.32]. The moment-curvature output from the program is illustrated in Figure 8.9 for the large-scale barrier wall test performed by the authors [8.23], which is discussed in detail in Chapter 14. The wall was a 230 mm thick cast in-situ RC barrier wall that was 1.5 m tall and 3 m long. A bilinear approximation of the moment-curvature response, highlighting the yield curvature, yield moment, ultimate curvature and ultimate moment capacity, is also presented in Figure 8.9.

The bilinear approximation is constructed by assuming the yield point corresponds to the point on the moment-curvature response curve when either the tensile strain in the extreme reinforcing bar reaches its yield stress or the extreme concrete compressive fibre reaches a strain of 0.002 (i.e. the strain value on the stress-strain curve corresponding to the maximum compressive stress being reached, for normal strength concrete). The ultimate point is assumed to be when the tensile strain in the extreme reinforcing bar reaches a strain of 0.04 or the extreme concrete compressive fibre reaches a strain of 0.004. A tensile strain limit of 0.04 is recommended by Menegon et al. [8.33] to prevent local bar buckling of the vertical reinforcement under reversed cyclic loading in walls without ligatures (i.e. cross-ties). A compressive strain limit of 0.004 is recommended by Priestley et al. [8.25] for columns or walls without ligatures confining the compressive region (which is the same value they recommend for the cover concrete in confined sections).

When undertaking advanced non-linear methods of analysis, the "likely" in-situ material properties should be adopted in the analysis rather than the minimum characteristic material properties that are commonly used in design. Detailed guidance with respect to the expected material properties for standard grades of reinforcement and concrete available in Australia is presented in Menegon et al. [8.7]. A sensitivity analysis should also be performed to assess how the expected variability in the input parameters can affect the design and overall behaviour of the barrier wall.

8.4.3 Empirical Equations

The bilinear moment-curvature response of barrier walls can also be calculated using empirical equations, such as those presented in Equations (8.16a) to (8.16h). These empirical equations were developed (and therefore validated for) for barriers walls with the following parameters:

- Wall thicknesses ranging from 200 to 800 mm;
- Concrete compressive strength (f_c) ranging from 30 to 55 MPa;
- Vertical reinforcement yield stress (f_{sy}) ranging from 400 to 600 MPa;
- Vertical reinforcement strain hardening ratio (f_{su}/f_{sy}) ranging from 1.1 to 1.3; and
- Vertical reinforcement ratio (p_{st}) ranging from 0.005 to 0.022.

$$M_y = 0.85 p_{st} f_{sy} b (d_o)^2 \tag{8.16a}$$

$$M_u = p_{st} f_{sy} b (d_o)^2 \tag{8.16b}$$

$$\phi_y = \left[\frac{(\varepsilon_{sy})^{1.5}}{(d_o)(f_c)^{0.25}} \right] (36 p_{st} - 860 p_{st}^2) \times 10^3 \tag{8.16c}$$

$$\phi_p = \min[\phi_{p,t}; \phi_{p,c}] \tag{8.16d}$$

$$\phi_{p,t} = \left[\frac{E_s'}{(d_o)(f_c)^{0.5}} \right] (0.1 + 200 p_{st}) \times 10^{-6} \tag{8.16e}$$

$$\phi_{p,c} = \left[\frac{(f_c)^{0.5}}{(d_o)^{0.5} (\varepsilon_{sy})^{0.75}} \right] (2.5 - 65 p_{st}) \times 10^{-6} \tag{8.16f}$$

$$\phi_u = \phi_y + \phi_p \tag{8.16g}$$

$$E_c I_{\text{eff}} = \frac{M_y}{\phi_y} \tag{8.16h}$$

where $\phi_{p,t}$ and $\phi_{p,c}$ are the plastic curvature governed by the tensile strain limit of 0.04 in the reinforcement being reached and the compressive strain limit of 0.004 in the concrete being reached, respectively; ε_{sy} is the yield stress of the vertical reinforcement; and E'_s is the post-elastic modulus of the vertical reinforcement, which may be taken to be equal to the difference between the ultimate and yield stress divided by the difference between the ultimate and yield strain, i.e. $E'_s = (f_{su} - f_{sy})/(\varepsilon_{su} - \varepsilon_{sy})$, since a bilinear stress-strain model for reinforcement has been assumed.

The empirical equations presented were developed using a parametric study that was undertaken using WHAM. The parametric study consisted of approximately 1,000 different wall cross sections, which comprised combinations of the wall parameters across the ranges stated in the summary above. The accuracy of the equations is presented in Figure 8.10, which plots a comparison of the actual value determined using WHAM and the predicted value using the respective equation for the 1,000 different wall cross-sections. The equations consistently predicted the yield point to a high level of accuracy, with the predictions being within ±15% of the actual value more than 99% of the time. The ultimate point was also predicted to a high level of accuracy, with the predictions being within ±15% of the actual value more than 90% of the time. The ±15% boundaries are shown in Figure 8.10.

Consideration also needs to be given to the fact that the properties of concrete can vary from mix-to-mix due to the inherent nature of the material and therefore affect the overall displacement behaviour of the element. For example, the water-cement ratio or type of aggregate used in the mix can affect the Young's modulus of the hardened concrete [8.34]. The Australian Standard for concrete structures, AS 3600 [8.17] specifies a blanket range of ±20% for the Young's modulus, whereas Eurocode 2 [8.8] specifies up to a 30% decrease or 20% increase depending on the specific type of aggregate used in the mix. For context, a ±30% change in the axial stiffness of the concrete can change the effective flexural stiffness ($E_c I_{\text{eff}}$) of the section by around ±10% (as an upper-bound limit). Designers should be aware of the inherent uncertainties associated with predicting the displacement behaviour of RC elements and undertake sensitive analyses as appropriate.

8.4.4 Tapered Barrier Walls

Barrier walls often have a tapered thickness across their height, such as those used on the side of roadways and bridges. To address such situations, it is proposed to assume two different typical thicknesses in the design. The

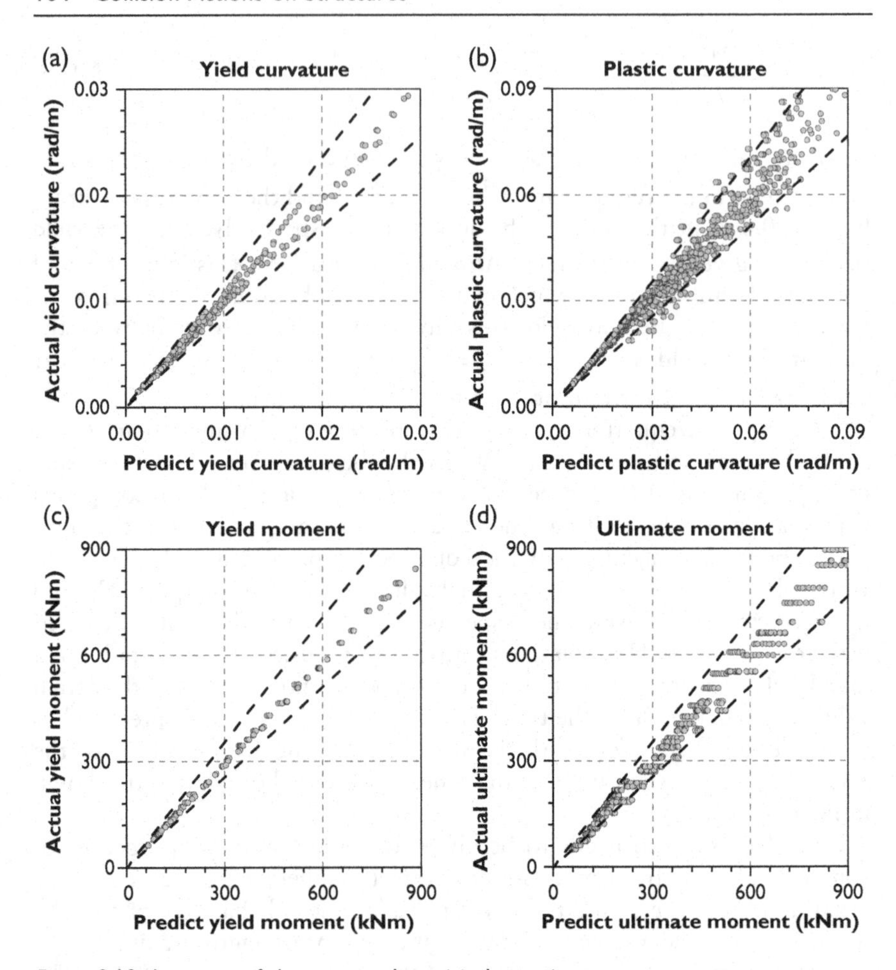

Figure 8.10 Accuracy of the proposed empirical equations.

first thickness, t_{eq}, which is the equivalent thickness of the tapered cross section assuming an elastic response (Figure 8.11(a)), can be used to calculate the yield point. Equation (8.17) for estimating the value of t_{eq} is derived from fundamental mechanics and is an approximation of the equivalent thickness of a cantilevered wall (with uniform thickness) that will have the same elastic stiffness (displacement) as a taper wall section. Equation (8.17) assumes the steepness of the taper is constant across the height of the barrier. The second thickness, t_{Lp}, which can be taken as the average thickness of the wall across the height of the plastic hinge length (Figure 8.11(b)), can be used to calculate the inelastic ultimate point. These two thicknesses, in conjunction with the equations presented previously, can be used to approximate the flexural displacement behaviour of a cantilever barrier with a tapered thickness.

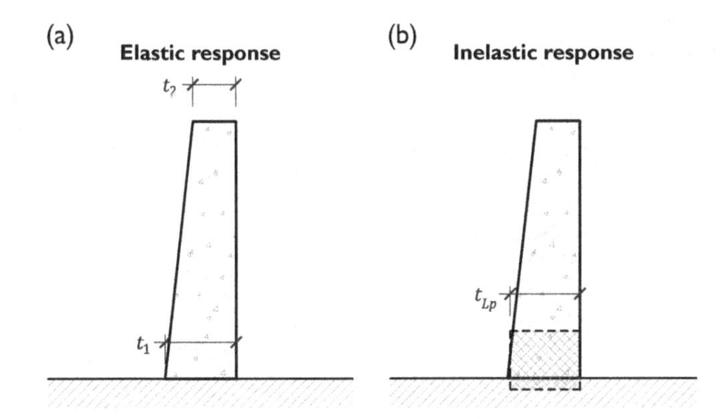

Figure 8.11 Taper barrier walls.

$$t_{eq} = t_1\left[0.6\left(\frac{t_2}{t_1}\right)^{0.4} - 0.4\right] \quad \text{Where: } 0.3 \le \frac{t_2}{t_1} \le 1.0 \tag{8.17}$$

8.5 WORKED EXAMPLE

Consider the example scenario of a spherical granite boulder specimen of 0.8 m in diameter (with density of 2,700 kg/m^3), striking an RC barrier at an impact velocity of 7 m/s. The barrier is 3 metres tall, 6 metres long and 800 mm thick. The longitudinal reinforcement consists of N24 steel bars at 200 mm spacing which have yield strength of 500 MPa and Young's modulus of 200,000 MPa. Concrete cover is 30 mm. The concrete has a characteristic cylinder compressive strength of 40 MPa and Young's modulus of 32,000 MPa. A Poisson's ratio of 0.2 is assumed for both the boulder and concrete. The impact occurs 500 mm from the top of the wall. Assume the barrier has a fully fixed (rigid) base support for the purpose of this example. The barrier wall is presented in Figure 8.12. For the barrier in question, determine the following:

1. The denting and spalling behaviour;
2. The punching behaviour; and
3. The overall flexural behaviour

Solution

The solution to each behaviour is presented in each of the following sub-sections.

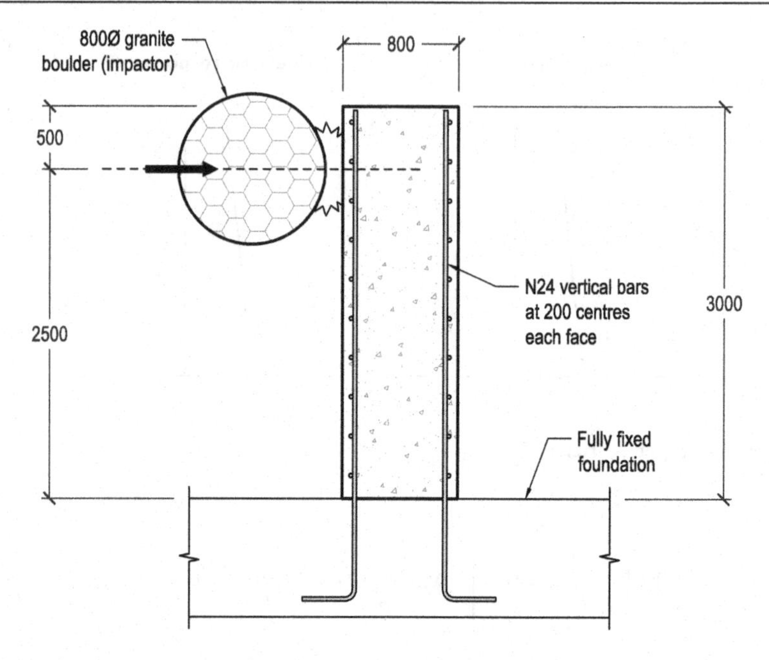

Figure 8.12 Barrier wall worked example.

8.5.1 Denting and Spalling

$$R_i = \frac{D_i}{2} = \frac{0.8}{2} = 0.4 \text{ m}$$

$$m = \frac{4}{3}\pi R_i^3 \rho_b = \frac{4}{3}\pi (0.4)^3 (2700) = 724 \text{ kg}$$

Total mass of stem wall:

$$m_w = \rho_c b D H_w = 2400(6)(0.8)(3) = 34560 \text{ kg}$$

From Table 2.1:

$$\lambda m = 0.25 m_w = 0.25(34560) = 8640 \text{ kg}$$

$$\lambda = \frac{\lambda m}{m} = \frac{8640}{724} = 11.9$$

By Equation (6.10b):

$$M_c = \left(\frac{\lambda}{1+\lambda}\right) m = \left(\frac{11.9}{1+11.9}\right)(724) = 668 \text{ kg}$$

Equations (6.39a) to (6.39d) are used to determine the modelling parameters of a granite boulder:

$$k_{n100} = 84.273v_0 + 160.86 = 84.273(7) + 160.86 = 750.8 \text{ MN/m}^p$$

$$k_n = k_{n100}\sqrt{\frac{D_i}{100}} = 750.8\sqrt{\frac{800}{100}} = 2124 \text{ MN/m}^p$$

$$p = 1.21 + 0.0175v_0 + 0.0005D_i = 1.21 + 0.0175(7) + 0.0005(800)$$
$$= 1.73$$

$$\text{KE}_0 = \frac{1}{2}mv_0^2 = \frac{1}{2}(724)(7)^2 = 17.7 \, k\text{J}$$

$$\text{COR} = 0.068\text{KE}_0^{-0.433} = 0.068(17.7)^{-0.433} = 0.02$$

By the use of the calculated values of p and COR alongside Figure 6.11(a), the value of $C_{F_{c,\max}}$ is estimated to be close to 2.8. The maximum contact force value can then be estimated using Equation (6.33):

$$F_{c,\max} = C_{F_{c,\max}} k_n \left[\left(\frac{p+1}{2k_n}\right) M_c v_0^2\right]^{\frac{p}{p+1}}$$

$$= 2.8(2124 \times 10^6)\left[\left(\frac{1.73+1}{2(2124 \times 10^6)}\right)(668)(7)^2\right]^{\frac{1.73}{1.73+1}} = 6.5 \text{ MN}$$

By the use of Equation (8.4),

$$r_a = \varphi\left[\frac{3}{4}R_i F_{c,\max}\left(\frac{1-\nu_1^2}{E_1} + \frac{1-\nu_2^2}{E_2}\right)\right]^{\frac{1}{3}} = 2$$

$$\times \left[\frac{3}{4}(0.4)(6.5)\left(\frac{1-0.2^2}{50000} + \frac{1-0.2^2}{32000}\right)\right]^{\frac{1}{3}} = 0.0916 \text{ m}$$

Mean contact stress is calculated from Equation (8.1):

$$\sigma_{vm} = P_m = \frac{F_{c,\max}}{\pi r_a^2} = \frac{6.5}{\pi(0.0916)^2} = 247 \text{ MPa}$$

$$f_c = 1.06f_c' = 1.06(40) = 42.4 \text{ MPa}$$

From Equations (8.5) and (8.6):

$$H_B = \frac{f_c - 10.829}{0.0684} = \frac{42.4 - 10.829}{0.0684} = 462 \text{ MPa}$$

$$Y = \frac{H_B}{1.5} = \frac{462}{1.5} = 308 \text{ MPa}$$

Limiting stress to cause denting:

$$\frac{Y}{\sqrt{3}} = \frac{308}{\sqrt{3}} = 178 \text{ MPa}$$

which is exceeded by $\sigma_{vm} = 247$ MPa.

Limiting stress to cause spalling:

$$\frac{H_B}{\sqrt{3}} = \frac{617}{\sqrt{3}} = 267 \text{ MPa}$$

which is not exceeded by $\sigma_{vm} = 247$ MPa.

Denting without spalling of the concrete is predicted to occur in the considered impact scenario.

8.5.2 Punching

$$d_{om} = 800 - 30 - \frac{24}{2} = 758 \text{ mm} = 0.758 \text{ m}$$

Equations (8.10a) to (8.10c) are employed to estimate the surface area of the shear plug:

$$A_G = \pi\sqrt{2}\,(d_{om}^2 + 2r_a d_{om}) = \pi\sqrt{2}\,(0.758^2 + 2(0.0916)(0.758)) = 3.17 \text{ m}^2$$

$$D = d_{om} + r_a = 0.64 + 0.0916 = 0.85 \text{ m}$$

As impact occurs at 500 mm from the top of the wall:

$$x = D - 0.5 = 0.85 - 0.5 = 0.35 \text{ m}$$

$$\frac{x}{D} = \frac{0.35}{0.85} = 0.41$$

$$A_x = D^2\left[-2.14\left(\frac{x}{D}\right)^3 + 3.37\left(\frac{x}{D}\right)^2 + \frac{x}{D} - 0.01\right]$$
$$= 0.85^2[-2.14(0.41)^3 + 3.37(0.41)^2 + 0.41 - 0.01] = 0.59 \text{ m}^2$$

$$A_E = A_G - A_x = 3.17 - 0.59 = 2.58 \text{ m}^2$$

$$f_{ct} = 0.36\sqrt{f'_c} = 0.36\sqrt{40} = 2.28 \text{ MPa}$$

Adopt $\dot{\varepsilon} = 1$ for rockfall impact.

$$\alpha_t = \frac{1}{1 + 0.8f'_c} = \frac{1}{1 + 0.8(40)} = 0.03$$

$$K_t = \left(\frac{\dot{\varepsilon}}{10^{-6}}\right)^{\alpha_t} = \left(\frac{1}{10^{-6}}\right)^{0.03} = 1.51$$

Contributions of $F_{I,p}$ to resistance against punching failure are equivalent to the acceleration of the shear plug multiplied by its mass. Acceleration response of the wall can be determined by dynamic simulations. Step-by-step implementation simulations of the cantilever wall are demonstrated in Section 13.5.6. Acceleration profile of the wall when contact force reaches its peak (t = 0.57 ms) is taken from Section 13.5.6, as shown in Figure 8.13.

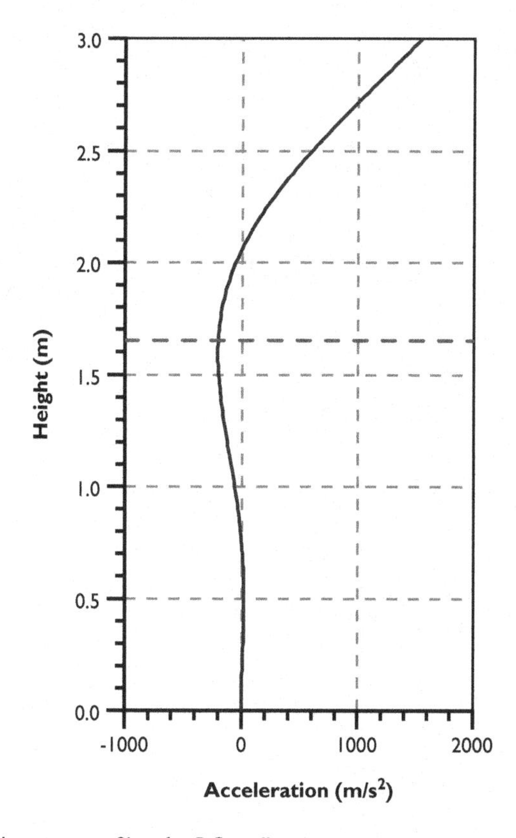

Figure 8.13 Acceleration profile of a RC wall.

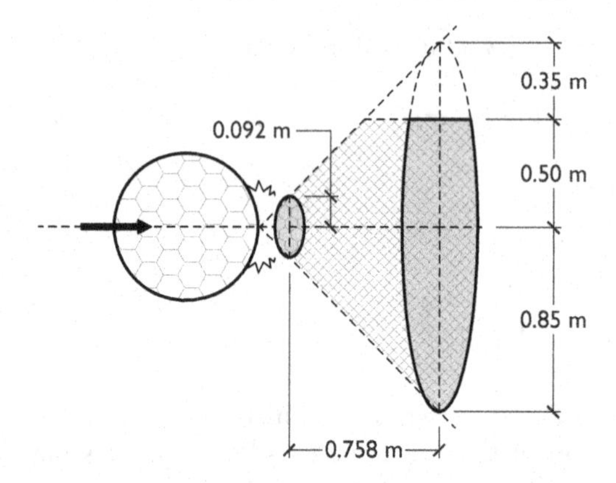

Figure 8.14 Cone representing shear plug of a RC wall.

A shear plug is defined as a cone, as shown in Figure 8.14. The region covered by the shear plug is above the dotted line shown in Figure 8.13. The shear plug is sliced into multiple pieces, which are then multiplied by their respective acceleration value to give inertial forces. Summing the forces contributed by all the slices gives $F_{I,p} = 1.54$ MN.

Thus,

$$F_{\text{capacity}} = K_t f_{ct} A_v \sin\alpha + F_{I,p} = 1.51(2.28)(2.58)\sin 45° + 1.54 = 7.8 \text{ MN}$$

In this example scenario, the predicted tensile resistant capacity of 7.8 MN is higher than the predicted peak contact force of 6.5 MN. Thus, no punching failure is expected.

8.5.3 Flexural Response Behaviour

$$A_{st} = \pi\left(\frac{24}{2}\right)^2 \times \frac{6000}{200} = 13571.7 \text{ mm}^2$$

$$p_{st} = \frac{A_{st}}{bd_o} = \frac{13571.7}{6000(758)} = 0.003$$

$$\varepsilon_{sy} = \frac{f_{sy}}{E_s} = \frac{500}{200000} = 0.0025$$

By the use of Equation (8.16a):

$$M_y = 0.85 p_{st} f_{sy} b (d_o)^2 = 0.85(0.003)(500 \times 10^3)(6)(0.758)^2 = 4395 \text{ kNm}$$

By the use of Equation (8.16c):

$$\phi_y = \left[\frac{(\varepsilon_{sy})^{1.5}}{(d_o)(f_c)^{0.25}} \right] (36 p_{st} - 860 p_{st}^2) \times 10^3 = \frac{0.0025^{1.5}}{0.758(40 \times 10^3)^{0.25}} (36(0.003) - 860(0.003)^2) \times 10^3 = 0.00117 \text{ rad/m}$$

By the use of Equation (8.16h):

$$E_c I_{\text{eff}} = \frac{M_y}{\phi_y} = \frac{4395}{0.00117} = 3.76 \times 10^6 \text{ kNm}^2$$

From Table 2.1:

$$k_{\text{eff}} = \frac{3 E_c I_{\text{eff}}}{H_w^3} = \frac{3(3.76 \times 10^6)}{3^3} = 4.18 \times 10^5 \text{ kN/m}$$

$$\Delta = \frac{m v_0}{\sqrt{m k_{\text{eff}}}} \sqrt{\lambda \left(\frac{1 + \text{COR}}{1 + \lambda} \right)^2} = \frac{724(7)}{\sqrt{724(4.18 \times 10^8)}} \sqrt{11.9 \left(\frac{1 + 0.02}{1 + 11.9} \right)^2} \times 10^3 = 2.5 \text{ mm}$$

By the use of Equation (8.13a):

$$\Delta_y = \frac{\phi_y H_w^2}{3} = \frac{0.00117(3)^2}{3} \times 10^3 = 3.5 \text{ mm}$$

As $\Delta < \Delta_y$, the wall is expected to respond within its elastic limit.

8.6 CLOSING REMARKS

In this chapter, new calculation procedures for assessing the likelihood of damage in the form of denting, spalling, punching and bending have been introduced in Sections 8.2 to 8.4 based on comparison of the calculated demand of the impact action against the calculated capacity. The demand parameters are the maximum contact force, along with geometrical and mechanical properties of the impactor, whereas the capacity is a function of the grade of concrete and its thickness. The section on bending as introduced in Section 8.4 is divided into sub-sections covering the modelling of the flexural displacement behaviour of an RC element and incorporating its non-linear moment-curvature behaviour.

REFERENCES

8.1 Hertz, H., 1882, "Uber die Beriihrung fester elastischer Korper (On the contact of elastic solids)", *Journal für die reine und angewandte Mathematik*, Vol. 92, pp. 156–171.
8.2 Popov, V.L., Heß, M., and Willert, E., 2019, *Handbook of Contact Mechanics: Exact Solutions of Axisymmetric Contact Problems*, Springer Nature, Berlin, Germany.
8.3 Li, H., Chen, W., and Hao, H., 2019, "Influence of drop weight geometry and interlayer on impact behavior of RC beams", *International Journal of Impact Engineering*, Vol. 131, pp. 222–237, doi:10.1016/j.ijimpeng.2019.04.02
8.4 Majeed, Z.Z.A., Lam, N.T.K., and Gad, E.F., 2021, "Predictions of localised damage to concrete caused by a low-velocity impact", *International Journal of Impact Engineering*, Vol. 149, p. 103799, doi:10.1016/j.ijimpeng.2020.103799
8.5 Fischer-Cripps, A.C., 2007, *Introduction to contact mechanics*, Springer, U.S.
8.6 Szilágyi, K., Borosynói, A., and Dobó, K., 2011, "Static indentation hardness testing of concrete: a long established method revived", *Epitoanyag-Journal of Silicate Based & Composite Materials*, Vol. 2011, pp. 2–8.
8.7 Menegon, S.J., Tsang, H.H., Wilson, J.L., and Lam, N.T.K., 2021, "Statistical analysis of material properties and recommended values for the assessment of RC structures in Australia", *Australian Journal of Structural Engineering*, Vol. 22(3), pp. 191–204. doi:10.1080/13287982.2021.1946993
8.8 European Committee for Standardization (CEN), 2004, *Eurocode 2: Design of concrete structures - Part 1-1: General rules and rules for buildings*, European Committee for Standardization (CEN), Brussels, Belgium.
8.9 Malvar, L.J. and Crawford, J.E., 1998, "Dynamic increase factors for concrete". Twenty-Eighth DDESB Seminar Orlando, FL, August 98.
8.10 Dinic, G. and Perry, S., 1990, "Shear plug formation in concrete slabs subjected to hard impact", *Engineering Fracture Mechanics*, Vol. 35(1–3), pp. 343–350, doi:10.1016/0013-7944(90)90213-Z
8.11 Saatci, S. and Vecchio, F.J., 2009, "Effects of shear mechanisms on impact behavior of reinforced concrete beams", *ACI Structural Journal*, Vol. 106(1), pp. 78–86.
8.12 Zhao, D.-B., Yi, W.-J., and Kunnath, S.K., 2017, "Shear mechanisms in reinforced concrete Beams under impact loading", *Journal of Structural Engineering*, Vol. 143(9), p. 04017089, doi:10.1061/(ASCE)ST.1943-541X.0001818
8.13 Fu, Y., Yu, X., Dong, X., Zhou, F., Ning, J., Li, P., and Zheng, Y., 2020, "Investigating the failure behaviors of RC beams without stirrups under impact loading", *International Journal of Impact Engineering*, Vol. 137, p. 103432, doi:10.1016/j.ijimpeng.2019.103432
8.14 Schmidt-Hurtienne, B., 2000, *Ein dreiaxiales Schädigungsmodell für Beton unter Einschluss des Dehnrateneffekts bei Hochgeschwindigkeitsbelastung*, Inst. für Massivbau und Baustofftechnologie, Karlsruhe, Germany.
8.15 Ngo, T., Mendis, P., Gupta, A., and Ramsay, J., 2007, "Blast loading and blast effects on structures–an overview", *Electronic Journal of Structural Engineering*, Vol. 7, pp. 76–91.

8.16 Othman, H. and Marzouk, H., 2016, "Strain rate sensitivity of fiber-reinforced cementitious composites", *ACI Materials Journal*, Vol. 113(2), pp. 143–150.
8.17 Standards Australia, 2018, *AS 3600 Concrete Structures*, Standard Australia Limited, New South Wales, Australia.
8.18 Standards New Zealand, 2006, *NZS 3101.1 Concrete structures standard: Part 1 – The design of concrete structures*, Standards New Zealand, Wellington, New Zealand.
8.19 International Federation for Structural Concrete (fib), 2013, *fib Model Code for Concrete Structures 2010*, Wilhelm Ernst & Sohn, Berlin, Germany.
8.20 American Concrete Institute, 2012, *ACI 352.1R-11 Guide for Design of Slab-Column Connections in Monolithic Concrete Structures*, American Concrete Institute, Farmington Hills, MI.
8.21 Moehle, J., 2015, *Seismic Design of Reinforced Concrete Buildings*, McGraw-Hill Education, New York, NY.
8.22 Munter, S., Lume, E., Woodside, J., and McBean, P., 2015, *Guide to Seismic Design and Detailing of Reinforced Concrete Building in Australia*, Steel Reinforcement Institute of Australia (SRIA), Roseville, NSW.
8.23 Yong, A.C.Y., Lam, N.T.K., Menegon, S.J., and Gad, E.F., 2020, "Experimental and analytical assessment of the flexural behaviour of cantilevered RC walls subjected to impact actions", *Journal of Structural Engineering*, Vol. 146(4), p. 04020034, doi:10.1061/(ASCE)ST.1943-541X.0002578
8.24 Menegon, S.J., Tsang, H.H., Wilson, J.L., and Lam, N.T.K., 2021, "RC walls in Australia: displacement-based seismic design in accordance with AS 1170.4 and AS 3600", *Australian Journal of Structural Engineering*, doi:10.1080/13287982.2021.1954306
8.25 Priestley, M.J.N., Calvi, G.M., and Kowalsky, M.J., 2007, *Displacement-Based Seismic Design of Structures*, IUSS Press, Pavia, Italy.
8.26 Yong, A.C.Y., 2019, *Impact-resistance of Reinforced Concrete Structures*, The University of Melbourne, Melbourne, Australia.
8.27 Standards Australia and Standards New Zealand, 2019, *AS/NZS 4671:2019 Steel for the reinforcement of concrete*, Standards Australia Limited and Standards New Zealand, Sydney and Wellington.
8.28 Menegon, S.J., Wilson, J.L., Lam, N.T.K., and McBean, P., 2018, "RC walls in Australia: seismic design and detailing to AS 1170.4 and AS 3600", *Australian Journal of Structural Engineering*, Vol. 19(1), pp. 67–84, doi: 10.1080/13287982.2017.1410309
8.29 Menegon, S.J., Wilson, J.L., Lam, N.T.K., and Gad, E.F., 2020, "Development of a user-friendly and transparent non-linear analysis program for RC walls", *Computers and Concrete*, Vol. 25(4), pp. 327–341, doi:10.12989/cac.2020.25.4.327
8.30 Menegon, S.J., 2019, "WHAM: a user-friendly and transparent non-linear analysis program for RC walls and building cores", Available from: downloads.menegon.com.au/1/20190901.
8.31 Mander, J.B., Priestley, M.J.N., and Park, R., 1988, "Theoretical stress-strain model for confined concrete", *ASCE Journal of Structural Engineering*, Vol. 114(8), pp. 1827–1849, doi:10.1061/(ASCE)0733-9445(1988)114:8(1804)

8.32 Menegon, S.J., Wilson, J.L., Lam, N.T.K., and Gad, E.F., 2021, "Tension stiffening model for limited ductile reinforced concrete walls", *Magazine of Concrete Research*, Vol. 73(7), pp. 366–378, doi:10.1680/jmacr.20.00211
8.33 Menegon, S.J., Wilson, J.L., Lam, N.T.K., and Gad, E.F., 2019, "Experimental testing of nonductile RC wall boundary elements", *ACI Structural Journal*, Vol. 116(6), pp. 213–225, doi:10.14359/51718008
8.34 Standards Australia, 2021, *AS 3600:2018 Supp 1:2021 Concrete Structures-Commentary*, Standard Australia Limited, New South Wales, Australia.

Chapter 9

Vehicle Collision on a Bridge Deck – Application

9.1 INTRODUCTION

Collision hazards on a bridge deck are common occurrences in metropolitan areas around the world. The typical design scenario is a vehicle colliding on a barrier, which is positioned on the edge of the bridge deck. The way the collision action is estimated can have significant implications on the design of both the superstructure and the sub-structure (including the foundation). The collision action of a moving vehicle is transient in nature. However, in the majority of design standards, the collision action is approximated by applying an equivalent static design force to the structure. The static analysis approach can result in over-designing of the bridge and its foundations (which is the focus of this chapter) depending on the typology and structural proportions of the bridge, since the dynamic properties of the bridge and transient nature of the design action have both not been taken into account. In situations where the codified (equivalent static) collision loads are of greater magnitude than the lateral seismic actions, the former can greatly influence and govern the sizing of the structural elements on the bridge, including the crosshead, the pier and the foundation. Cost implications stemming from over-designing the foundations in particular, can be significant depending on site and subsoil conditions. While this chapter is limited in scope to vehicular collision scenarios on bridge decks, the behaviour described equally applies to rail bridges. Collision actions on rail bridges (from derailed trains) are outside the scope of this book.

Current approaches of designing for vehicular collisions on a bridge deck include: (1) equal energy method, (2) prescriptive equivalent static design forces and (3) physical crash testing. Important parameters such as energy losses and inertial resistances are neglected in the first two approaches, whereas the third approach can be costly and is essentially limited to testing barriers in isolation (i.e. the bridge structure itself is not tested), thereby limiting its utility in the day-to-day structural design. All three approaches are presented and reviewed in Section 9.2.

DOI: 10.1201/9781003133032-9

In current practice, numerical simulations are rarely conducted to assess the dynamic response behaviour of a bridge structure to determine the internal design actions (e.g. shear force or bending moments) that are generated by a collision scenario. The implementation of the simulations is filled with challenges, even if the required software is available. For example, defining the vehicular forcing function (i.e. the time-history of the impact force that is applied to the barrier in a collision) can be filled with ambiguities and uncertainties. Normalised vehicular forcing functions developed using the Hunt and Crossley (H&C) model [9.1], which was first introduced in Chapter 6 (refer Section 6.6), are presented in this chapter for a series of different typical vehicular collision scenarios (Section 9.3). A case study, which utilises two different types of numerical modelling for the bridge, is then presented to illustrate how the normalised forcing functions can be applied (Section 9.4). A simplified modelling methodology based on the use of closed-form expressions is then introduced to estimate the structural response behaviour of the bridge in a much simpler, and expedient, manner (Section 9.5). Details of the derivations and validation of the presented analytical model is also presented. It is noted that the simplified methodology based on the use of closed-form expressions only provides approximations to facilitate speedy evaluation. Designers should rely on numerical modelling for achieving better accuracy and for undertaking a final design. The closed-form expressions are used to calculate the response of the case study (as introduced in Section 9.4) to illustrate how the expressions are applied in practice and also compare their accuracy to results from the numerical simulations (Section 9.6).

It is noted that Chapter 9 is primarily focused on the application of the two proposed methodologies: numerical simulation and closed-form solution. Validations of these methodologies to establish their credibility are covered in Chapter 10.

9.2 CURRENT PRACTICE

9.2.1 Equal Energy Method

The equal energy method, as first introduced in Chapter 3 (e.g. Equation (3.14b)), is based on well-recognised equal energy principles [9.2] and is represented here as Equation (9.1a). When multiplying the result obtained from Equation (9.1a) by the stiffness of the barrier (k), the required strength of the barrier or the considered structural element to resist the collision is presented; refer to Equation (9.1b), which is identical to Equation (3.13). Note that the term V_b represents base shear in Equation (3.13) and is replaced by a general term F^* to represent the equivalent static design force. The force calculated using Equation (9.1b) is the collision force applied at the top of the barrier (or at the designated "impact point").

$$\Delta = \frac{mv_0}{\sqrt{mk}} \tag{9.1a}$$

$$F^* = v_0\sqrt{mk} \tag{9.1b}$$

The displacement and strength demand as predicted by the use of Equations (9.1a) and (9.1b) can be very high as the implicit assumption is that the total amount of kinetic energy delivered by the moving vehicle is wholly absorbed by the target. The sharing of energy on impact as described in the earlier chapters of the book has not been taken into account by either Equations (9.1a) or (9.1b). The over-conservatism of this modelling approach has resulted in a very limited uptake of these expressions in the design of vehicular barriers and the supporting bridge structure in design practices.

9.2.2 Equivalent Static Design Force Method

The design approach of treating the collision action on a vehicular barrier as an equivalent static force is straightforward and easy to apply, and is widely adopted in contemporary highway standards of practices around the globe [9.3–9.5]. The magnitude of the stipulated design force is based on the barrier performance level, which is in turn dependent on a number of factors such as traffic volumes, volumes of heavy vehicles and location of the bridge. Design forces stipulated by the American Association of State Highway and Transportation Officials (AASHTO) [9.3] and the Australian Standard for bridge design actions, AS 5200.2 [9.5], are listed in Table 9.1 as an example. The forces listed in the table are lateral forces to be applied to the barrier. Prescriptive vertical and longitudinal forces, which are also stipulated by these standards, are not presented herein.

The design collision forces stipulated by AASHTO [9.3] and AS 5100.2 [9.5] are based on a mathematical model presented in Hirsch [9.6], as shown by Equation (9.2). The basis of the analytical model represented by Equation (9.2) is presented in the schematic diagram of Figure 9.1.

$$F = \frac{1.25mv_\theta^2\sin^2\theta}{AL\sin\theta - B(1 - \cos\theta) + \Delta} \tag{9.2}$$

The model was developed initially in Olson and Post [9.7] and was later calibrated against empirical information that was derived from numerically simulated and experimental data [9.6]. Input parameters into the model include mass of the colliding vehicle (m), velocity of impact (v_θ), angle of impact (θ), dimensions of the vehicle (AL and B) and displacement of the barrier (Δ). For a rigid barrier, zero displacement is assumed; thus, the

Table 9.1 Lateral static design forces stipulated in major codes of practice

Test Level	*Performance Level (AS 5100.2)*	*Static Design Force (kN)*		*Height Static Load Is Applied (AS 5100.2)*
		AASHTO [9.3]	*AS 5100.2 [9.5]*	
TL-1	N/A	60	N/A	N/A
TL-2	Low	120	150	600 mm
TL-3	N/A	240	N/A	N/A
TL-4	Regular	240	300	900 mm
TL-5	Medium	552	600	1200 mm
TL-6	Special	778	750	1800 mm
> TL-6	> Special	N/A	1200	1500 mm

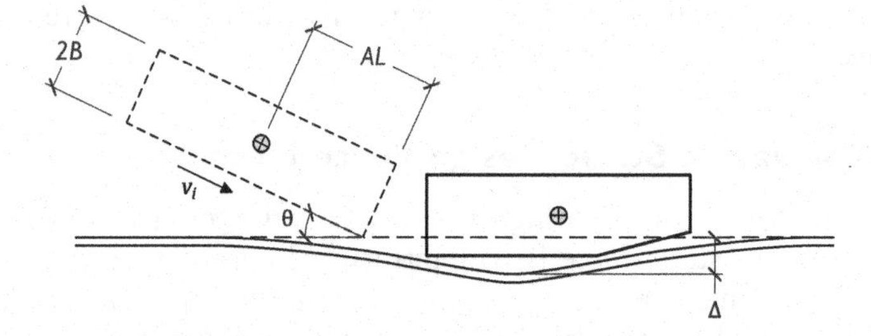

Figure 9.1 Mathematical model derived by Olson and Post [9.7].

design forces so calculated from the model are solely dependent on the deformation of the vehicle.

The mass and stiffness of the bridge, which are influential on the outcome of the collision with respect to the behaviour of the overall structure, have not been incorporated into the calculation. Applying an equivalent static design force at the barrier on the bridge deck may give results that exaggerate the amount of bending moment transmitted to the base of the pier and its foundation because the mitigating effects of the inertial resistance developed within the bridge are neglected.

9.2.3 Physical Crash Testing

An alternative approach to the design of vehicular barriers (in isolation) is by conducting physical crash tests for evaluating their potential performance in terms of providing containment. For each performance level of the barrier, the respective standards of practice (as cited previously in Section 9.2.2) provide specific collision-resistant criteria that need to be fulfilled; refer to Table 9.2 for the controlling strength criteria. The main limitation with physical crash testing is that only the barrier in isolation is tested. The transmission of the collision action to the rest of the bridge and its foundation is not assessed by the experimental investigation.

Table 9.2 Crash test criteria

Test Level	*Vehicle Type*	*Vehicle Mass (kg)*	v_θ *(km/h)*		θ *(°)*
			AASHTO [9.3]	*AS 5100.1 [9.8]*	
TL-1	Pickup Truck	2,270	48	N/A	25
TL-2	Pickup Truck	2,270	72	70	25
TL-3	Pickup Truck	2,270	97	N/A	25
TL-4	Single-Unit Truck	10,000	89	90	15
TL-5	Van-Type Semi-Trailer	36,000	80	90	15
TL-6	Tractor-Tanker Trailer	36,000	80	100	15
> TL-6	Articulated Van	44,000	N/A	100	15

9.3 NORMALISED VEHICULAR FORCING FUNCTIONS

An alterative procedure for assessing the overall bridge structure to the vehicular collision scenario is to build a numerical model of the bridge and apply a forcing function that simulates the collision action on the barrier. A series of normalised forcing functions have been generated using the H&C model [9.1] based on physical crash testing data, which is presented in Chapter 10. The calibrations have been undertaken for both the semi-trailer and single-unit truck barrier collisions, i.e. TL-5 and TL-4 performance levels, respectively. The collision consists of two impacts, i.e. the front mass (m_f) impact from the engine and the rear mass (m_r) impact of the rear axle. The calibrated dynamic factors are summarised in Table 9.3; however, the reader is directed to Chapter 10 for further details of how these numbers were derived.

A summary of the various normalised forcing functions is shown in Table 9.4. The mass and impact velocity of the semi-trailer are varied from 32 t to 40 t and 70 km/h to 90 km/h, respectively, whilst the mass and impact velocity of a single-unit truck vary from 7.5 t to 12.5 t, and 80 km/h to 100 km/h, respectively. An angle of impact of 15° is adopted in accordance with the requirements outlined in Table 9.2. The front-to-rear

Table 9.3 Calibrated dynamic factors

Vehicle	*Front/Rear*	k_n *(kN/m^p)*	*p*	*COR*
Semi-Trailer (TL-5 scenario)	Front	2,300	2	0.2
	Rear	9,000	1.5	0.5
Single-Unit Truck (TL-4 scenario)	Front	400	2	0.2
	Rear	1,500	2	0.5

Table 9.4 Summary of impact scenarios

Vehicle	m (t)	v_θ (km/h)	m_f (t)	m_r (t)
Semi-Trailer	32	70	17.6	14.4
		80		
		90		
	36	70	19.8	16.2
		80		
		90		
	40	70	22	18
		80		
		90		
Single-Unit Truck	7.5	80	3	4.5
		90		
		100		
	10	80	4	6
		90		
		100		
	12.5	80	5	7.5
		90		
		100		

mass ratio of a semi-trailer and single-unit truck is taken to be 0.55:0.45 and 0.4:0.6, respectively, which is in line with results from physical crash tests that were used to develop the forcing functions (refer Chapter 10). Each forcing function is simplified into 30 data points, which provide sufficient accuracy to represent the forcing function, as shown in Figures 9.2(a) and 9.2(b). These simplified functions (comprised 30 points) have been tabulated for each collision scenario, which designers can duplicate and adopt in their own numerical modelling. Tables 9.5 and 9.6 are for a semi-trailer and single-unit truck collision scenario, respectively.

9.4 MODELLING OVERALL BRIDGE RESPONSE TO VEHICULAR COLLISION ON THE BRIDGE DECK

9.4.1 General

This section will show how the normalised forcing functions developed previously can be used to model a collision scenario on a case study bridge. The case study bridge is presented in Figure 9.3. The collision being considered is a 36 t semi-trailer at a velocity of 90 km/h and an impact angle of 15°.

Two different approaches have been adopted by the authors for modelling the overall behaviour of the bridge to the collision actions applied to

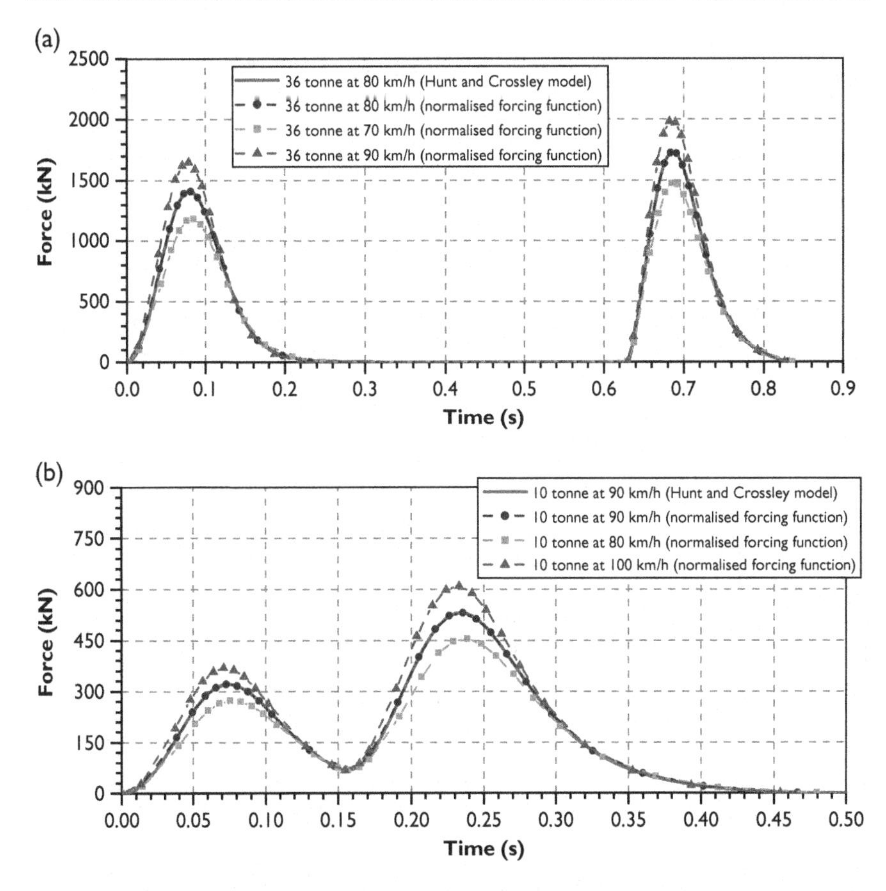

Figure 9.2 Example of forcing function simplified into 30 data points for (a) 36 tonne semi-trailer impact and (b) 10 tonne single-unit truck impact.

the barrier. The first approach relies on the use of a commercial software package (referred hereafter as the "detailed model"), whereas the second approach only involves programming a simplified "three-legged frame" model of the bridge in *MATLAB* (referred hereafter as the "simplified model"). The simplified model has been adopted extensively throughout Chapter 10 for developing the normalised forcing functions.

The case study bridge is a highway structure located in Melbourne and was a concrete viaduct. The superstructure of the viaduct was a couple of kilometres long, 12 metres above the ground and the form of construction was post-tensioned concrete box girders that typically had average span lengths of 40 metres between piers. The self-weight of the bridge was estimated to be 196.5 kN/m. A cracked effective stiffness ratio of 0.35 was assumed for the bridge deck and the barrier. A higher value of 0.7 was assumed for the box girder and pier, as they were expected to experience much less cracking than the deck.

Table 9.5 Forcing function for a semi-trailer weighting 32 to 40 tonnes and at speed of 70 to 90 km/h

32 Tonnes						*36 Tonnes*						*40 Tonnes*					
70 km/h		*80 km/h*		*90 km/h*		*70 km/h*		*80 km/h*		*90 km/h*		*70 km/h*		*80 km/h*		*90 km/h*	
t (s)	F_c *(kN)*	*t (s)*	F_c *(kN)*	*t (s)*	F_c *(kN)*	*t (s)*	F_c *(kN)*	*t (s)*	F_c *(kN)*	*t (s)*	F_c *(kN)*	*t (s)*	F_c *(kN)*	*t (s)*	F_c *(kN)*	*t (s)*	F_c *(kN)*
0.000	0.0	0.000	0.0	0.000	0.0	0.000	0.0	0.000	0.0	0.000	0.0	0.000	0.0	0.000	0.0	0.000	0.0
0.015	90.9	0.014	108.6	0.014	127.0	0.015	98.5	0.015	118.0	0.014	136.5	0.016	106.3	0.015	126.1	0.015	148.3
0.042	599.9	0.040	713.1	0.039	828.2	0.043	647.2	0.042	768.5	0.040	894.3	0.045	692.7	0.043	824.6	0.042	958.1
0.054	857.1	0.052	1019.4	0.050	1186.4	0.057	924.2	0.054	1098.7	0.052	1280.4	0.059	989.6	0.056	1178.3	0.054	1372.1
0.064	1005.2	0.062	1196.5	0.059	1394.7	0.067	1085.8	0.064	1292.4	0.062	1506.0	0.069	1163.2	0.067	1384.2	0.064	1614.3
0.073	1078.9	0.070	1286.5	0.068	1501.7	0.076	1166.0	0.073	1389.7	0.070	1621.9	0.079	1249.4	0.076	1488.9	0.073	1738.2
0.081	1090.7	0.078	1302.8	0.075	1523.4	0.085	1179.3	0.081	1408.6	0.079	1646.8	0.088	1264.7	0.084	1510.2	0.081	1765.6
0.090	1046.7	0.086	1253.5	0.083	1469.1	0.094	1133.1	0.090	1356.4	0.087	1589.4	0.097	1215.9	0.093	1456.1	0.090	1706.0
0.099	949.5	0.095	1140.4	0.092	1339.6	0.103	1028.8	0.099	1236.4	0.096	1451.8	0.107	1105.6	0.103	1327.9	0.099	1560.8
0.110	797.2	0.105	963.3	0.101	1137.2	0.114	865.3	0.109	1046.1	0.105	1233.9	0.118	932.6	0.113	1126.2	0.109	1330.2
0.123	589.2	0.118	716.4	0.113	850.7	0.128	640.6	0.123	779.0	0.118	926.2	0.133	691.3	0.127	841.2	0.122	1001.5
0.144	316.4	0.137	390.1	0.132	469.8	0.150	347.1	0.143	427.6	0.137	515.3	0.155	376.1	0.148	463.2	0.141	560.5
0.168	134.3	0.160	167.0	0.153	203.6	0.175	147.6	0.166	183.6	0.158	225.4	0.180	160.9	0.171	201.2	0.163	247.0
0.202	40.6	0.191	50.6	0.181	62.0	0.209	44.6	0.197	55.9	0.187	68.9	0.216	48.7	0.203	61.3	0.192	76.1
0.239	0.0	0.225	0.0	0.213	0.0	0.247	0.0	0.232	0.0	0.219	0.0	0.254	0.0	0.238	0.0	0.224	0.0
0.630	0.0	0.630	0.0	0.630	0.0	0.630	0.0	0.630	0.0	0.630	0.0	0.630	0.0	0.630	0.0	0.630	0.0
0.638	149.1	0.637	174.3	0.637	202.6	0.638	160.1	0.638	187.3	0.638	214.0	0.638	171.4	0.638	200.7	0.638	229.5
0.657	838.6	0.656	980.6	0.656	1120.7	0.658	897.5	0.658	1051.7	0.657	1204.6	0.659	954.8	0.659	1116.4	0.658	1280.9
0.667	1140.4	0.666	1331.8	0.665	1529.8	0.669	1222.3	0.668	1428.8	0.667	1642.1	0.670	1300.4	0.669	1521.4	0.669	1745.9
0.675	1304.5	0.674	1526.0	0.673	1754.9	0.677	1397.6	0.676	1635.6	0.675	1879.6	0.679	1487.2	0.678	1741.2	0.677	1999.3
0.682	1371.7	0.681	1607.2	0.680	1848.2	0.685	1470.8	0.683	1723.5	0.682	1981.5	0.687	1565.3	0.686	1833.8	0.685	2108.5
0.689	1362.6	0.688	1598.6	0.687	1839.4	0.692	1461.9	0.691	1714.5	0.690	1973.1	0.695	1556.2	0.693	1825.4	0.692	2100.2
0.697	1286.2	0.695	1511.6	0.694	1741.3	0.700	1380.5	0.698	1621.1	0.697	1868.2	0.703	1470.6	0.701	1727.4	0.700	1989.5
0.705	1146.8	0.703	1349.9	0.701	1557.6	0.708	1231.6	0.707	1450.1	0.705	1671.1	0.712	1312.1	0.710	1545.3	0.708	1781.9
0.714	947.6	0.712	1116.0	0.710	1292.1	0.718	1019.5	0.716	1201.0	0.714	1387.8	0.722	1087.1	0.720	1281.0	0.718	1477.8
0.726	689.8	0.724	814.9	0.722	942.3	0.731	742.4	0.728	876.7	0.726	1014.1	0.735	792.8	0.733	936.2	0.731	1083.4
0.745	381.4	0.742	449.6	0.740	521.9	0.751	410.3	0.748	484.7	0.745	560.4	0.756	437.1	0.753	517.6	0.750	598.4
0.767	177.8	0.764	209.3	0.761	241.9	0.774	191.3	0.770	225.5	0.767	260.1	0.780	204.1	0.776	240.9	0.773	277.5
0.793	66.0	0.789	77.5	0.785	89.3	0.801	70.8	0.796	83.1	0.793	95.8	0.808	75.3	0.803	89.9	0.800	102.1
0.826	0.0	0.821	0.0	0.817	0.0	0.836	0.0	0.830	0.0	0.826	0.0	0.845	0.0	0.835	0.0	0.834	0.0

Table 9.6 Forcing function for a single-unit truck weighting 7.5 to 12.5 tonnes and at speed of 80 to 100 km/h

7.5 Tonnes						*10 Tonnes*						*12.5 Tonnes*					
80 km/h		*90 km/h*		*100 km/h*		*80 km/h*		*90 km/h*		*100 km/h*		*80 km/h*		*90 km/h*		*100 km/h*	
t (s)	*F_c (kN)*	*t (s)*	*F_c (kN)*	*t (s)*	*F_c (kN)*	*t (s)*	*F_c (kN)*	*t (s)*	*F_c (kN)*	*t (s)*	*F_c (kN)*	*t (s)*	*F_c (kN)*	*t (s)*	*F_c (kN)*	*t (s)*	*F_c (kN)*
0.000	0.0	0.000	0.0	0.000	0.0	0.000	0.0	0.000	0.0	0.000	0.0	0.000	0.0	0.000	0.0	0.000	0.0
0.013	17.4	0.013	20.3	0.012	23.5	0.014	20.7	0.014	24.4	0.013	27.9	0.015	23.4	0.015	27.7	0.014	32.3
0.037	117.9	0.035	138.5	0.034	160.1	0.040	140.4	0.038	164.9	0.037	191.3	0.043	160.5	0.041	189.2	0.040	[illegible]19.1
0.047	171.0	0.046	201.1	0.044	232.3	0.051	204.3	0.050	239.8	0.048	277.2	0.055	233.7	0.053	274.9	0.052	[illegible]18.4
0.055	202.9	0.053	237.9	0.052	274.6	0.060	243.4	0.058	286.0	0.057	329.8	0.064	279.3	0.062	328.3	0.060	[illegible]79.5
0.063	220.4	0.061	258.5	0.059	297.9	0.068	265.2	0.066	311.1	0.064	358.5	0.073	305.9	0.070	359.0	0.068	[illegible]14.1
0.070	226.3	0.067	264.7	0.065	304.7	0.076	273.8	0.073	320.6	0.071	369.1	0.080	317.1	0.078	371.4	0.076	[illegible]27.7
0.076	222.0	0.074	259.6	0.072	298.3	0.083	270.6	0.080	316.0	0.078	363.0	0.088	315.5	0.085	368.5	0.083	[illegible]23.2
0.084	209.1	0.081	243.5	0.078	279.2	0.091	257.1	0.088	299.2	0.085	342.9	0.096	302.2	0.093	351.4	0.091	[illegible]02.4
0.091	187.9	0.088	218.0	0.086	248.9	0.099	234.2	0.096	271.1	0.093	309.2	0.104	278.3	0.101	322.2	0.099	[illegible]67.0
0.100	159.3	0.097	183.3	0.095	208.2	0.108	202.4	0.105	232.8	0.102	263.7	0.114	244.8	0.111	280.6	0.108	[illegible]17.4
0.126	81.9	0.124	90.5	0.121	99.3	0.133	115.2	0.131	127.0	0.128	138.8	0.139	150.6	0.136	165.5	0.134	[illegible]80.7
0.145	47.4	0.143	49.3	0.141	51.1	0.149	75.6	0.148	79.3	0.146	82.3	0.151	113.1	0.149	120.2	0.148	[illegible]22.5
0.154	37.8	0.153	37.0	0.153	36.1	0.157	67.7	0.155	67.8	0.155	67.5	0.159	102.7	0.158	104.8	0.157	[illegible]05.8
0.164	52.5	0.164	59.1	0.160	46.6	0.165	77.3	0.164	82.2	0.164	88.4	0.169	117.0	0.168	126.2	0.165	[illegible]21.0
0.171	78.1	0.171	91.8	0.171	106.5	0.171	98.5	0.171	112.0	0.171	127.0	0.181	172.2	0.181	199.7	0.174	[illegible]72.7
0.189	189.8	0.188	223.7	0.187	259.3	0.192	226.8	0.191	266.3	0.190	307.7	0.194	261.8	0.193	306.8	0.192	[illegible]53.2
0.203	284.6	0.201	334.2	0.200	385.0	0.207	342.2	0.205	400.9	0.204	462.0	0.210	394.6	0.209	462.4	0.207	[illegible]33.1
0.213	340.1	0.211	398.2	0.209	458.6	0.219	411.3	0.216	481.0	0.215	553.6	0.223	476.7	0.221	557.2	0.219	[illegible]40.7
0.222	368.4	0.220	430.4	0.218	494.6	0.229	446.7	0.226	521.7	0.224	599.6	0.234	519.4	0.231	606.2	0.229	[illegible]96.3
0.231	373.6	0.228	436.4	0.226	501.5	0.238	454.8	0.235	530.9	0.233	609.6	0.245	530.2	0.242	618.4	0.239	[illegible]09.9
0.240	360.3	0.237	420.7	0.234	483.5	0.248	439.7	0.245	512.8	0.242	588.8	0.255	513.6	0.252	598.7	0.249	[illegible]87.1
0.249	330.4	0.246	385.7	0.242	443.1	0.259	404.0	0.255	471.4	0.251	540.9	0.267	472.9	0.262	551.2	0.259	[illegible]32.1
0.260	286.0	0.256	333.8	0.252	383.8	0.270	350.3	0.266	408.3	0.262	469.1	0.279	410.3	0.275	478.2	0.271	[illegible]48.7
0.272	228.5	0.268	266.7	0.264	307.1	0.284	280.3	0.279	326.9	0.275	375.8	0.294	329.2	0.289	383.6	0.284	[illegible]40.0
0.289	161.0	0.284	188.0	0.279	216.1	0.303	197.6	0.297	230.7	0.292	264.9	0.314	232.0	0.308	270.7	0.303	[illegible]10.5
0.316	86.2	0.310	100.6	0.305	115.5	0.333	105.7	0.326	123.5	0.320	141.7	0.346	124.1	0.339	144.5	0.333	[illegible]66.0
0.349	39.4	0.341	46.0	0.335	52.8	0.368	48.2	0.360	56.2	0.353	64.5	0.385	56.4	0.376	65.8	0.368	75.5
0.388	13.8	0.379	16.1	0.371	18.5	0.412	16.8	0.402	19.6	0.393	22.5	0.432	19.6	0.421	22.9	0.412	26.3
0.448	0.0	0.436	0.0	0.425	0.0	0.480	0.0	0.466	0.0	0.454	0.0	0.507	0.0	0.492	0.0	0.479	0.0

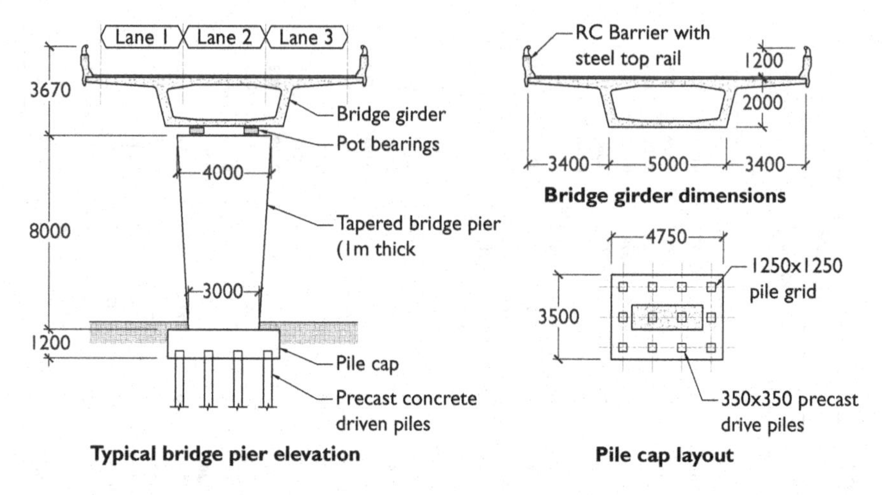

Figure 9.3 Case study bridge.

The viaduct was found on soft soils and the sub-structure was built of precast concrete driven piles, which were arranged in 3 × 4 grid pile group connected to a 1 m thick pile cap. The piles were driven to refusal, which occurred at a depth of about 30 metres. The rotational stiffness (k_R) of the pile group was calculated to be approximately 2,000 MNm/rad.

9.4.2 Detailed Model

The case study bridge was modelled in commercial software *Strand7* [9.9]. The beam elements are as shown in the schematic diagram of Figure 9.4. The beam elements simulating the box girder and the barrier were modelled with a width (into the page) equal to twice the height of the barrier, similar to what was assumed for the reinforced concrete (RC) wall example presented in Chapter 8. The tapered sections were modelled as discrete constant thickness sections that stepped down across the length of the taper. The collision was assumed to occur directly in line with the pier and as such, the total mass of the structure in the model was taken as the mass of the pier plus 40 metres' length of the bridge box girder, asphalt and barrier (i.e. corresponding to 20 metres of viaduct on either side of the pier). Additional lumped masses were used (at locations shown in Figure 9.4) to simulate the 40-metre portion of the bridge girder that has not been modelled.

The contact force time-history proposed in the previous section (for the collision scenario being considered) was applied to the model at the top of the barrier. Linear transient dynamic analyses were conducted to assess the response behaviour of the structure when subjected to the transient forcing function. The linear transient dynamic analysis was a "modal analysis" and a damping ratio of 5% (of critical damping) was adopted

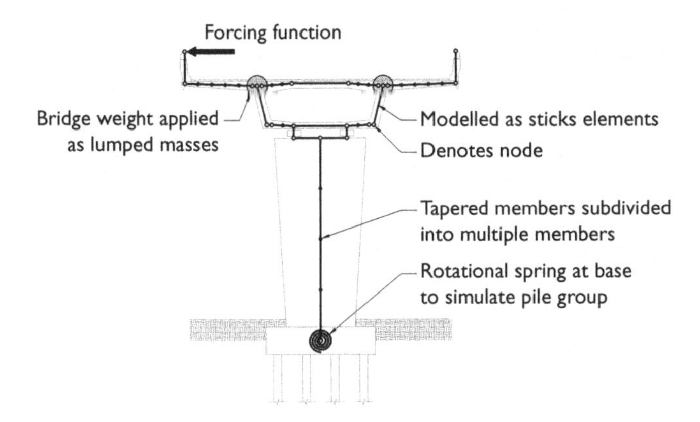

Figure 9.4 Detailed model of case study bridge in Strand 7.

based on recommendations in the well-known bridge design textbook of Priestley et al. [9.10]. The results of the analysis are presented in Section 9.4.4.

9.4.3 Simplified Model

A simplified model was developed where the numerical model presented above has been simplified into a "three-legged frame" with only four members forming half of the bridge structure, as shown in Figure 9.5. Members #1, #2, #3 and #4, as annotated in Figure 9.5, represent the barrier, bridge deck, box girder and pier, respectively. The simplified frame is modelled as a multi-degree-of-freedom (MDOF) system and discretised into 250 lumped masses, with an additional lumped mass (m_{add}) at the intersection of members 3 and 4 to approximate the portion of the bridge that has not been included in the model explicitly (via the self-weight of the members modelled). The contact force time-history for the collision scenario was applied to the model by use of the *Duhamel Integration Method* (as introduced in Chapter 13, which shows the implementation of the method on an Excel spreadsheet, or on MATLAB). As for the detailed model, the simplified model was subject to dynamic modal analysis, assuming a damping ratio of 5%.

The barrier (member #1) was modelled as a 0.25 metre thick element as was consistent with the detailed model. The bridge deck (member #2) was modelled as a 0.3 metre thick element, which was the average thickness of the tapered cantilevered deck slab. The box girder (member #3) was modelled as a 2 m deep cross section to act as a rigid offset between the bridge deck and the top of the pier. The bridge pier (member #4) was modelled as a 3.5 m long member based on averaging the length of the pier.

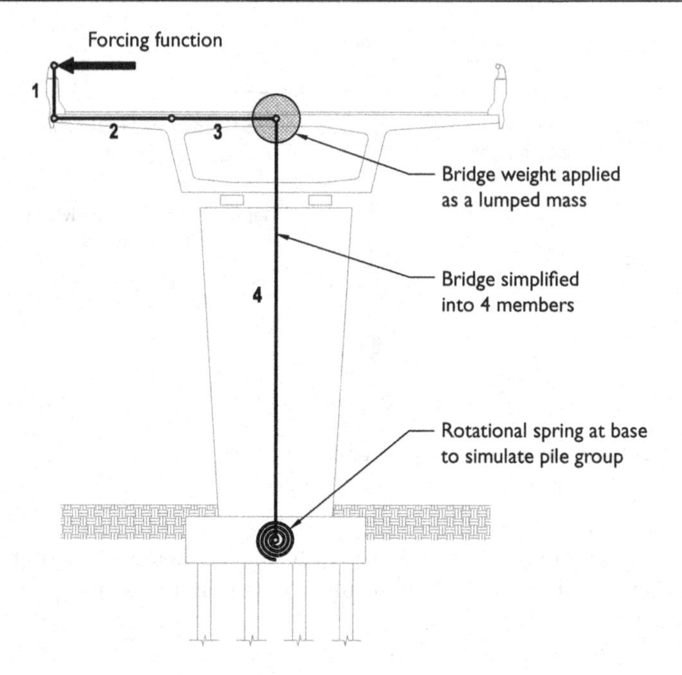

Figure 9.5 Simplification of numerical model.

All the RC members as described have the same stiffness modification factors as presented earlier (0.35 for the barrier and the bridge deck, and 0.7 for the box girder and pier).

9.4.4 Comparison of Results

The results of the detailed and simulated analysis models are presented in Figure 9.6(a) and 9.6(b) for the total barrier displacement and base bending moment at the base of the pier, respectively.

9.5 CLOSED-FORM EXPRESSION

9.5.1 General

This section presents a procedure for the rapid assessment of a bridge structure subject to a collision scenario on the bridge deck. A design methodology based on the use of closed-form expression is presented herein for estimating the maximum bending moment that is transmitted to the base of the bridge pier. However, it is important to verify preliminary sizing estimated by the closed-form expression using results from numerical modelling when developing the final design solution.

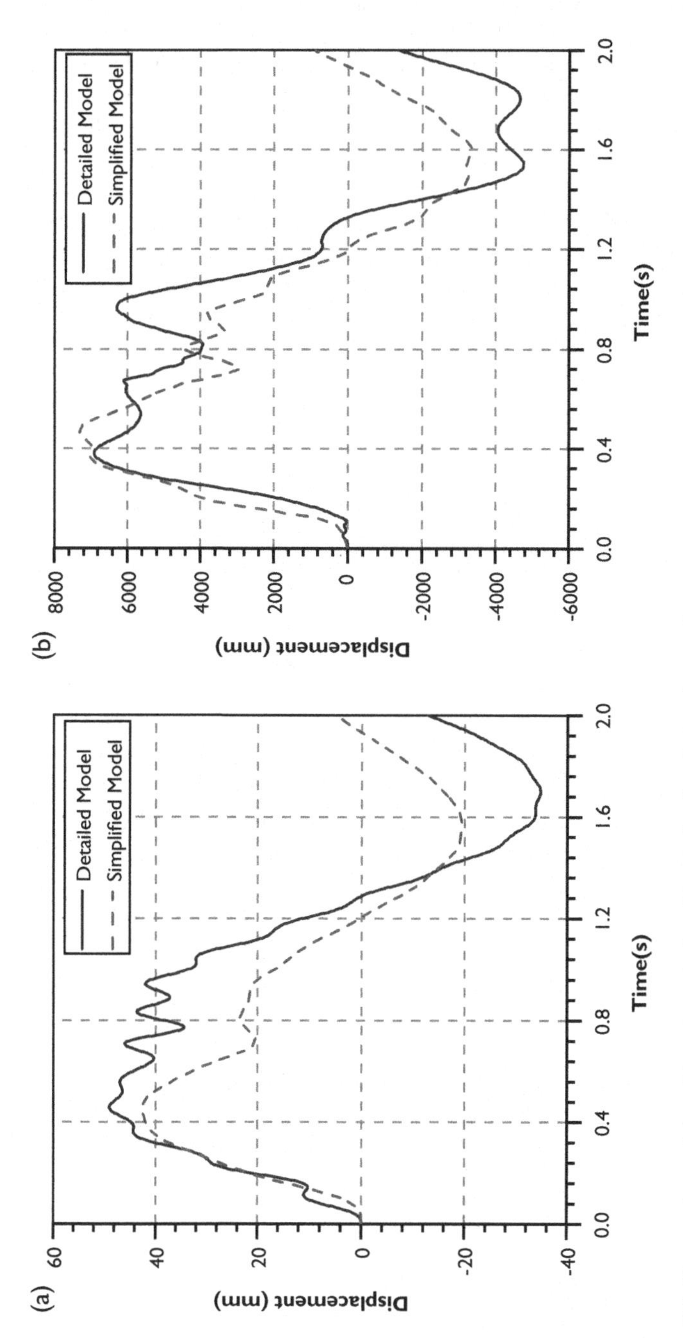

Figure 9.6 Comparison of time-histories calculated from the detailed and simplified numerical models when subjected to the forcing function for the collision scenario being considered: (a) displacement at the top of the bridge pier and (b) bending moment transmitted to the base of the bridge pier.

Equation (3.9a) in Chapter 3 is modified into Equation (9.3) by incorporating an additional parameter γ to take into account the effects of a delay in the transfer of momentum following the collision. A COR value of 0.35 is adopted, and the value of γ is recommended to be 0.5 and 0.9 for a semi-trailer and single-unit truck, respectively. The COR and γ values are developed in Chapter 10 (the reader is directed to Figure 10.14 specifically). Substituting the recommended value of COR = 0.35 into Equation (9.3) and multiplying it by the height dimension (L_6 defined in Figure 9.7) gives Equation (9.4) for estimating the maximum bending moment transmitted to the base of the bridge pier. The generalised mass and stiffness to be adopted, for use in Equation (9.4), are presented in the following section for a typical bridge structure.

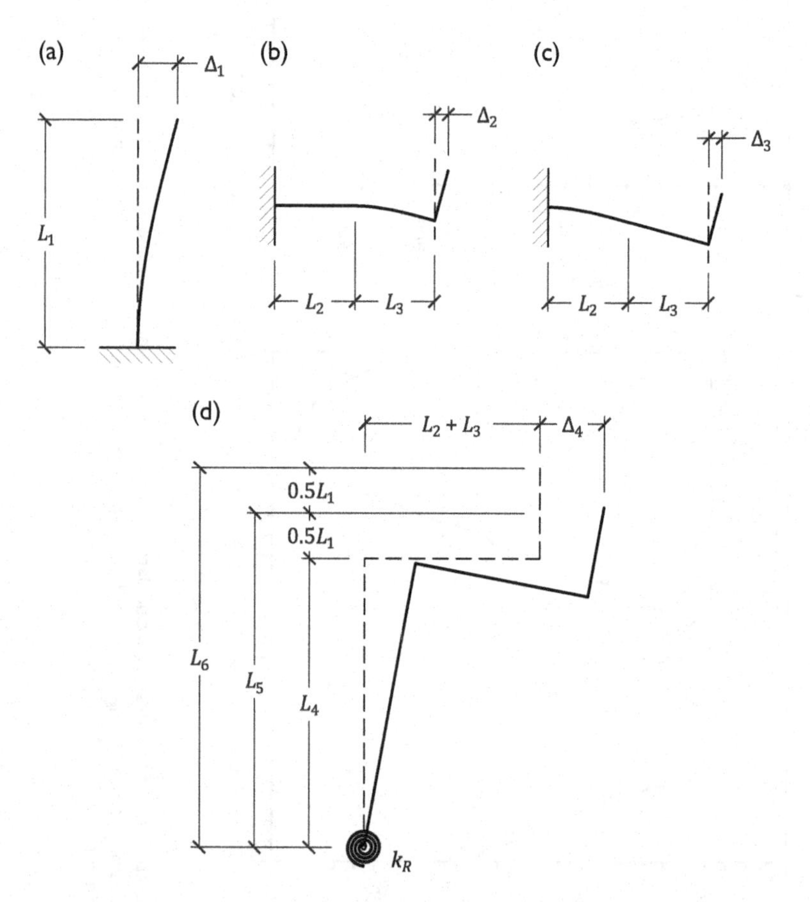

Figure 9.7 Displacement of (a) member #1, (b) member #2, (c) member #3 and (d) member #4.

$$F^* = v_0\sqrt{mk}\sqrt{\lambda\left(\frac{1 + \text{COR}}{1 + \lambda}\right)^2} \times \gamma \tag{9.3}$$

$$M^* = 1.35v_0\sqrt{mk}\left(\frac{\sqrt{\lambda}}{1 + \lambda}\right)L_6 \times \gamma \tag{9.4}$$

9.5.2 Derivations of Generalised Mass and Stiffness

Expressions for estimating the generalised mass and stiffness of the bridge structure as a whole are first derived. Consider the simplified four-member frame shown in Figure 9.5. Note the member identification numbers (IDs) that are annotated on the diagram. These member IDs will be frequently referred to in the rest of this section. The displacement contribution of each member (to the overall lateral displacement of the bridge) is considered separately, as shown in Figure 9.7.

Values for the generalised mass and stiffness of a cantilevered wall (i.e. member #1) of Figure 9.7(a) can be found using Equations (9.5) and (9.6), respectively. Their derivations can be found in Chapter 2.

$$m_1^* = 0.25M_{t1} \tag{9.5}$$

$$k_1^* = \frac{3EI_1}{L_1^3} \tag{9.6}$$

where $M_{t1} = \rho A_1 L_1$ is the total mass of the barrier (or member #1).

As for member #2, as shown in Figure 9.7(b), the flexing of the member causes base rotation of the barrier (member #1). The displacement of the barrier from this rotation can be represented by a rigid rod, which is fitted with a rotational spring at the base (with rotational stiffness k_{R2}), as shown in Figure 9.8.

The generalised mass of a structural element can be calculated using Equation (2.14b), which is re-written here as Equation (9.7). Derivation of the equation can be found in Chapter 2.

$$m^* = \int_0^L \rho A\psi(x)^2 dx \tag{9.7}$$

where $\psi(x)$ is the shape function, which has been normalised to unity at the point of contact, i.e. at the free (top) end of the cantilever. $\psi(x)$ of the rigid rod is as defined by Equation (9.8), which results in a linear displacement profile along the height of the rod.

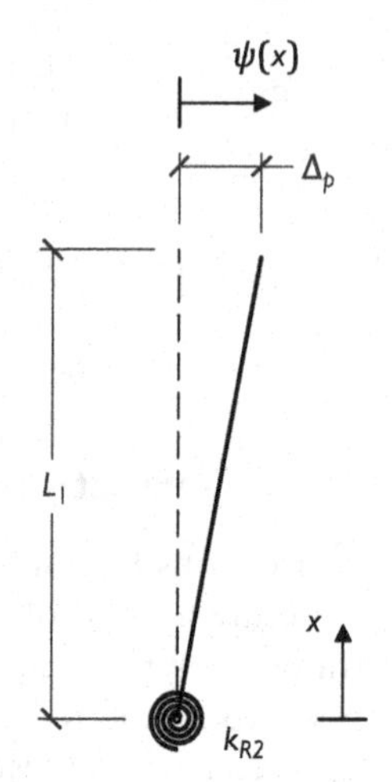

Figure 9.8 Sketch showing shape function of rigid rod.

$$\psi(x) = \frac{x}{L_1} \tag{9.8}$$

Substituting Equation (9.8) into Equation (9.7) and solving for the integral gives Equation (9.9), which provides an estimate for the generalised mass of the barrier associated with rigid body rotation.

$$m_2^* = \int_0^{L_1} \rho A_1 \left(\frac{x}{L_1}\right)^2 dx = \frac{M_{t1}}{3} \tag{9.9}$$

The generalised stiffness of the barrier associated with the same rigid body rotation is given by Equation (9.10):

$$k_2^* = \frac{P}{\Delta_P} \tag{9.10}$$

The bending moment (M_b) and angle of rotation (θ) at the base of the rod are as defined by Equations (9.11) and (9.12), respectively:

$$M_b = PL_1 \tag{9.11}$$

$$\theta = \frac{\Delta_P}{L_1} \tag{9.12}$$

The bending moment at the base of the rod is a function of the rotational stiffness and angle of rotation, as shown by Equation (9.13). Parameter k_{R2}

may in turn be determined using Equation (9.14), which was derived by applying the unit load method.

$$M_b = k_{R2}\theta \tag{9.13}$$

$$k_{R2} = \frac{EI_2}{L_2} \tag{9.14}$$

Substituting Equations (9.11) and (9.12) into Equation (9.13) and re-arranging gives Equation (9.15):

$$P = \frac{k_{R2}\Delta_P}{L_1^2} \tag{9.15}$$

Substituting Equation (9.15) into Equation (9.10) gives Equation (9.16):

$$k_2^* = \frac{k_{R2}}{L_1^2} \tag{9.16}$$

Substituting Equation (9.14) into Equation (9.16) gives Equation (9.17), which provides estimates for the contribution of member #2 to its generalised stiffness:

$$k_2^* = \frac{EI_2}{L_1^2 L_2} \tag{9.17}$$

The contribution by member #3 to its own generalised mass and stiffness as shown in Figure 9.7(c) can be derived in a similar manner as for member #2, as represented by Equations (9.18) and (9.19):

$$m_3^* = \frac{M_{t1}}{3} \tag{9.18}$$

$$k_3^* = \frac{EI_3}{L_1^2 L_3} \tag{9.19}$$

As for the displacement of member #4, two additional geometrical parameters are introduced in Figure 9.7(d), as defined by Equations (9.20) and (9.21). The rotation of member #4 representing the bending of the bridge pier is typically insignificant for bridges that are supported on piles like the example design shown in Figure 9.3.

$$L_5 = \frac{L_1}{2} + L_4 \tag{9.20}$$

$$L_6 = L_1 + L_4 \tag{9.21}$$

Contributions to the generalised mass of the pier can be resolved into three components, which are all associated with its own rotational motion: (a) contributions by the motion of member #1, (b) contributions by the motion of members #2 and #3 and the additional bridge mass m_{add} and (c) contributions by the motion of the pier itself (member #4). These three component conbtributions are denoted as m_a, m_b and m_c, respectively. Every contribution to the generalised mass is dependent on both the mass, which is engaged into motion, and the relative amplitude of the respective motions. With the assumption of a linear displacement profile, as shown in Figure 9.7(d), the values of m_a and m_b are each scaled as per their respective height dimensions L_5 and L_4, as shown by Equations (9.22) and (9.23):

$$m_a = M_{t1}\left(\frac{L_5}{L_6}\right) \tag{9.22}$$

$$m_b = (M_{t2} + M_{t3} + m_{\text{add}})\left(\frac{L_4}{L_6}\right) \tag{9.23}$$

The remaining component, m_c, is derived by the use of Equation (9.7). The mode shape of member #4 is as defined by Equation (9.24):

$$\psi(x) = \frac{x}{L_6} \tag{9.24}$$

Equation (9.24) is then substituted into Equation (9.7) and the integral is solved, as shown by Equation (9.25):

$$m_c = \int_0^{L_4} \rho A_4\left(\frac{x}{L_6}\right)^2 dx = \frac{M_{t4}}{3}\left(\frac{L_4}{L_6}\right)^2 \tag{9.25}$$

where $M_{t4} = \rho A_4 L_4$ is the total mass of the pier (member #4).

Taking the summation of Equations (9.22), (9.23) and (9.25) gives the total contributions by member #4 to its own generalised mass, as shown by Equation (9.26):

$$m_4^* = M_{t1}\left(\frac{L_5}{L_6}\right) + (M_{t2} + M_{t3} + m_{\text{add}})\left(\frac{L_4}{L_6}\right) + \frac{M_{t4}}{3}\left(\frac{L_4}{L_6}\right)^2 \tag{9.26}$$

Equation (9.16), which was derived for a rigid rod is also applicable for the modelling of the generalised stiffness for member #4, and is re-written as Equation (9.27):

$$k_4^* = \frac{k_R}{L_6^2} \tag{9.27}$$

Stiffness parameters k_1^*, k_2^* , k_3^* and k_4^* essentially represent four springs in series, which can be combined into one spring by the use of Equation (9.28). Taking the reciprocal of Equation (9.28) gives Equation (9.29) for determining the generalised stiffness of the part of the bridge structure that is engaged in responding to the vehicular collision. The generalised stiffness components in Equation (9.29) can be determined using Equations (9.6), (9.17), (9.19) and (9.27), respectively.

$$\frac{1}{k} = \frac{1}{k_1^*} + \frac{1}{k_2^*} + \frac{1}{k_3^*} + \frac{1}{k_4^*} \tag{9.28}$$

$$k = \left(\frac{1}{k_1^*} + \frac{1}{k_2^*} + \frac{1}{k_3^*} + \frac{1}{k_4^*}\right)^{-1} \tag{9.29}$$

Contributions by the four members to the generalised mass of the structure (engaged in responding to the vehicular collision) may be combined by the use of their respective flexibilities (i.e. reciprocal of their stiffness), as shown by Equation (9.30). The weighting factor attributed to each member is defined as their respective flexibility ratio. The generalised mass components in Equation (9.30) can be determined using Equations (9.5), (9.9), (9.18) and (9.26), respectively.

$$\lambda m = \left(\frac{k}{k_1^*}\right) m_1^* + \left(\frac{k}{k_2^*}\right) m_2^* + \left(\frac{k}{k_3^*}\right) m_3^* + \left(\frac{k}{k_4^*}\right) m_4^* \tag{9.30}$$

In summary, this section presents the development of Equations (9.29) and (9.30), which are used as input parameters into Equation (9.4) for determining the maximum bending moment (M^*) transmitted to the base of the bridge pier.

9.6 APPLICATION OF THE CLOSED-FORM EXPRESSION

Consider the example box girder viaduct featured in the case study (Figure 9.3). By the use of closed-form expression presented in the chapter, determine the amount of bending moment transmitted to the base of the pier and its foundation when subjected to a collision of a 36 t semi-trailer at the barrier (on the bridge deck) at a velocity of 90 km/h and an angle of 15°. Check the accuracy of the estimate by comparison with the same estimate made by numerical simulations.

Solution

Total mass of members #1 and #2:

$$M_{t1} = \rho L_1 B_1 D_1 = 2400(1.37)(2.74)(0.25) = 2252.3 \text{ kg}$$

$$M_{t2} = \rho L_2 B_2 D_2 = 2400(3.16)(2.74)(0.3) = 6234 \text{ kg}$$

Area of half of a box girder forming member #3 that has a sectional area: $A_3 = 1.951 \text{ m}^2$. Total mass of member #3:

$$M_{t3} = \rho A_3 B_3 = 2400(1.951)(2.74) = 12830 \text{ kg}$$

From Section 9.4.1, the self-weight of the bridge is 196.5 kN/m with span length of 40 m. Additional mass that has not been modelled:

$$m_{\text{add}} = 196.5(40) \times \frac{1000}{9.81} - 2252.3 - 6234 - 12830 \approx 780000 \text{ kg}$$

Total mass of member #4:

$$M_{t4} = \rho L_4 B_4 D_4 = 2400(10.9)(1)(3.5) = 91560 \text{ kg}$$

Contributions by each member to the respective generalised mass:

$$m_1^* = 0.25 M_{t1} = 0.25(2252.3) = 563 \text{ kg}$$

$$m_2^* = m_3^* = \frac{M_{t1}}{3} = \frac{2252.3}{3} = 751 \text{ kg}$$

$$L_5 = \frac{L_1}{2} + L_4 = \frac{1.37}{2} + 10.9 = 11.6 \text{ m}$$

$$L_6 = L_1 + L_4 = 1.37 + 10.9 = 12.3 \text{ m}$$

$$m_4^* = M_{t1}\left(\frac{L_5}{L_6}\right) + (M_{t2} + M_{t3} + m_{\text{add}})\left(\frac{L_4}{L_6}\right) + \frac{M_{t4}}{3}\left(\frac{L_4}{L_6}\right)^2 = 2252.3\left(\frac{11.6}{12.3}\right) + (6234 + 12830 + 780000)\left(\frac{10.9}{12.3}\right) + \frac{91560}{3}\left(\frac{10.9}{12.3}\right)^2 = 734000 \text{ kg}$$

Second moment of area of each member:

$$I_1 = \frac{B_1 D_1^3}{12} = \frac{2.74(0.25)^3}{12} = 3.57 \times 10^{-3} \text{ m}^4$$

$$I_2 = \frac{B_2 D_2^3}{12} = \frac{2.74(0.3)^3}{12} = 6.17 \times 10^{-3}\ \text{m}^4$$

$$I_3 = \frac{B_3 D_3^3}{12} = \frac{2.74(2)^3}{12} = 1.\ 83\ \text{m}^4$$

Contributions by each member to the generalised stiffness:

$$k_1^* = \frac{3EI_1}{L_1^3} = \frac{3(37100 \times 10^3)(0.35 \times 3.57 \times 10^{-3})}{1.37^3} = 54147\ \text{kN/m}$$

$$k_2^* = \frac{EI_2}{L_1^2 L_2} = \frac{37100 \times 10^3(0.35 \times 6.17 \times 10^{-3})}{1.37^2(3.16)} = 13524\ \text{kN/m}$$

$$k_3^* = \frac{EI_3}{L_1^2 L_3} = \frac{37100 \times 10^3(0.7 \times 1.83)}{1.37^2(2.74)} = 8896 \times 10^3\ \text{kN/m}$$

$$k_4^* = \frac{k_R}{L_6^2} = \frac{2000 \times 10^3}{12.3^2} = 13220\ \text{kN/m}$$

Given the excessively high value of k_3^*, contribution of member #3 to the generalised mass and stiffness may be neglected.

Generalised stiffness of the structure as a whole:

$$k = \left(\frac{1}{k_1^*} + \frac{1}{k_2^*} + \frac{1}{k_4^*}\right)^{-1} = \left(\frac{1}{54147} + \frac{1}{13524} + \frac{1}{13220}\right)^{-1} = 5951\ \text{kN/m}$$

Generalised mass of the structure as a whole:

$$\lambda m = \left(\frac{k}{k_1^*}\right)m_1^* + \left(\frac{k}{k_2^*}\right)m_2^* + \left(\frac{k}{k_4^*}\right)m_4^* = 5951\left(\frac{563}{54147} + \frac{751}{13524} + \frac{734000}{13220}\right)$$
$$= 330800\ \text{kg}$$

$$\lambda = \frac{\lambda m}{m} = \frac{330800}{36000} = 9.2$$

Impact velocity perpendicular to the barrier:

$$v_0 = v_\theta \sin\theta = \left(\frac{90}{3.6}\right)\sin 15° = 6.47\ \text{m/s}$$

Applying Equation (9.4) for determining bending moment at the base of the pier (adopting $\gamma = 0.5$ for semi-trailer collision):

$$M^* = 1.35 v_0 \sqrt{mk}\left(\frac{\sqrt{\lambda}}{1+\lambda}\right) L_6 \times \gamma$$

$$= 1.35(6.47)\sqrt{36000(5951 \times 10^3)}\left(\frac{\sqrt{9.2}}{1+9.2}\right)(12.3) \times 0.5 = 7394 \text{ kNm}$$

The peak bending moment value as obtained from the numerical model (as shown in Figure 9.6(b)) is 7,322 kNm, which is about 1% different to the approximate estimate of 7,394 kNm, as found using the closed-form expressions presented in Section 9.5. The distribution of the γ factor (in the closed-form expression) based on calibration against results from numerical simulations is presented in Chapter 10.

9.7 CLOSING REMARKS

Current methods that are stipulated in codes of practice for guiding the design of bridges for countering vehicular collision are first reviewed to identify their limitations, which include the equal energy method, the equivalent static force method and physical crash testing. The common limitation across these methods lies in the accuracy of predicting the transmission of the collision action from the barrier to the rest of the bridge, including the base of the bridge pier and its foundation, since the inertial response of the bridge is not accounted for. Forcing functions for a range of vehicular collision scenarios have been derived and digitised in tabulated forms for input into the numerical models as an alternative procedure for calculated the design actions associated with collision scenario. Numerical modelling is then introduced using an RC viaduct as an example to illustrate how the force functions can be adopted. Separate to numerical modelling is the development of analytical modelling, which involves the use of closed-form expression for making predictions of internal forces (including bending moments) that are transmitted to the base of the bridge pier. The analytical model, which is relatively easy to use, is more suited to guiding the preliminary design of the bridge. Numerical modelling, which requires the computer model of the structure to be available (in the detailed design phase of the project), can be used for evaluating the accuracy of predictions made by the analytical model. The chapter concludes by presenting the application of the closed-form expressions using the case study bridge presented earlier in the chapter.

REFERENCES

9.1 Hunt, K.H. and Crossley, F.R.E., 1975, "Coefficient of restitution interpreted as damping in vibroimpact", *Journal of Applied Mechanics*, Vol. 42(2), p. 440, doi:10.1115/1.3423596

9.2 European Committee for Standardization (CEN), 2006, *Eurocode 1: Actions on structures - Part 1-7: General actions - Accidental actions*, European Committee for Standardization (CEN), Brussels, Belgium.

9.3 American Association of State Highway and Transportation Officials, 2012, *AASHTO LRFD Bridge Design Specifications, 6th edition*, American Association of State Highway and Transportation Officials, Washington DC, U.S.

9.4 Austroads, 2013, *Standardised Bridge Barrier Design*, Austroads Ltd, Sydney.

9.5 Standards Australia, 2017, *AS 5100.2 Bridge Design Part 2: Design Loads*, Standard Australia Limited, New South Wales, Australia.

9.6 Hirsch, T.J., 1978, *Analytical Evaluation of Texas Bridge Rails to Contain Buses and Trucks*, Texas Transportation Institute, Texas, U.S.

9.7 Olson, R.M. and Post, E.R., 1970, "Tentative service requirements for bridge rail systems", *NCHRP Report* (86).

9.8 Standards Australia, 2004, *AS 5100.1 Bridge Design Part 1: Scope and General Principles*, Standard Australia Limited, New South Wales, Australia.

9.9 Strand7 Pty Ltd, 2004, *Strand7 (R2.3)*, software developed by Strand7 Pty Ltd, Sydney, Australia.

9.10 Priestley, M.J.N., Seible, F., and Calvi, G.M., 1996, *Seismic Design and Retrofit of Bridges*, John Wiley & Sons, U.S.

Chapter 10

Vehicle Collision on a Bridge Deck – Model Validation

10.1 INTRODUCTION

This chapter presents the development and verifications of the methodologies proposed in Chapter 9. A series of full-scale crash test data is used to develop forcing functions to simulate the localised collision actions that occur from a vehicular collision on a bridge desk's barrier (Section 10.2). The developed forcing functions are then applied to a case study bridge to assess the overall bridge structure to the collision scenario. The overall bridge structure was modelled using a simplified four-member model, which was validated against a detailed analysis model of the bridge (Section 10.3). The simplified model was then utilised to derive normalised forcing functions using the Hunt and Crossley (H&C) [10.1] model for different collision scenarios (Section 10.4). This chapter concludes with the development of closed-formed expressions, featuring the use of a cushion reduction factor (γ) to account for the crumbling action of the vehicle, that can be used to quickly estimate the transmission of the collision actions to the rest of the structure (Section 10.5).

10.2 CRASH TEST DATA

Four full-scale crash tests were considered in the numerical modelling, as listed in Table 10.1, to determine an equivalent generalised forcing function to represent certain benchmark collision scenarios. The presented benchmarks correspond to the Regular and Medium performance levels, or TL-4 and TL-5, respectively. The TL-4 vehicle is a single-unit truck, which can be idealised as a single mass m_1 (as shown in Figure 10.1(a)). The TL-5 vehicle is a semi-trailer, and can be idealised as two separate masses m_1 and m_2, where m_1 was the total mass supported on the axles of the prime mover and m_2 was the total mass supported on the axles of the trailer (as shown in Figure 10.1(b)). The vehicle collides with the vehicular barrier at a velocity of v_θ and at an angle θ in accordance with crash test criteria as presented in Table 9.2 [10.2]. Each of the crash tests was conducted on a reinforced concrete (RC) barrier that was supported at the base with a fixed condition.

DOI: 10.1201/9781003133032-10

Table 10.1 Crash test

Test #	*Test Level*	*Type of Vehicle*	m_l *(kg)*	m_2 *(kg)*	v_θ *(m/s)*	θ *(deg)*	*Ref.*
1	TL-5	Semi-Trailer	20,462	15,903	21.9	15.0	[10.3]
2	TL-4	Single-Unit Truck	10,048	N/A	25.6	16.1	[10.4]
3	TL-4	Single-Unit Truck	10,133	N/A	25.4	14.6	[10.5]
4	TL-4	Single-Unit Truck	10,146	N/A	26.1	15.6	[10.6]

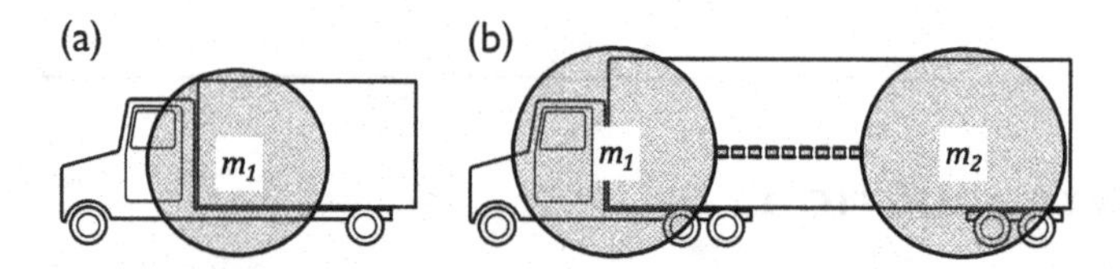

Figure 10.1 (a) TL-4 and (b) TL-5 crash test vehicles.

The time-histories recorded from these tests were used to generate the contact forcing functions (time-histories) for the respective test.

10.2.1 TL-5 Crash Test

Acceleration time-histories in all three directions (lateral, longitudinal and vertical directions, with respect to the vehicle) were recorded in the crash tests with the use of accelerometers. The accelerations were measured at both the centre of gravity of the prime mover (i.e. mass m_1) and at the rear axle (i.e. mass m_2). Prior to the occurrence of the contact, the axis of the vehicle is at an angle θ relative to the alignment of the barrier (i.e. Figure 10.2(a)). This angle varies with time in the course of the impact, as recorded in the respective test report.

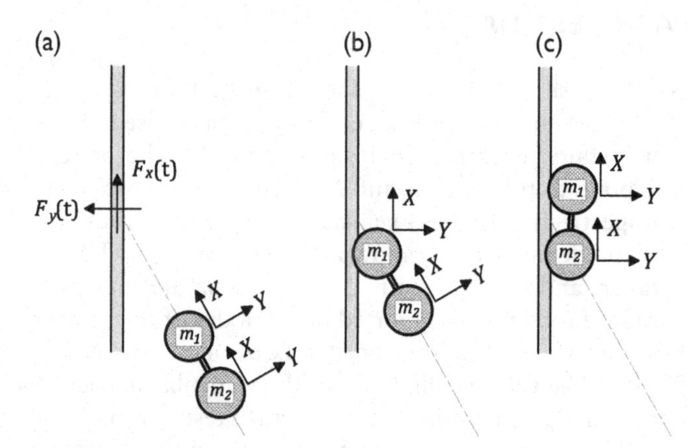

Figure 10.2 TL-5 crash behaviour.

Collision of a TL-5 truck on a barrier has two occurrences. The initial collision refers to the prime mover colliding with the barrier at an angle θ. During this phase of the collision, the prime mover rotates towards the barrier until the axis of the prime mover aligns with the barrier (i.e. Figure 10.2(b)). As the prime mover continues to move forward the trailer rotates (i.e. pivots about the fifth-wheel on the prime mover) until the rear axle collides laterally onto the barrier in a second collision. The axles of the trailer essentially collide with the barrier in a normal direction (i.e. Figure 10.2(c)). This behaviour is documented and clearly observable in a series of photos presented in test reports, which were taken from the "bird's eye view" during the crash test using the high-speed camera [10.3].

The contact force time-history generated by the first and second collisions (i.e. $F_{y,1}(t)$ and $F_{y,2}(t)$, respectively) were determined by multiplying the mass of the respective part of the truck engaged in the collision (i.e. m_1 and m_2, respectively) by its acceleration time-history in the direction normal to the barrier. The lateral acceleration of the first collision needs to be calculated by taking the lateral component from each of the x-axis and y-axis acceleration time-histories (i.e. $a_{x,1}(t)$ and $a_{y,1}(t)$, respectively), while also accounting for the rotation of the prime mover in the course of the collision, as described previously. The calculation has to be conducted with the use of Equation (10.1a), where θ is the initial angle that the axis of the prime mover makes with the barrier and $\theta(t)$ is the rotation time-history during the collision. Given that the second collision is the pounding of the rear axle onto the barrier, which essentially occurs perpendicular to the barrier, Equation (10.1b) is to be used. The time-histories of the contact force generated by the collision are presented in Figures 10.3(a) and 10.3(b) for the first and second collisions, respectively.

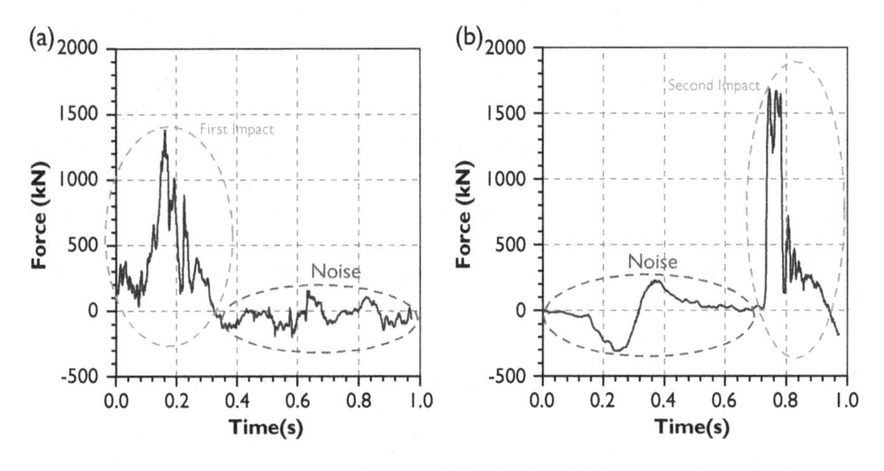

Figure 10.3 Force time-history of Test #1 at (a) front and (b) rear of the vehicle.

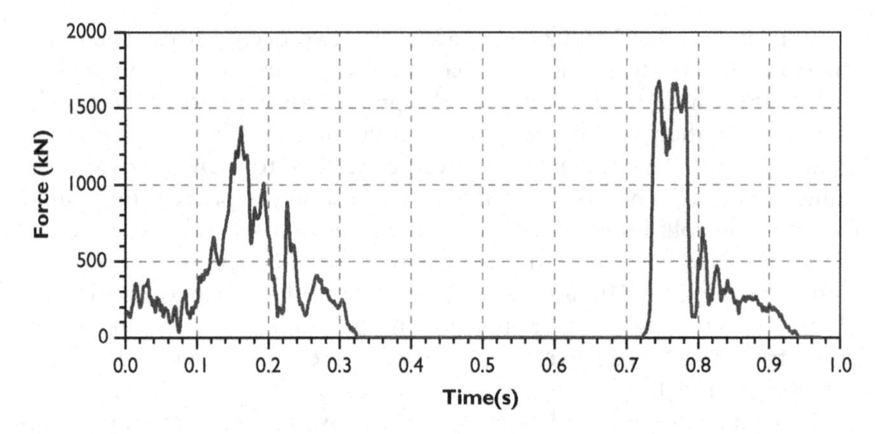

Figure 10.4 Contact force time-history for the TL-5 impact scenario.

$$F_{y,1}(t) = [a_{x,1}(t)sin(\theta + \theta(t)) + a_{y,1}(t)cos(\theta + \theta(t))] \times m_1 \tag{10.1a}$$

$$F_{y,2}(t) = a_{y,2}(t) \times m_2 \tag{10.1b}$$

The two contact force time-histories can be combined into one time-history for modelling the total response of the structure. There are noises preceding, and succeeding, each collision when neither the prime mover, nor the trailer, is "in contact" with the barrier, and are only resulted from vibration of the colliding objects. The total contact force time-history representing the two collisions collectively is obtained by overlaying/enveloping the two time-histories and removing all the noise. The result is shown in Figure 10.4.

10.2.2 TL-4 Crash Tests

The contact force time-history for the TL-4 collision can be obtained in the same manner as described for the modelling of the TL-5 collision; Equation (10.1a) is adopted. The resulting contact force time-histories for the three TL-4 crash tests is presented in Figure 10.5. A similar behaviour occurs for the single-unit truck impact where the engine collides with the barrier and the truck rotates and then the rear axle collides, causing two peaks in the force time-history (as is observed in Figure 10.5). The duration between these two collision points is significantly less for the single-unit truck compared to the semi-trailer in Section 10.2.1.

10.2.3 Limitation of the Contact Force Time-Histories

The contact force time-histories presented in this section were all recorded from crash tests conducted on RC barriers with a fixed base connection and an

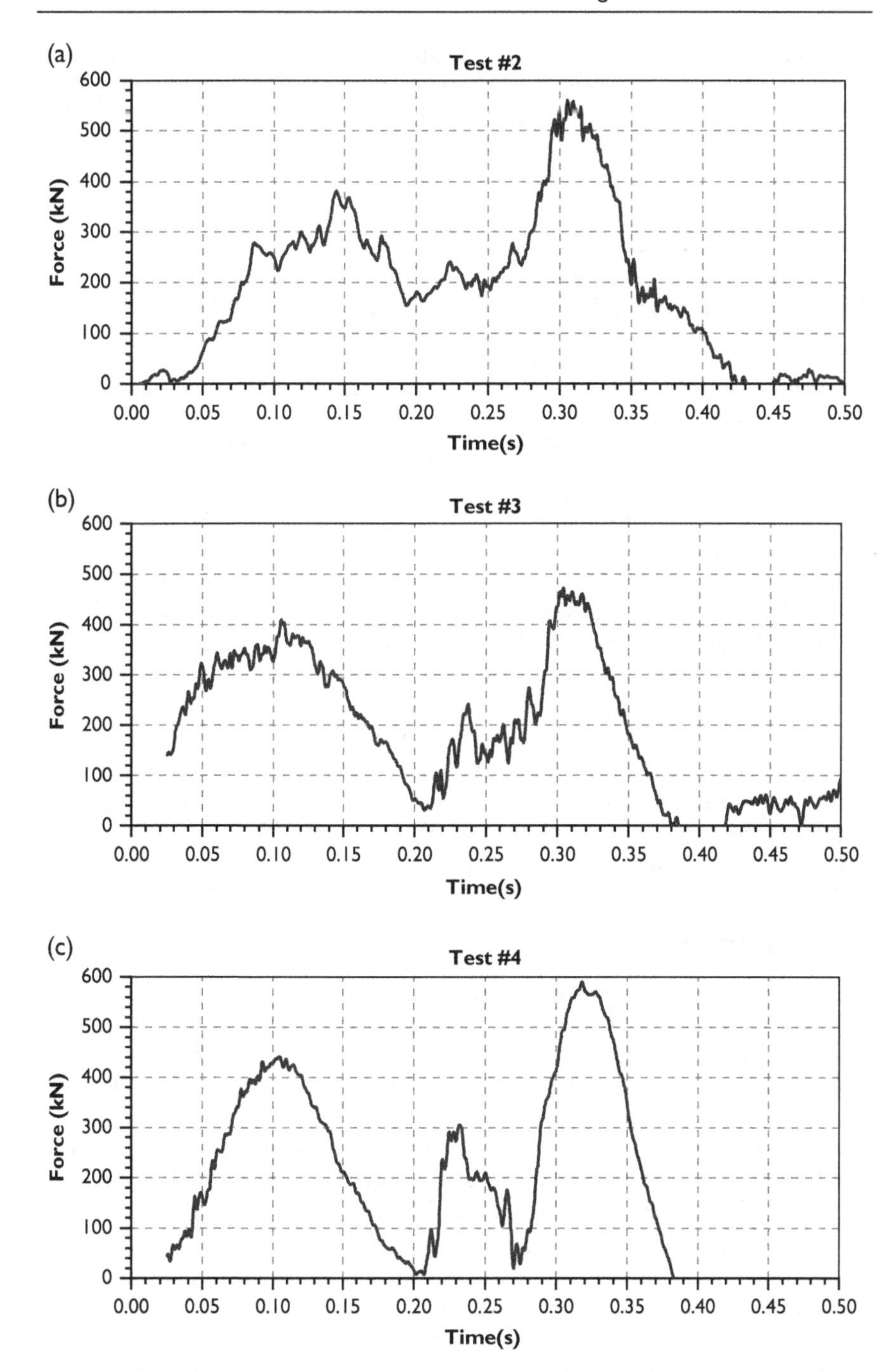

Figure 10.5 Contact force time-history for the TL-4 impact scenario.

overall height of 1,067 mm (42 inches) and 914 mm (36 inches) for the TL-5 and TL-4 scenarios, respectively. Barriers with a significantly different lateral stiffness (e.g. structural steel post/beam barriers or barriers with very flexible connection) would likely respond with different contact force time-histories.

10.3 NUMERICAL MODELLING

10.3.1 Validation of the Simplified Numerical Model

Validation of the simplified numerical model as introduced in Section 9.4.3 is presented herein. The validation is performed by comparing its response to the detailed model presented in Section 9.4.2, using the case study bridge presented in Section 9.4.1. The contact force time-histories recorded from both Tests #1 and #2 (Figures 10.4 and 10.5(a)) represent scenarios where the top of the barrier of the case study bridge (presented in Section 9.4.1) is subject to the collision action. The response behaviour of the bridge structure as calculated from the detailed and simplified numerical models to each of the contact force time-histories is presented in Figures 10.6 and 10.7 for Tests #1 and #2, respectively. These figures present (a) displacement of the top of the barrier relative to the base of the bridge, (b) bending moment at the base of the barrier, (c) bending moment transmitted to the bridge deck, (d) displacement at the top of the box girder and (e) bending moment transmitted to the base of the bridge pier. Good consistencies between results calculated from the detailed and simplified numerical models are shown. The satisfactory performance of the simplified numerical model is the basis of its adoption in studies presented in the rest of this chapter.

10.3.2 Parametric Study

A parametric study has been undertaken to understand how the case study bridge (as presented in Figure 9.3) responds to the contact force time-histories when different parameters of the case study bridge are modified. The results from the simplified numerical model are compared against the results derived from the prescriptive equivalent static design forces as stipulated in the bridge design standard AS 5100.2 [10.7] (refer Chapter 9, Table 9.1). The parametric study was conducted using the contact force time-histories derived from Tests #1 and #2 (for TL-5 and TL-4 performance levels, respectively). The requirements of AS 5100.2 for TL-5 and TL-4 are to apply at the top of the barrier an equivalent horizontal static point force of 600 kN and 300 kN, respectively.

The parametric study involved halving and doubling the mass (m_{add}) and base rotational stiffness (k_R) of the base model. Time-histories of the bending moment transmitted to the base of the bridge pier are presented in Figures 10.8 and 10.9 for Tests #1 (TL-5) and 2 (TL-4), respectively.

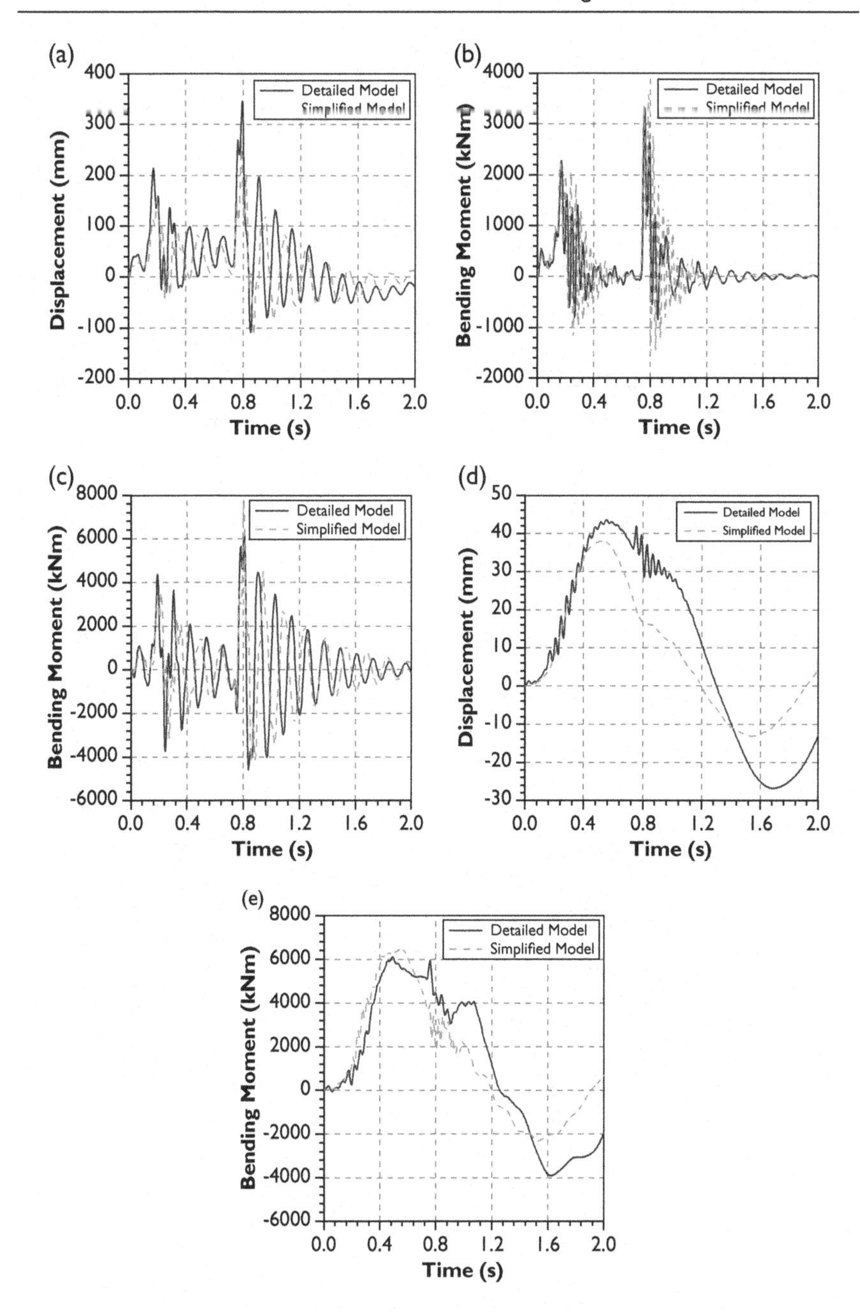

Figure 10.6 Comparison of time-histories calculated from the detailed and simplified numerical models when subjected to the forcing function of Crash Test #1: (a) Displacement at the top of the barrier, (b) bending moment at the base of the barrier, (c) bending moment transmitted to the fixed-end of the bridge deck, (d) displacement at the top of the bridge pier and (e) bending moment transmitted to the base of the bridge pier.

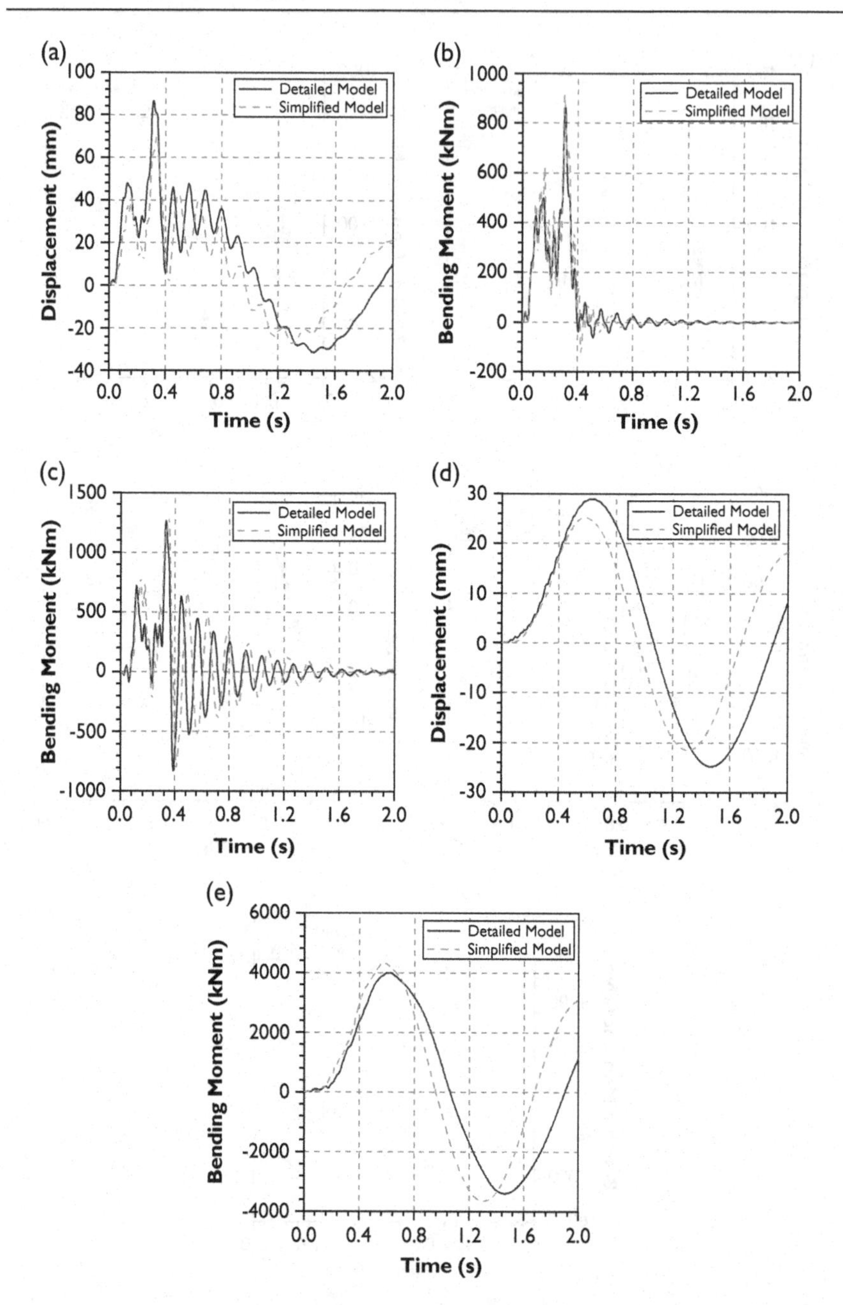

Figure 10.7 Comparison of time-histories calculated from the detailed and simplified numerical models when subjected to the forcing function of Crash Test #2: (a) Displacement at the top of the barrier, (b) bending moment at the base of the barrier, (c) bending moment transmitted to the fixed-end of the bridge deck, (d) displacement at the top of the bridge pier and (e) bending moment transmitted to the base of the bridge pier.

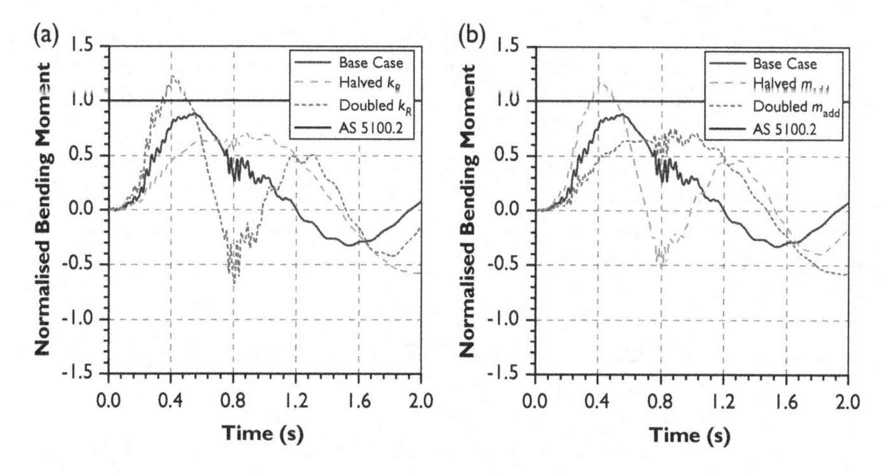

Figure 10.8 Results of parametric study of (a) k_R and (b) mass of bridge for Test #1 (TL-5).

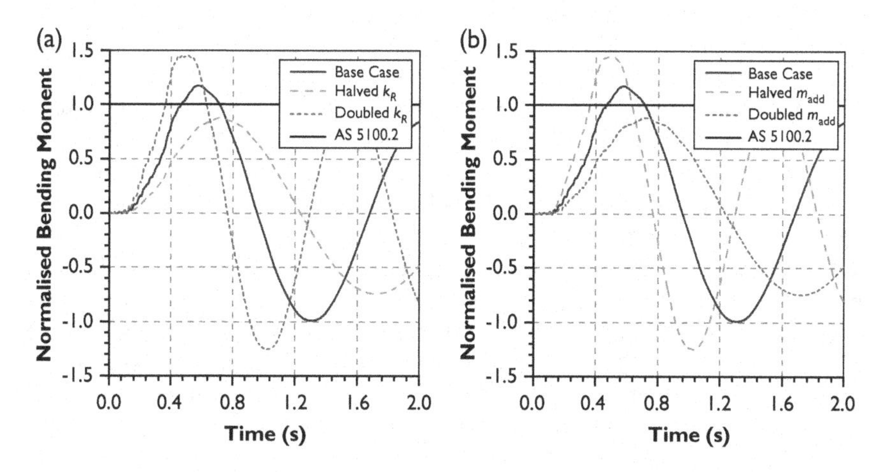

Figure 10.9 Results of parametric study of (a) k_R and (b) mass of bridge for Test #2 (TL-4).

The results of analyses are presented as ratios to the base bending moments calculated using the code prescribed equivalent horizontal static forces (i.e. 600 and 300 kN for TL-5 and TL-4, respectively). The base bending moments using the AS 5100.2 equivalent static loads are 7,362 kNm and 3,681 kNm for Tests #1 and 2, respectively.

It is shown in results from the comparative analyses that the maximum bending moment as obtained from numerical simulations can be significantly different to results derived from calculations as per the stipulations in AS 5100.2. Differences between the two sets of predictions depend on the configuration and sectional properties of the bridge structure. The amount of bending moment transmitted to the base of the bridge pier is shown to reduce with a decrease in the value of k_R or an increase in the total mass of the bridge.

The bending moment is also shown to increase as the value of k_R increases (i.e. the lateral stiffness of the bridge increases) or when the mass of the bridge decreases. In other words, the transmitted bending moment would decrease with increasing natural period of the bridge; and vice versa. These findings have significant implications to the design actions on the bridge bearings, the bridge pier (and its crosshead, if applicable) and the bridge foundation. The bending moment behaviour in the barrier and the bridge deck is not as sensitive to the dynamic properties of the bridge as a whole.

10.4 CALIBRATION OF PARAMETERS FOR HUNT AND CROSSLEY MODEL

The H&C model introduced in Chapter 6 can be used to generate a normalised vehicular forcing function. To fulfil this objective, parameters in the H&C model need to be calibrated to achieve a good match between the simulated and experimentally recorded contact force time-histories, as presented in the previous section. Step-by-step illustration of the use of the H&C model is presented in Chapter 13 (refer Section 13.5.8). Two peaks of the forcing functions are typically observed from data recorded from crash tests involving the semi-trailer (TL-5), and the single-unit truck (TL-4), colliding at the barrier with the front and rear of the vehicle. As described earlier, the contact forcing time-history for the collision event was obtained by the superposition of two contact force time-histories for the TL-5 collision. Whereas for the TL-4 collision the contact force time-history was obtained purely from the centre of gravity accelerogram in the vehicle. Input parameters to the model include: mass at the front of the vehicle (m_f); mass at the rear of the vehicle (m_r); resultant impact velocity (v_0); target mass (λm); target stiffness (k); contact stiffness (k_n); exponent p; coefficient of restitution (COR); and the time lag between the first and second collision (t_g).

10.4.1 Front Mass (m_f) and Rear Mass (m_r)

For the single-unit truck (TL-4 impact), m_f is taken as the weight that is applied through the front axle and m_r is taken as the weight that is applied through the rear axle/s. This is in contrast to the semi-trailer (TL-5 impact) where the m_f and m_r are the weight of the prime mover and weight applied through the rear axles of the trailer, respectively, as shown in Figure 10.1(b). Masses measured at each wheel of the vehicle taken from crash test reports [10.3–10.6,10.8] are summarised in Table 10.2.

10.4.2 Resultant Impact Velocity (v_0)

The impact velocity (v_0) is the component of velocity perpendicular to the barrier, i.e. $v_0 = v_\theta sin\ \theta$, where values of v_θ and θ are the velocity of collision

Table 10.2 Impactor mass and velocity

Test #	m_f (kg)	m_r (kg)	v_0 (m/s)
1	20,462	15,903	5.66
2	4,482	5,566	7.09
3	3,733	6,400	6.41
4	3,533	6,613	7.01

and the angle of the axis of the vehicle with respect to the barrier immediately prior to the initial contact, respectively, as previously presented in Table 10.1. The values of v_0 are listed in Table 10.2 for ease of reference.

10.4.3 Target Mass (λm)

The target mass, which was taken as 400 kg, was approximately the weight of the barrier of a length equals to twice its height (as is consistent with the assumption of an angle of dispersion of 45 degrees in the transmission of the collision action). The target mass is shown to not significantly affect the magnitude of the contact force, as is evident from the results obtained from a parametric study in which the mass of the target (λm) was varied in the range of 100 to 10,000 kg. The change in the target mass by two orders of magnitude has been found to result in negligible differences to the maximum contact force, the duration of contact and the amount of transmitted impulse (i.e. area under the force time-history graph). Thus, errors in the estimated value of λm should not have any bearings on the applied forcing function of the impact.

10.4.4 Target Stiffness (k)

The lateral stiffness of the barrier was approximated to be 82,800 kN/m. A parametric investigation was conducted to determine the extent of the influence of the lateral stiffness on the contact forcing function. It involved varying the value of k in the range 1,000 to 1,000,000 kN/m. The change in the peak contact force, contact duration and transmitted impulse with an increasing value of k are presented in Figure 10.10. It is shown that the modelled forcing function would converge as the value of k exceeds the 40,000 kN/m threshold. The contact force is shown to remain unchanged, assuming everything else is kept the same, as the stiffness of the barrier is varied provided that its stiffness is above the minimum threshold. Therefore, the contact force time-histories developed should be applicable to most RC barriers walls as their stiffness should not vary that drastically and within these limits. However, barriers of a completely different material or structural typology may need their own normalised forcing functions derived.

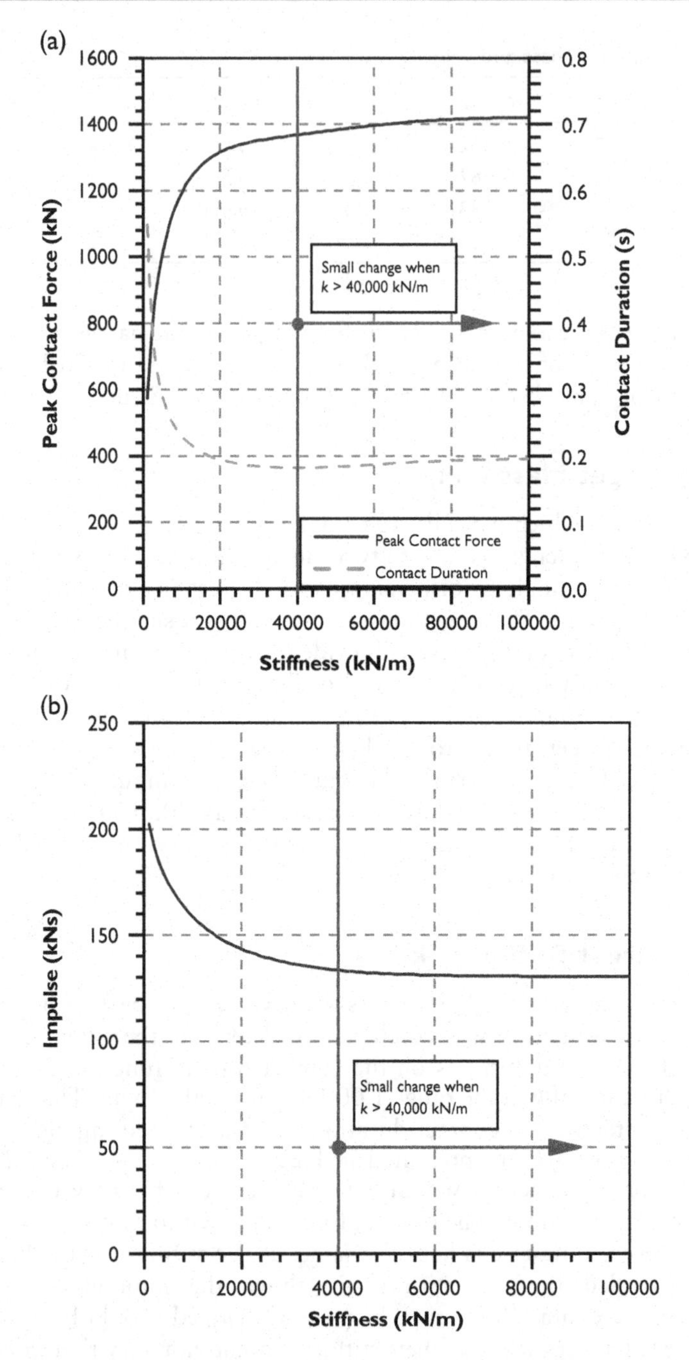

Figure 10.10 (a) Peak contact force, contact duration and (b) impulse versus target stiffness.

10.4.5 Time Gap (t_g)

The value of t_g is observed from the time-histories of Figures 10.4 and 10.5 to be approximately 0.63 s and 0.15 s for semi-trailer (TL-5) and single-unit truck (TL-4) collisions, respectively.

10.4.6 Contact Stiffness (k_n), Exponent (p) and Coefficient of Restitution (COR)

With all the parameters pre-determined, the dynamic factors k_n, p and COR were the remaining parameters requiring calibration. Calibrations that have been undertaken were based on matching the numerically simulated, and the recorded, contact force time-histories associated with Tests #1 and #2, which involved the semi-trailer (TL-5) and single-unit truck (TL-4), respectively. Information obtained from the calibration as listed in Table 10.3 was used for generating the contact force time-histories, which were employed subsequently in the analyses of the case study bridge.

The contact force time-history generated by the calibrated H&C model for Test #1 (semi-trailer) is shown in Figure 10.11(a). This forcing function was then inputted into the simplified numerical model (which was first introduced in Section 9.4.3) to simulate the response of the case study bridge. The simulated deflection of the barrier, displacement at the top of the pier and bending moment transmitted to the base of the bridge pier are presented in Figures 10.11(b), 10.11(c) and 10.11(d), respectively. The accuracy of the simulations is verified in the figures, which present comparison with the contact force time-histories derived from physical crash testing.

Similarly, the calibrated dynamic parameters for the single-unit truck in Table 10.3 have also been used for generating the contact forcing function, as shown in Figure 10.12(a). The generated forcing function has been used subsequently for analysing the response behaviour of the case study bridge. The comparison of the simulated contact force time-history and the original contact force time-history derived from the physical crash testing associated with the TL-4 scenario are presented in Figures 10.12(b) to 10.12(d). These comparisons demonstrate satisfactory performance of the simulated contact force time-history.

Table 10.3 Calibrated dynamic factors

Vehicle	*Front/Rear*	k_n *(kN/m*^{p}*)*	*p*	*COR*
Semi-Trailer (TL-5 scenario)	Front	2,300	2	0.2
	Rear	9,000	1.5	0.5
Single-Unit Truck (TL-4 scenario)	Front	400	2	0.2
	Rear	1,500	2	0.5

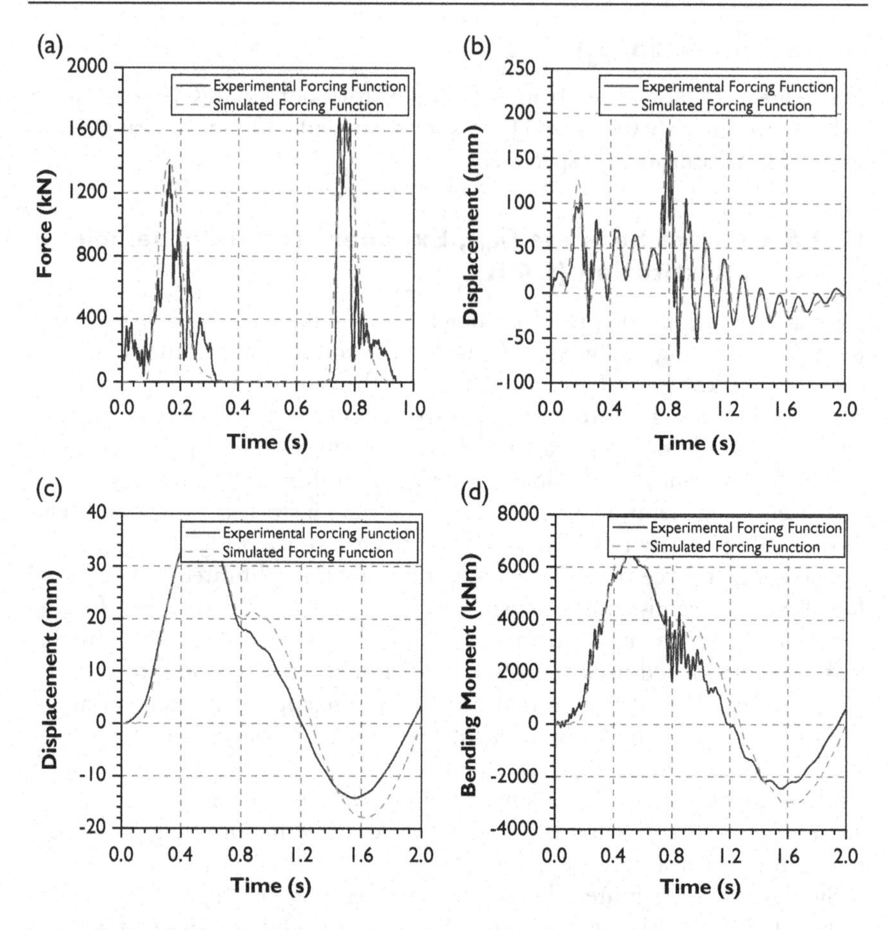

Figure 10.11 (a) Forcing function, (b) displacement time-history of barrier, (c) displacement time-history of pier and (d) bending moment time-history at foundation for Test #1.

The verification studies as described have been carried out subsequently for Tests #3 and #4 by the use of the H&C model, along with the calibrated dynamic factors as shown in Table 10.3. The bar charts in Figures 10.13(a) and 10.13(b) present the comparison of the maximum displacement at the top of the pier and bending moment transmitted to its base, respectively. The results of these two comparisons add further to the credibility of the H&C model for simulating the contact force time-histories.

10.5 CLOSED-FORM EXPRESSION

Equation (3.9a) presented in Chapter 3 is re-written here as Equation (10.2). As explained in Chapter 3, Equation (10.2) is only valid for hard-impact

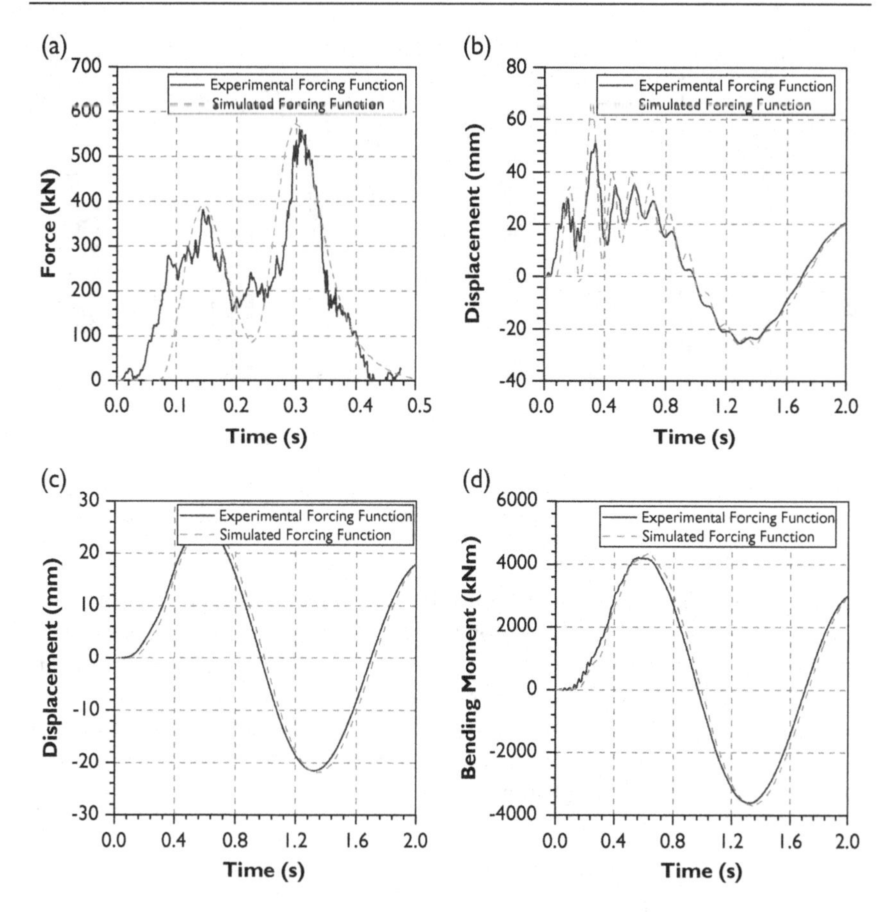

Figure 10.12 (a) Forcing function, (b) displacement time-history of barrier, (c) displacement time-history of pier and (d) bending moment time-history at foundation for Test #2.

scenarios and is not applicable when there is a delay in the transfer of momentum (i.e. a cushioning effect occurs). In the case of a vehicular collision scenario, the front part of the vehicle would typically be experiencing crumbling upon the initial collision. As the crumbing takes time to complete, the transfer of momentum is delayed (in the same manner as the cushioning of the impact by fallen rocks by gabion cells that are placed in front of the rockfall barrier, as presented in Chapter 7). An additional factor γ is introduced in Chapter 7 to take into account the reduction in target displacement and reaction force due to the effect of cushioning. However, the design chart of Figure 7.5 in Chapter 7 showing the value of γ as a function of the period ratio is not applicable for modelling vehicular collision because of a number of reasons: (1) there can be two consecutive collisions as opposed to the one-off collision by a fallen rock; (2) taking p = 1 and COR = 0.01 is only valid in modelling the cushioning action of gabion cells that are place in front

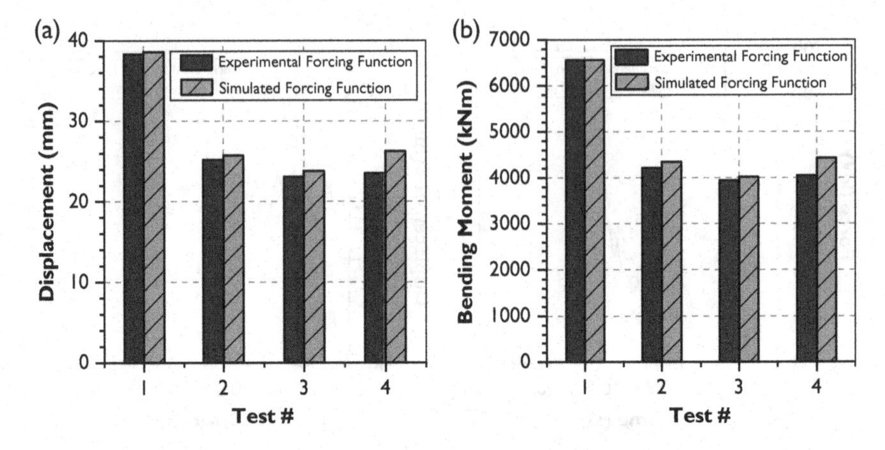

Figure 10.13 Comparison of maximum (a) pier displacement and (b) bending moment at foundation simulated by experimental forcing function and Hunt & Crossley model.

of a rockfall barrier, and is not applicable for modelling vehicular collisions as the two types of forcing functions are not comparable; and (3) a bridge structure with multiple components (barrier, girder and pier) is a multi-degree-of-freedom (MDOF) structural system unlike a rockfall barrier that is a cantilevered wall that can be modelled as a single-degree-of-freedom (SDOF) structural system. The value of γ needs to be specifically determined for a vehicular collision on a bridge.

$$F_{\text{hard}} = v_0 \sqrt{mk} \sqrt{\lambda \left(\frac{1 + \text{COR}}{1 + \lambda} \right)^2} \tag{10.2}$$

The simplified numerical model, which has been validated in Section 10.3.1, is used herein for modelling the bending moment that is transmitted to the base of the bridge pier in a vehicular collision scenario. Some 140,000 simulations have been carried out in accordance with parameters (i.e. contact force time-history) that have been calibrated against recordings from Tests #1 and #2, with each simulation having different input parameters to define the structural model of the bridge. The parametric study was guided by Table 10.4, which presents the range of values for each parameter for input into the numerical model. In the collation of results generated by the study, the maximum bending moment transmitted to the base of the bridge pier (denoted herein as $M_{\text{simulation}}$) was taken from each simulation.

Meanwhile, the closed-form expression of Equation (10.2) was used to determine the base shear of the bridge pier (denoted herein as F_{hard}). Values of k and λm required for input into Equation (10.2) had been found by using Equations (9.29) and (9.30), respectively, and COR = 0.35 was adopted in view of results obtained from calibrations as presented in Section 10.4.6. Implicit in the closed-form expression is the assumption of a

Table 10.4 Range of input parameters

Parameters	*Lower Bound*	*Upper Bound*	*Units*
m_{add}	500	1,200	t
k_R	1,750	5,000	MNm/rad
L_1	1.07	1.67	m
D_1	0.2	0.3	m
L_2	1.5	4	m
D_2	0.3	0.4	m
L_3	2	4	m
L_4	8.5	15	m
D_4	2	5	m
B_4	0.8	2	m

"hard impact," which refers to the instantaneous transfer of momentum generated by the collision of a hard impactor (without crumbling). The base shear so calculated from the use of Equation (10.2) is accordingly denoted as F_{hard}, which is then multiplied by the height dimension (L_6) to give M_{hard}, which refers to the maximum bending moment transmitted to the base of the bridge pier in a hard impact, as shown by Equation (10.3). It is noted that the mitigating action of crumbling has only been taken into account by the numerical model, and not by the closed-form expression. Thus, differences between the simulated bending moment ($M_{simulation}$) and the calculated bending moment by use of the closed-form expression (M_{hard}) are essentially reflective of the crumbling (or cushioning) action of the colliding vehicle. The cushioning factor (γ) is accordingly taken as the ratio of the two bending moment values, as shown by Equation (10.4).

$$M_{hard} = v_0\sqrt{mk}\sqrt{\lambda\left(\frac{1 + COR}{1 + \lambda}\right)^2} L_6 \tag{10.3}$$

$$\gamma = \frac{M_{simulation}}{M_{hard}} \tag{10.4}$$

The values of γ so derived from the parametric investigation have been used for modelling its probabilistic distribution $P(\gamma)$, as shown in Figure 10.14(a) and (b) for semi-trailer and single-unit truck, respectively. The value of γ where the $P(\gamma)$ function peaks is shown to be 0.5 (rounded up from 0.48) and 0.9 (rounded up from 0.88) for a semi-trailer and single-unit truck, respectively. The area under the probabilistic functions bound by two limiting γ values (as shown by the vertical lines overlaid on the graph) represent approximately 80% of the total area under each curve. Thus, the two limits define the range of γ values that 80% of the results fall within. These ranges

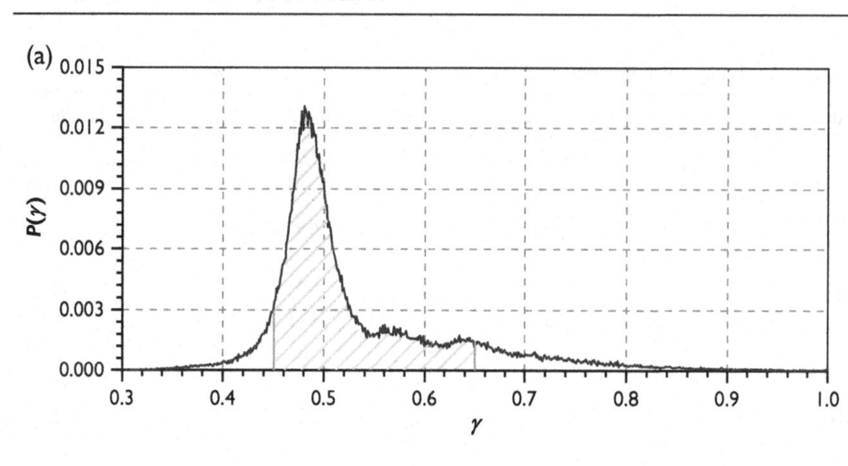

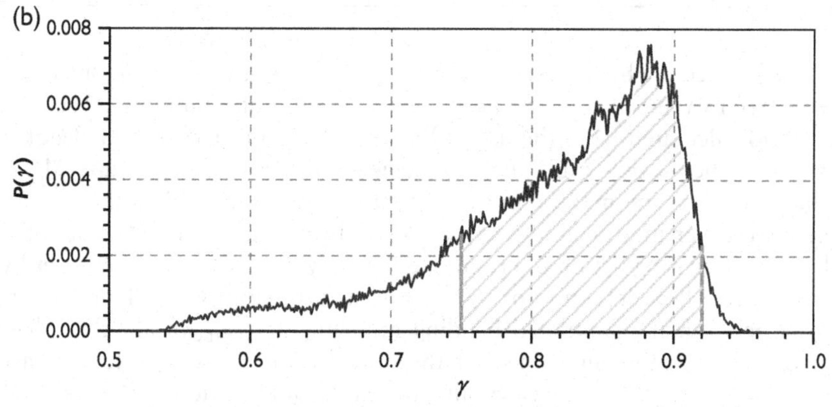

Figure 10.14 Probability distribution of γ for (a) semi-trailer and (b) single-unit truck.

are 0.45 to 0.65 for semi-trailer, and 0.75 to 0.92 for a single-unit truck. As for all recommendations based on empirical modelling, the solutions for γ as presented are not exact solutions. The closed-form expressions as introduced in Chapter 9 and evaluated in this chapter can be used to guide the configuration and sizing of members in the preliminary design of the bridge structure for withstanding the collision action. However, for determining the final design solution, numerical simulations should also be employed for assuring satisfactory performance once the computer model of the bridge becomes available.

10.6 CLOSING REMARKS

Physical crash test data was utilised to develop contact force time-histories for the collision behaviour of various vehicles against a cast in-situ barrier wall. The simplified numerical model, which was first introduced in

Chapter 9, is then verified against a more detailed numerical model that was implemented on a commercial software. The good consistency of results from analyses of the same structure responding to the same forcing functions taken from crash test data gives credibility to the simplified model. The Hunt and Crossley model has been used to approximate parameters that can be used to generate normalised forcing functions for a collision scenario, which were used to develop the suite of normalised forcing functions in Chapter 9. A closed-form expression for predicting the maximum base shear and base bending moment to the collision scenario is presented. The methodology, which was introduced initially in Chapter 3 for modelling a hard-collision scenario, is modified by a cushion reduction factor (γ) for emulating the crumbling of the colliding vehicle. Some 140,000 simulations have been carried out to develop recommendations for the value of γ for a semi-trailer truck, and single-unit truck, based on calibrating closed-form solutions against results from numerical modelling.

REFERENCES

10.1 Hunt, K.H. and Crossley, F.R.E., 1975, "Coefficient of restitution interpreted as damping in vibroimpact", *Journal of Applied Mechanics*, Vol. 42(2), p. 440, doi:10.1115/1.3423596

10.2 American Association of State Highway and Transportation Officials, 2012, *AASHTO LRFD Bridge Design Specifications*, 6th edition, American Association of State Highway and Transportation Officials, Washington DC, U.S.

10.3 Sheikh, N.M., Kovar, J.C., Cakalli, S., Menges, W.L., Schroeder, G.E., and Kuhn, D.L., 2019, *Analysis of 54-Inch Tall Single-Slope Concrete Barrier on a Structurally Independent Foundation*, Texas A&M Transportation Institute, Texas, U.S.

10.4 Sheikh, N.M., Bligh, R.P., and Menges, W.L., 2011, *Determination of minimum height and lateral design load for MASH test level 4 bridge rails*, Texas A&M Transportation Institute, Texas, U.S.

10.5 Moran, S.M., Bligh, R.P., Menges, W.L., Schroeder, W., and Kuhn, D.L., 2020, *Mash Test 4-12 of Shallow Anchorage Single Slope Traffic Rail (SSTR)*, Texas A&M Transportation Institute, Texas, U.S.

10.6 Williams, W.F., Sheikh, N.M., Menges, W.L., Kuhn, D.L., and Bligh, R.P., 2018, *Crash Test and Evaluation of Restrained Safety-Shape Concrete Barriers on Concrete Bridge Deck*, Texas A&M Transportation Institute.

10.7 Standards Australia, 2017, *AS 5100.2 Bridge Design Part 2: Design Loads*, Standard Australia Limited, New South Wales, Australia.

10.8 Sheikh, N.M., Bligh, R.P., Kovar, J.C., Cakalli, S., Menges, W.L., Schroeder, G.E., and Kuhn, D.L., 2020, *Development of Structurally Independent Foundations for 36-Inch Tall Single Slope Traffic Rail (SSTR) for MASH TL-4*, Texas A&M Transportation Institute, Texas, U.S.

Chapter 11

Collision on a Pole Found in Soil

11.1 INTRODUCTION

The common way of firmly securing a pole into the ground is having it embedded into the soil down to a certain depth. The normal way of constructing this type of pole that is made of steel is to have the embedded segment cast into a concrete jacket forming a composite section which is then put into a pre-formed socket to be buried in the ground. The construction is complete as the socket is backfilled. This type of pole is referred to here as a soil-embedded pole (left of Figure 11.1). Determining the required depth of embedment is an important part of the design and is treated in Section 11.2. Algebraic expressions for making predictions of the response of the pole in a collision scenario are presented in Section 11.3. The alternative way of securing the pole into the ground is to have the base of the pole cast into a concrete slab forming a free-standing assemblage, which is then put into a shallow pit (right of Figure 11.1). The assemblage derives its stability to resist overturning by virtue of the self-weight of the concrete slab. As the pit is backfilled, this type of pole that performs the same function as a soil-embedded pole is suited to locations where there are constraints at shallow depths such as underground utilities that are not to be disturbed or the occurrence of rock outcrops/boulders. In these situations, designing a pole that only involves excavating a shallow pit is preferred.

Algebraic expressions for predicting the capacity of a free-standing pole-slab assemblage to resist a collision are presented in Section 11.4. Illustrations by worked examples of the calculation methodology for both types of poles are presented in Section 11.5.

The authors originally gained experience on this topic through conducting investigations into the design of steel baffles that were installed on a hill slope for containing fallen boulders to protect lives and property in the event of a landslide or a rockfall. The main design objective for those installations was to ensure that a fallen boulder striking the top end of the baffle would rebound on impact. The amount of maximum displacement and slope of deflection sustained by the pole can be checked using

DOI: 10.1201/9781003133032-11

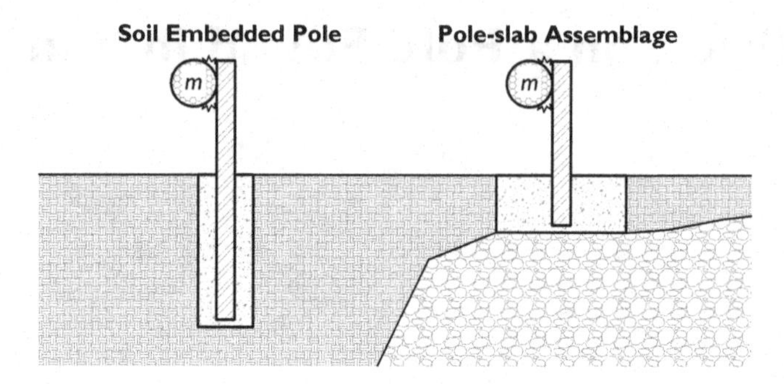

Figure 11.1 Two types of pole construction.

calculations presented in this chapter. With either form of baffle construction, the ability of the deflected pole to bounce back a flying object is controlled by the maximum slope of deflection.

The introduced design methodology can be adapted for the design of poles carrying other protective functions when exposed to collision hazards. In designing a pole to withstand collision by a vehicle, the height of contact (i.e. point of load application) would be much closer to the ground surface than the top of the pole. In applying expressions presented in this chapter for analysis of the impact action, the part of the pole above the point of contact may be ignored in the calculation. In other words, taking the top of the pole as the point of contact would result in a conservative design. Choosing a compact tubular section as the cross section of the pole serves to safeguard against premature failure by local buckling, or by shear.

11.2 EMBEDMENT CONSIDERATIONS IN A LATERALLY LOADED POLE

The first step in the design of a soil-embedded pole is the assessment of the ability of the surrounding soil to restrain rotation when collided upon. The pole is classified as *flexible* if it responds to lateral load, or collision action, as if it was a vertical cantilever found on a fixed base at the depth of virtual fixity (C), and there is no whole-body rotation of the pole caused by deformation of the surrounding soil (refer Figure 11.2). In contrast, the impact-resistant behaviour of a *rigid* soil-embedded pole (with a shallow embedment depth) is dominated by whole body rotation. The assessment of the lateral impact resistant capacity of this type of pole, which is very dependent on the type of subgrade, requires detailed investigation of the backfill material. A *semi-rigid* soil-embedded pole, which responds to an impact with a of whole-body rotation and deformation of the pole itself has the benefits of having both the pole and the surrounding soil to share the absorption of energy. However, the robustness of a *rigid*, or *semi-rigid*, pole

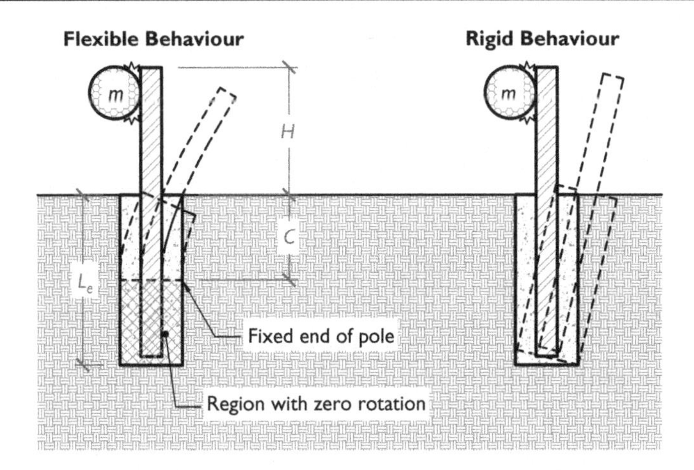

Figure 11.2 Flexible and rigid lateral behaviour of a soil-embedded pole.

is more difficult to predict than a *flexible* pole because of the need to model soil deformation behaviour in highly transient conditions.

In the absence of a rational and transparent design procedure for the design of a *rigid pole* or *semi-rigid pole*, full-scale field trials are warranted to ensure satisfactory performance in a collision scenario. Design recommendations presented in this chapter may be used for designing the test specimens for such field trials.

The behaviour of a flexible pole subject to a collision would not be as sensitive to changes in the conditions of the soil and is therefore more robust. Predictions of the response behaviour of the pole are also more straightforward. For poles to be installed in a diversity of site conditions, the flexible pole design approach is preferred because of the higher degree of robustness as the response behaviour of the pole is not as sensitive to changes in conditions of the soil. Designing a flexible pole also has merits from the maintenance perspectives as the foundation is more likely to remain intact in situations where the pole has sustained damage in a collision incident. The need to reconstruct the foundation may therefore be spared.

Investigations undertaken by the authors and their collaborators involving physical and computer-simulated collision testing have transpired into recommendations as presented by Equation (11.1) and Figure 11.3, which was developed by Perera and Lam [11.1]:

$$C = 0.45\,[L_e/T_s]_{\text{lim}}\,T_s \tag{11.1}$$

where $[L_e/T_s]_{\text{lim}}$ is determined using Figure 11.3; L_e is the embedment length of the pole; and T_s is the soil stiffness parameter.

The limiting L_e to T_s ratio, referred to as $[L_e/T_s]_{\text{lim}}$, defines the point of transition from rigid to flexible behaviour of the pole. In other words, the

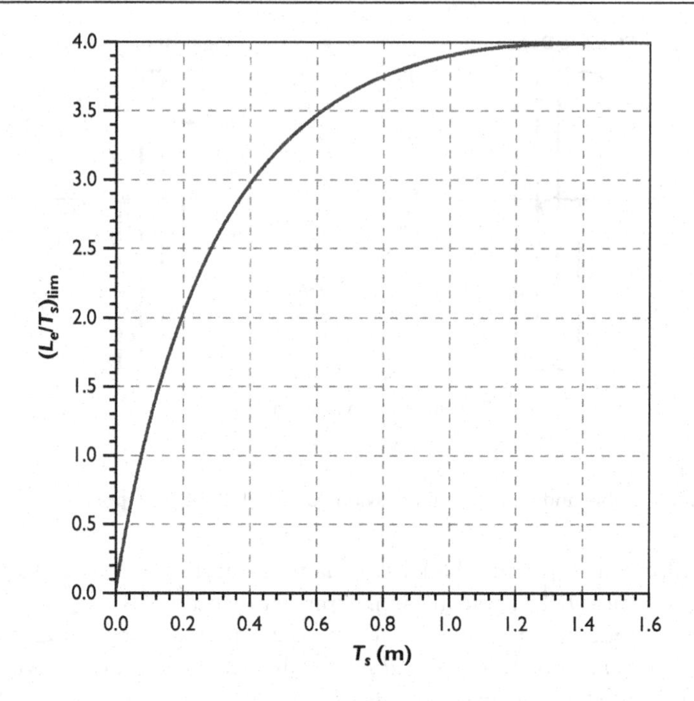

Figure 11.3 Limiting $\mathbf{L_e}$ to $\mathbf{T_s}$ ratio versus $\mathbf{T_s}$.

actual L_e/T_s ratio needs to be greater than $[L_e/T_s]_{\text{lim}}$ to ensure that the pole behaves in a flexible manner. As shown in Figure 11.3, the required length of L_e to achieve flexible behaviour for high T_s values (i.e. $T_s \geq 1.6$ m) is equal to $4T_s$, which corresponds to $C = 1.8T_s$. A trend of decreasing values of $[L_e/T_s]_{\text{lim}}$ (hence reducing the required length L_e) with the lowering of the value of T_s is evident. Thus, reducing the stiffness of the pole has the benefit of reducing the depth of the socket that needs to be excavated to install the pole. The design of the pole must also consider the required yield resistance at the position of the "weak link" where the flexural stress is critical.

The soil stiffness parameter (T_s), which has the dimension of length, is defined by Equation (11.2). The controlling parameters for characterising the soil stiffness are flexural rigidity of the embedded segment of the pole (EI) and the coefficient of subgrade modulus variation (η_h). If the soil-embedded pole is surrounded by a concrete encasement, such as a bored pier, the EI value for use in Equation (11.2) would be that of the concrete-bored pier and not the pole itself.

$$T_s = \sqrt[5]{\frac{EI}{\eta_h}} \qquad (11.2)$$

By calculating the value of T_s from Equation (11.2) and reading $[L_e/T_s]_{\text{lim}}$ from Figure 11.3, the depth of virtual fixity (C) can be estimated by the use

Table 11.1 Typical values of η_h in cohesionless soil

Subgrade Classification	η_h (MN/m^3)
Soft Organic Silt	0.15
Soft Clay	0.35–0.70
Loose Sand	5–10
Medium Dense Sand	10–25
Dense Sand	25– 40
Very Dense Sand	40–100
Dense Sandy Gravel	100–150

of Equation (11.1), which provides the minimum depth to which the pole needs to be embedded into the soil. A worked example is presented in Section 11.5.1 to illustrate how the estimate is made.

Details of correlations between the stiffness parameters as described and the relative density of the soil can be found in Garassino et al. [11.2] and Tomlinson and Woodward [11.3]. Table 11.1 presents the correlation in a simplified format for broad classifications of soil types based on information provided in these cited references.

With cohesionless soils, the value of η_h is shown to vary by an order of magnitude from loose sand to very dense sand (Table 11.1). There are also considerable uncertainties in the value of η_h within a soil classification. For example, the value of η_h may vary from 40 to 100 MN/m^3 for very dense sand. If a mid-range value of say 70 MN/m^3 is assumed the actual value may differ from the design assumption by as much as 30 MN/m^3. However, this amount of error is translated to less than 10% difference in the value of T_s according to Equation (11.2). Thus, the value of η_h may be taken as 70 MN/m^3 (the average value in the range) in design calculations when dealing with installations in very dense sand. The value of η_h for cohesive soil is shown to be much lower than cohesionless soil.

11.3 DESIGN OF A SOIL-EMBEDDED POLE TO RESIST COLLISION

A manual and easy to operate design procedure, which does not require the use of complex computer analysis, has been developed by the authors and their colleagues [11.1] and is presented herein for determining the deflection demand of a soil-embedded pole subject to a collision. The simplified assumption to make in the impact-resistant design is that the pole would behave as a vertical cantilever, which is supported at the virtual depth of fixity (C), as shown in the schematic diagram of Figure 11.4(a). The displacement of the pole is resolved into three components, as listed in the following:

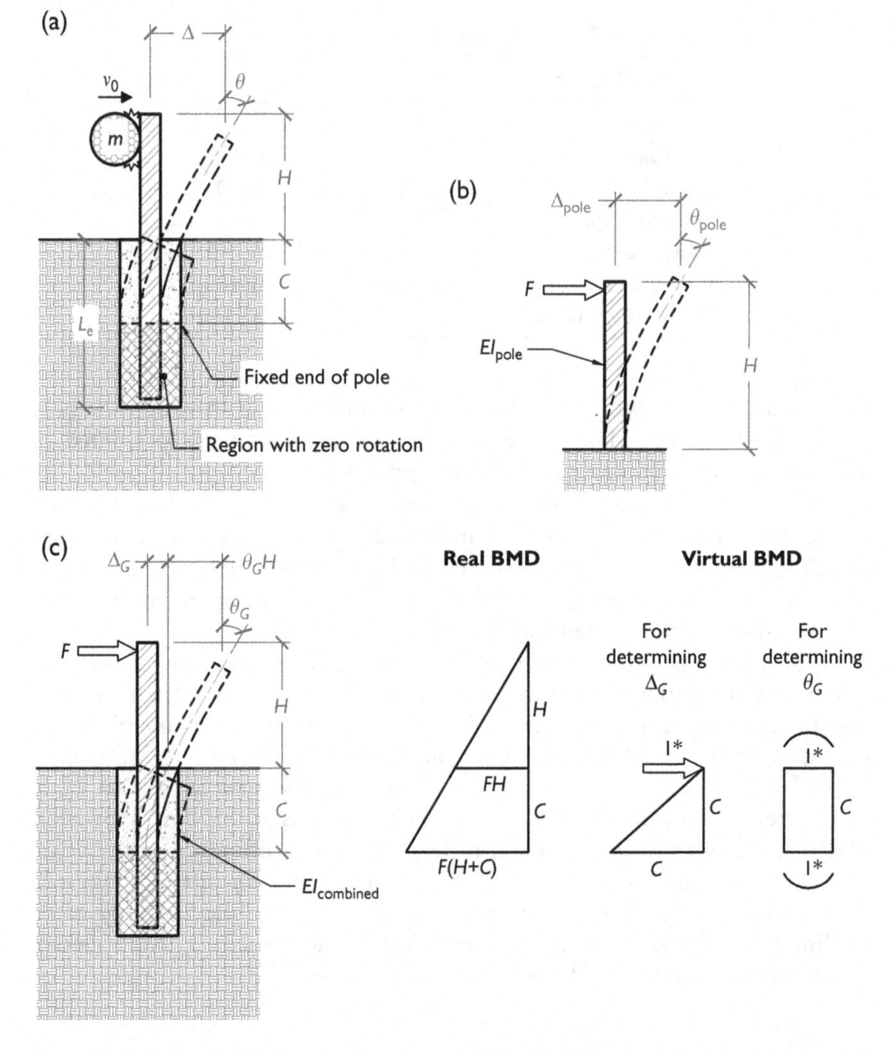

Figure 11.4 Soil-embedded pole: (a) schematic diagram, (b) calculation methodology of Δ_{pole}, and (c) calculation methodology of Δ_G and $\theta_G H$.

i. Δ_{pole} = deflection of the above ground segment of the baffle as if it was a pole fixed at the ground level, as shown in Figure 11.4(b);
ii. Δ_G = deflection of the sub-surface segment of the baffle at the ground level; and
iii. $\theta_G H$ = deflection of the pole at height H above ground caused by rotation of the pole at the ground level.

The calculation methodology for finding deflection components (ii) and (iii) is based on the virtual force method, as illustrated in Figure 11.4(c). The sum of the three components gives an estimate for the value of Δ_{top}, which is the total deflection of the pole at the top when subject to a given lateral force F.

The deflection components listed previously and the associated rotation (slope of deflection) can be found using Equations (11.3) to (11.5). Parameter C in Equations (11.4a) and (11.4b) is determined from Equation (11.1), which is to be read in conjunction with Figure 11.3.

$$\Delta_{pole} = \frac{FH^3}{3EI_{pole}} \tag{11.3a}$$

$$\theta_{pole} = \frac{FH^2}{2EI_{pole}} \tag{11.3b}$$

$$\Delta_G = \frac{C(2F(H+C)+FH)(C)}{6EI_{combined}} = \frac{F(2C^3+3HC^2)}{6EI_{combined}} \tag{11.4a}$$

$$\theta_G = \frac{1 \times (FH + F(H+C))(C)}{2EI_{combined}} = \frac{F(2HC+C^2)}{2EI_{combined}} \tag{11.4b}$$

$$\Delta_{top} = \Delta_G + \theta_G H + \Delta_{pole} \tag{11.5a}$$

$$\theta_{top} = \theta_G + \theta_{pole} \tag{11.5b}$$

The lateral stiffness of the pole (k_e) can be found accordingly using Equation (11.6). Substituting Equations (11.3a), (11.4a) and (11.4b) into Equation (11.5a) gives an equation for estimating Δ_{top}, which can then be substituted into Equation (11.6a) to give Equation (11.6b) for estimating k_e:

$$k_e = \frac{F}{\Delta_{top}} \tag{11.6a}$$

$$k_e = \left[\frac{(H+C)^3 - H^3}{3EI_{combined}} + \frac{H^3}{3EI_{pole}}\right]^{-1} \tag{11.6b}$$

Calculations for predicting the amount of displacement resulting from the collision may then be undertaken. This part of the calculation requires the following parameters to be identified:

i. mass of impactor (m);
ii. velocity of impact (v_o);
iii. lateral stiffness of the pole (k_e), which can be found using Equation (11.6b); and
iv. mass ratio (λ) which can be found using expressions presented in Chapter 2.

The maximum amount of deflection of the pole (Δ) generated by the impact can be found by either Equation (3.7a) or (3.7b) for scenario Types (1) and (2), respectively, as presented in Chapter 3. Should the weight of the boulder be much heavier than the pole, the boulder is assumed to remain attached to the pole up to the instance when the deflection reaches the maximum limit. Thus, Equation (3.7b) may be adopted. This is a simplified representation of real behaviour, as explained in Chapter 3. Equation (3.7b) is re-written as Equation (11.7); and the generic stiffness term k is replaced by k_e, which represents the lateral stiffness of the pole:

$$\Delta = \frac{mv_o}{\sqrt{mk_e}}\sqrt{\frac{1}{1+\lambda}} \tag{11.7}$$

The amount of rotation (θ) at the top of the pole that is generated by the collision can be found using Equation (11.8a). Substituting Equations (11.5a) and (11.5b) into Equation (11.8a) gives Equation (11.8b):

$$\theta = \frac{\theta_{\text{top}}}{\Delta_{\text{top}}}\Delta \tag{11.8a}$$

$$\theta = \frac{3}{2}\left[\frac{(H+C)^2 EI_{\text{pole}} + H^2 EI_{\text{concrete}}}{(H+C)^3 EI_{\text{pole}} + H^3 EI_{\text{concrete}}}\right]\Delta \tag{11.8b}$$

A step-by-step illustration of the use of the analytical model as described is presented in the form of a worked example in Section 11.5.1.

11.4 DESIGN OF A POLE-SLAB ASSEMBLAGE TO RESIST COLLISION

This section presents a procedure for assessing a pole-slab assemblage to collision actions. The calculation procedure is based on the simplified (and conservative) assumption that the pole together with the base slab (refer to the schematic diagram of Figure 11.5) is free-standing and does not have any restraint from the backfill material surrounding it.

The basis of the calculation procedure for predicting rotation of the assemblage and horizontal deflection at the top (Δ_H) is based on the diagrams of Figure 11.6. Critical locations are marked by points A, B and C, which represent the pivot point of overturning, centre of gravity of base slab and top of pole, respectively. Points B' and C' are their respective positions when the pole-slab assemblage is at its maximum displacement/rotation.

The use of equal angular momentum and equal energy principles in predicting the lifting of the centre of gravity of a free-standing, and rigid,

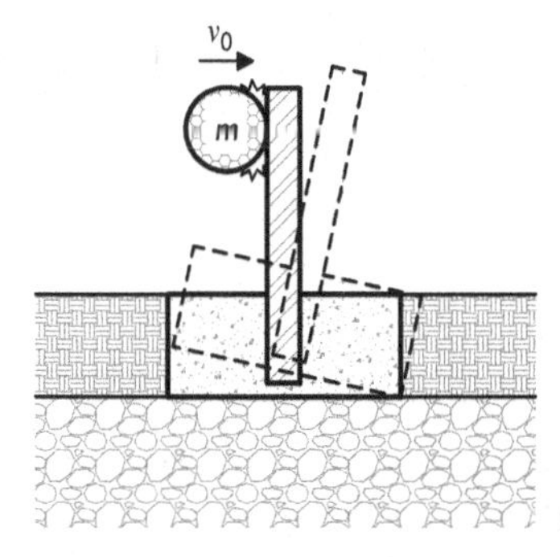

Figure 11.5 Free-standing pole-slab assemblage.

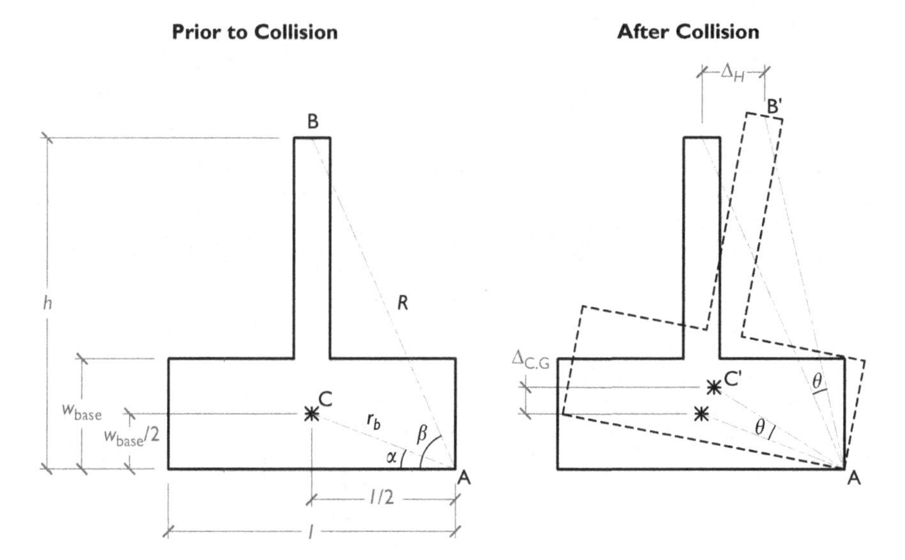

Figure 11.6 Kinematics of the overturning of the baffle-base slab assemblage.

object was first introduced in Chapter 4 where free-standing installations of various cross sections were used as examples. The lifting of the centre of gravity of these objects was predicted using Equation (4.15a) alongside Equations (4.2) and (4.6), which are reproduced here as Equations (11.9) to (11.11). These expressions can be adapted for use in predicting the displacement of the pole-slab assemblage provided that the pole is sufficiently rigid (that there is negligible deformation of the pole) as the assemblage is lifted when collided upon. The self-weight of the pole may be neglected as it is typically negligible in comparison with the weight of the base slab, which may be estimated by the use of Equation (11.12).

$$\Delta_{C.G.} = \frac{mv_0^2}{2Mg}\frac{\kappa h}{R}\left(\frac{1 + \mathrm{COR}}{1 + \kappa}\right)^2 \tag{11.9}$$

$$\kappa = \frac{I_\theta}{mhR} \tag{11.10}$$

$$I_\theta = M\left(\frac{w_{\text{base}}^2 + l^2}{3}\right) \tag{11.11}$$

$$M = \rho_c w_{\text{base}} l^2 \tag{11.12}$$

Should the pole be of a slender design and hence flexes when collided upon, the use of Equation (11.13) for the prediction of the value of $\Delta_{C.G.}$ is expected to provide a conservative prediction given that no energy loss occurring on contact is assumed.

$$\Delta_{C.G.} = \frac{mv_0^2}{2Mg} \tag{11.13}$$

Once the value of $\Delta_{C.G.}$ is known, the displacement of the assemblage as represented by the value of θ and Δ_H can be found by applying kinematics principles either by graphical means or by the use of Equations (11.14) to (11.17). Refer to Figures 11.7(a) and 11.7(b), which show the lines extracted from Figure 11.6 for the derivations of Equations (11.14) and (11.16), respectively. It should be noted that length $AC = AC' = r_b = \sqrt{(w_{\text{base}}/2)^2 + (l/2)^2}$ and length $AB = AB' = R$.

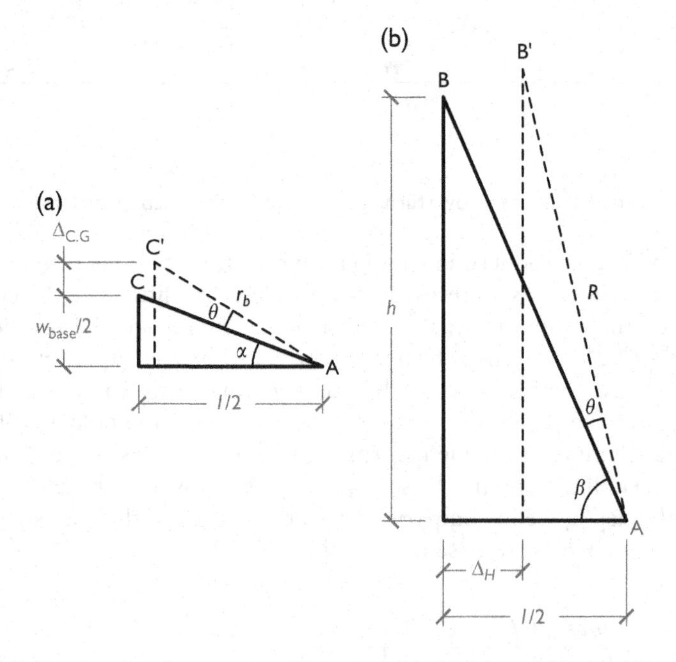

Figure 11.7 Sketches for derivations of Equations (a) (11.14) and (b) (11.16).

$$\theta = \sin^{-1}\left(\frac{\Delta_{C.G.} + \frac{w_{base}}{2}}{\sqrt{\frac{w_{base}^2 + l^2}{4}}}\right) - \alpha \tag{11.14}$$

$$\alpha = \tan^{-1}\left(\frac{w_{base}}{l}\right) \tag{11.15}$$

$$\Delta_H = \frac{l}{2} - R\cos(\theta + \beta) \tag{11.16}$$

$$\beta = \tan^{-1}\left(\frac{2h}{l}\right) \tag{11.17}$$

11.5 WORKED EXAMPLES

11.5.1 Design of a Pole Embedded in Very Dense Sand

Consider a 150 mm × 150 mm square hollow section with a section thickness of 10 mm. The steel section has a Young's modulus of 200 GPa and a second moment of area of 16.5×10^6 mm^4. The pole is embedded in very dense sand as shown in Figure 11.8. The section is estimated to have mass of 41.3 kg/m. The embedded segment of the pole is cast in concrete with a cover of 50 mm, forming a jacket of square cross section and of side lengths measuring 250 mm. The pole (in the jacket) is subjected to the collision of a boulder with diameter of 1 m and impact velocity of 6 m/s. A reduced Young's modulus value of 15,000 MPa may be assumed of the concrete to allow for the effects of crack formation. Estimate the maximum amount of displacement and rotation of the pole when collided upon by the boulder.

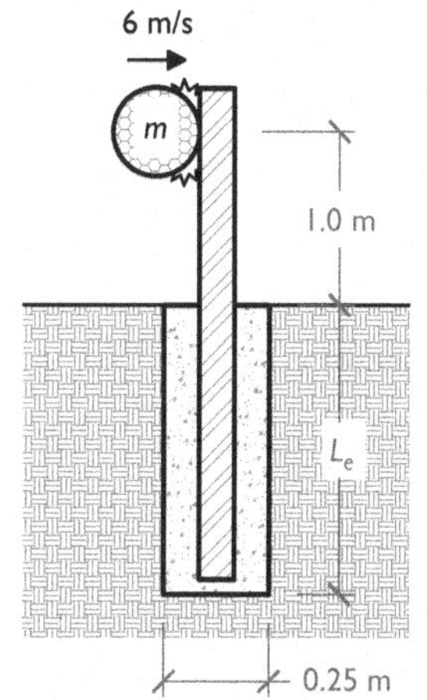

Figure 11.8 Worked example of a soil-embedded pole.

Solution

$$m = 2300 \times \frac{4}{3}\pi\left(\frac{1}{2}\right)^3 = 1204 \text{ kg}$$

$$EI_{\text{pole}} = 200000(16.5 \times 10^6) \times 10^{-9} = 3300 \text{ kNm}^2$$

$$I_{\text{concrete}} = \frac{250(250)^3 - 150(150)^3}{12} = 283.3 \times 10^6 \text{ mm}^4$$

$$EI_{\text{concrete}} = 15000(283.3 \times 10^6) \times 10^{-9} = 4250 \text{ kNm}^2$$

$$EI_{\text{combined}} = EI_{\text{pole}} + EI_{\text{concrete}} = 3300 + 4250 = 7550 \text{ kNm}^2$$

From Table 11.1, the value of η_h may be taken as 70 MN/m^3 for very dense sand.

Calculations based on the use of Equation (11.2) are shown as follows:

$$T_s = \sqrt[5]{\frac{EI_{\text{combined}}}{\eta_h}} = \sqrt[5]{\frac{7550}{70 \times 10^3}} = 0.64 \text{ m}$$

By reading off the design chart of Figure 11.3: $[L_e/T_s]_{\text{lim}} = 3.6$ approximately for $T_s = 0.64$ m. The embedment length (L_e) is accordingly to be at least equal to 2.3 m (= 3.6 × 0.64 m) in order to ensure flexible behaviour of the soil-embedded pole.

By the use of Equation (11.1):

$$C = 0.45[L_e/T_s]_{\text{lim}}\, T_s = 0.45(3.6)(0.64) = 1.04 \text{ m}$$

By the use of Equation (11.6b):

$$k_e = \left[\frac{(H + C)^3 - H^3}{3EI_{\text{combined}}} + \frac{H^3}{3EI_{\text{pole}}}\right]^{-1} = \left[\frac{(1 + 1.04)^3 - 1^3}{3(7550)} + \frac{1^3}{3(3300)}\right]^{-1}$$
$$= 2317 \text{ kN/m}$$

The pole is considered as a cantilevered member with height $H + C = 2.04$ m. Thus, the total mass is:

$$m_{\text{pole}} = 2.04(41.3) = 84.3 \text{ kg}$$

The generalised mass of a cantilever member is calculated as follows as per recommendations presented in Table 2.1 (of Chapter 2):

$$\lambda m = 0.25 m_{\text{pole}} = 0.25(84.3) = 21.1 \text{ kg}$$

$$\lambda = \frac{\lambda m}{m} = \frac{21.1}{1204} = 0.02$$

By the use of Equation (11.7):

$$\Delta = \frac{m v_o}{\sqrt{m k_e}} \sqrt{\frac{1}{1 + \lambda}} = \frac{1204(6)}{\sqrt{1204(2317 \times 10^3)}} \sqrt{\frac{1}{1 + 0.02}}$$

$$= 0.135 \text{ m or } 135 \text{ mm}$$

By the use of Equation (11.8b):

$$\theta = \frac{3}{2}\left[\frac{(H + C)^2 EI_{\text{pole}} + H^2 EI_{\text{concrete}}}{(H + C)^3 EI_{\text{pole}} + H^3 EI_{\text{concrete}}}\right]\Delta$$

$$= \frac{3}{2}\left[\frac{2.04^2(3300) + 1^2(4250)}{2.04^3(3300) + 1^3(4250)}\right](0.135) = 0.113 \text{ rad or } 6.5°$$

11.5.2 Design of a Pole-Slab Assemblage

Consider a pole-slab assemblage that is installed in a shallow pit, as shown in Figure 11.9. The pole is subjected to the collision of a boulder of 0.75 m in diameter at an impact velocity of 10 m/s. A COR value of 0.4 is taken. The pole is assumed to be sufficiently rigid that its deformation may be neglected. Estimate the maximum amount of displacement and rotation of the pole in the collision scenario.

Solution

$$m = 2300 \times \frac{4}{3}\pi\left(\frac{0.75}{2}\right)^3 = 508 \text{ kg}$$

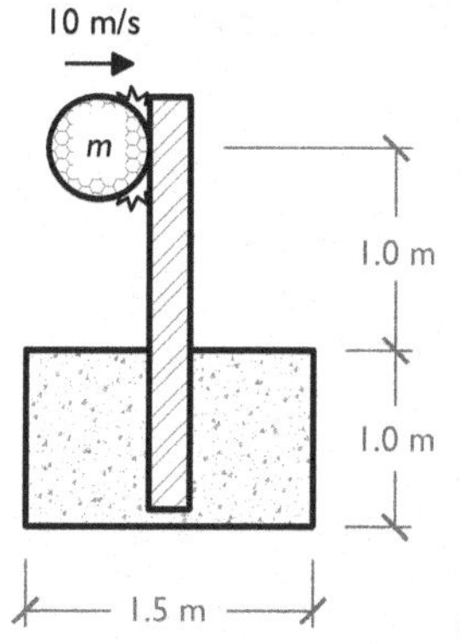

Figure 11.9 Worked example of a pole-slab assemblage.

$$\text{KE}_0 = \frac{1}{2}mv_0^2 = \frac{1}{2}(508)(10)^2 = 25.4 \text{ kJ}$$

By the use of Equation (11.12):

$$M = \rho_c w_{\text{base}} l^2 = 2400(1)(1.5)^2 = 5400 \text{ kg}$$

By the use of Equation (11.11):

$$I_\theta = M\left(\frac{w_{\text{base}}^2 + l^2}{3}\right) = 5400\left(\frac{1^2 + 1.5^2}{3}\right) = 5400 \times 1.083 = 5850 \text{ kgm}^2$$

$$R = \sqrt{\left(\frac{1.5}{2}\right)^2 + 2^2} = 2.14 \text{ m}$$

By the use of Equation (11.10):

$$\kappa = \frac{I_\theta}{mhR} = \frac{5850}{508(2)(2.14)} = 2.7$$

The lifting of the centre of gravity of the assemblage may be estimated by the use of Equation (11.9):

$$\Delta_{\text{C.G.}} = \frac{mv_0^2}{2Mg}\frac{\kappa h}{R}\left(\frac{1 + \text{COR}}{1 + \kappa}\right)^2 = \frac{508(10)^2}{2(5400)(9.81)}\left(\frac{2.7(2)}{2.14}\right)\left(\frac{1 + 0.4}{1 + 2.7}\right)^2$$
$$= 0.17 \text{ m}$$

By the use of Equations (11.14) to (11.17):

$$\alpha = \tan^{-1}\left(\frac{h_b}{w_b}\right) = \tan^{-1}\left(\frac{1}{1.5}\right) = 0.59 \text{ rad}$$

$$\theta = \sin^{-1}\left(\frac{\Delta_{\text{C.G.}} + \frac{w_{\text{base}}}{2}}{\sqrt{\frac{w_{\text{base}}^2 + l^2}{4}}}\right) - \alpha = \sin^{-1}\left(\frac{0.17 + \frac{1}{2}}{\sqrt{\frac{1^2 + 1.5^2}{4}}}\right) - 0.59 = 0.25 \text{ rad or } 14.3°$$

$$\beta = \tan^{-1}\left(\frac{2h}{l}\right) = \tan^{-1}\left(\frac{2(2)}{1.5}\right) = 1.21 \text{ rad}$$

$$\Delta_H = \frac{l}{2} - R\cos(\theta + \beta) = \frac{1.5}{2} - 2.14\cos(0.25 + 1.21) = 0.75 - 0.45$$
$$= 0.51 \text{ m or } 510 \text{ mm}$$

11.6 CLOSING REMARKS

A pole can be secured to the ground by having it embedded into soil, or by casting the pole into a concrete base slab forming a pole-slab assemblage that is then placed in a shallow pit. The mechanism of the two types of poles in withstanding a collision is very different. With a soil-embedded pole, the design approach is to treat the pole as a flexible element that responds to a collision action mainly by flexure. A compact section can be chosen in order to safeguard against premature failure by local buckling, or by shear. Guidelines in relation to the required depth of embedment of the pole into the subgrade have been provided. With a free-standing pole-slab assemblage, the stability to resist overturning is by virtue of the self-weight of the concrete slab. With both design approaches, algebraic expressions for predicting the maximum displacement and slope of deflection of the pole for a given collision scenario have been presented. Both design methodologies are illustrated with worked examples.

REFERENCES

11.1 Perera, J.S. and Lam, N., 2022, "Soil-embedded steel baffle with concrete footing responding to collision by a fallen or flying object", *International Journal of Geomechanics*, Vol. 22(3), p. 04021311, doi:10.1061/(ASCE)GM.1943-5622.0002299

11.2 Garassino, A., Jamiolkowski, M., and Pasqualini, E. 1976. *Soil modulus for laterally loaded piles in sands and NC clays.* In *Sechste Europaeische Konferenz Fuer Bodenmechanik und Grundbau*, Vol. 1.2, 429–434, INSTITUT FUER GRUNDBAU UND BODENMECHANIK, TU WIEN.

11.3 Tomlinson, M. and Woodward, J., 2014, *Pile Design and Construction Practice*, CRC Press, Abingdon, Oxon.

Chapter 12

Debris and Hail Impact on Aluminium Roofing and Cladding

12.1 INTRODUCTION

This chapter deals with denting (requiring the panel to be replaced) and fracture (leading to leaks of rainwater in a storm), as shown in Figure 12.1(a) and 12.1(b), respectively. These are the common forms of damage to aluminium roofing and cladding when impacted by a flying, or fallen, debris.

The analytical model to be introduced is for estimating permanent deformation caused to the surface of the aluminium and associated tensile strain. The commonly known controlling parameters are the mass of the impactor, incident velocity of impact, thickness of the targeted plate and the grade of the aluminium alloy. Another controlling parameter that is not as well known is the location of strike within the panel. This location-sensitive phenomenon can be explained by the significant change in participating mass activated by the impact with change in the impact location. Closed-form expressions for predicting the indentation are presented (Section 12.2). An inexpensive method involving non-destructive testing conducted on a plate specimen for the accurate determination of the participating mass is then presented. The expressions presented are applicable to the whole range of windborne debris that can cause damage to roofing and claddings (Section 12.3). In the next section, hail damage is given special attention given the high terminal velocity that can be developed in a large hailstone under free fall. In the context of hail impact, the correlation of the predicted impact velocity value with the size of hailstone has been developed by considering the terminal velocity in combination with horizontal velocity in windborne conditions (Section 12.4). The presented calculation methodology is illustrated with numerous worked examples, as is consistent with the format of earlier chapters (Section 12.5).

12.2 INDENTATION INTO ALUMINIUM PANEL

An expression based on energy principles for predicting the amount of permanent indentation (w_o) into the surface of an aluminium panel was first

DOI: 10.1201/9781003133032-12

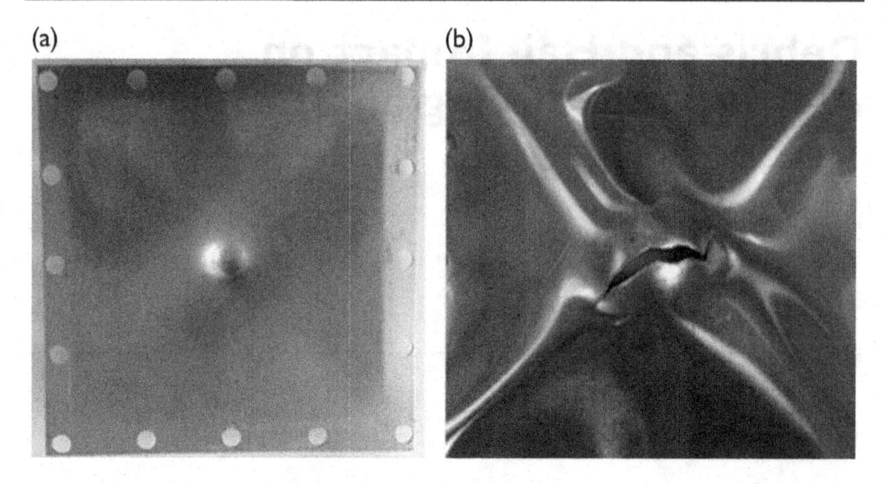

Figure 12.1 Damage to aluminium panel by debris impact: (a) denting, (b) fracture.

derived by Duffey [12.1], and subsequently modified by Calder and Goldsmith [12.2] and Mohotti et al. [12.3] to incorporate the effects of strain hardening of materials. Expressions that have been developed can be simplified into a form as shown by Equation (12.1), which expresses w_o as a function of the energy of absorption by the panel in an impact (ΔE), thickness of the panel (t_p) and yield strength (f_y) of the alloy. Note that Poisson's ratio (ν), which is an elastic phenomenon, is not present in Equation (12.1), which deals with permanent indentation in the plastic state.

$$w_o = \sqrt{\frac{4\Delta E}{\pi t_p f_y}} \tag{12.1}$$

For aluminium alloys, the value of f_y varies with the type of alloy.

The energy term in Equation (12.1), which is denoted as ΔE, represents the reduction in kinetic energy in the system from the level at the time immediately prior to the impact to the total kinetic energy remaining in the impactor and the target combined.

A convenient way of calculating ΔE is to employ the concept of the combinational mass (M_c), which was introduced in Chapter 6. The calculation involves the use of Equation (6.10b), which solves for M_c (to replace m) for substitution into Equation (6.10a) in the calculation for the value of ΔE. These two expressions are re-written as Equations (12.2) and (12.3), respectively:

$$M_c = \left(\frac{\lambda}{1+\lambda}\right)m \tag{12.2}$$

$$\Delta E = \frac{1}{2}M_c v_0^2 \tag{12.3}$$

where M_c is the combinational mass, λm is the generalised mass of the target participating in response to the impact action (and is also known as the participating mass), m is mass of the impactor and λ is the ratio of the participating mass and impactor mass.

Equation (12.3) applies to situations where the impactor does not rebound from the surface of the target but instead remains attached to the target at the conclusion of the impact. In situations where the impactor rebounds from the surface of the target (i.e. COR > 0), there can be significant additional energy losses, as some of the energy will be carried away by the impactor in the form of kinetic energy. Thus, the amount of energy remained on the targeted plate (ΔE), which is responsible for causing damage, is less than that calculated from Equation (12.3). The energy balance needs to be redefined as per Equation (12.4):

$$\Delta E = \mathrm{KE}_0 - (\mathrm{KE}_1 + \mathrm{KE}_2) \tag{12.4}$$

where KE_0 is the kinetic energy delivered by the impact, KE_1 is the kinetic energy carried by the impactor on rebound and KE_2 is the kinetic energy transferred to the target (the aluminium panel).

Equation (12.5), which can be used for estimating ΔE (taking into account energy consumed in the rebound), can be derived by making use of Equation (12.4), as shown in the appendix to this chapter (Section 12.7). Note that Equation (12.5) reverts to Equation (12.3) when COR $= 0$.

$$\Delta E = \frac{1}{2} M_c v_0^2 (1 - \mathrm{COR}^2) \tag{12.5}$$

The combinational mass (M_c), which is an input parameter into Equation (12.5), is a function of the λ factor. Note that recommendations given in Chapter 2 in relation to the participating mass are based on conditions of the element experiencing (free) natural vibration. They are only applicable when dealing with global response of the element (e.g. flexural bending of a beam), as its duration is typically much longer than the contact duration. On the other hand, permanent indentation at the point of contact occurs within a short period of time (in the range of milliseconds) after initial contact is made, and should be considered as a localised phenomenon. For this scenario, λm has been found to vary with the impact location and size of the impactor (as opposed to the size of the plate as one would expect). An inexpensive, and non-destructive, experimental method for accurate determination of the value of λm, as recommended in Shi et al. [12.4], is introduced in Section (12.3). Substituting Equation (12.5) into (12.1) gives Equation (12.6a), which can be simplified into Equation (12.6b):

$$w_o = \sqrt{\frac{2M_c v_0^2 (1 - \text{COR}^2)}{\pi t_p f_y}} \tag{12.6a}$$

$$w_o = \sqrt{\frac{0.64 M_c v_0^2}{t_p f_y} (1 - \text{COR}^2)} \tag{12.6b}$$

Given the maximum permanent indentation at the point of contact (w_o), the deformation profile of the panel can be estimated by the use of Equation (12.7) [12.5]:

$$W(r_c) = w_o e^{-0.023 r_c} \tag{12.7}$$

where $e^{-0.023 r_c}$ is the normalised permanent deformation and r_c is the distance from the position of contact.

Once the deformation profile has been found, the radial tensile strain in the aluminium $\varepsilon_s(r_c)$ can be estimated using Equation (12.8) [12.5]. As the maximum strain in the aluminium ($\varepsilon_{s,\max}$) occurs at the point of contact (i.e. $r_c = 0$), Equation (12.9) for estimating $\varepsilon_{s,\max}$ is obtained readily.

$$\varepsilon_s(r_c) = 2.65 \times 10^{-4} w_o^2 e^{-0.046 r_c} \tag{12.8}$$

$$\varepsilon_{s,\max} = 2.65 \times 10^{-4} w_o^2 \tag{12.9}$$

where w_o in the two equations are expressed in units of mm.

The utility of Equation (12.9) is for comparison against the limiting strains of damage. For example, if a tensile strain exceeding 0.12 is predicted, the panel is expected to experience fracture in the projected impact scenario (refer to Figure 12.1(b)) [12.6].

The balance of energy, as calculated from Equation (12.5), may be assumed to be totally expended to cause indentation into the aluminium panel. Making this assumption might result in giving conservative predictions of the value of w_o, given that some of the energy is dissipated by friction, and some is absorbed by the impactor, which experiences significant permanent deformation, or becomes disintegrated.

The degree of conservatism of the predictions obtained from the use of the presented method of calculation depends on the type of impactor. For example, the conservatism is minor if the impactor is rigid in comparison with the target, in which case only a very limited amount of energy is absorbed by the deformation of the impactor. In contrasts, the amount of energy dissipation through compressive deformation, or crushing, of cementitious-based debris particles, or hail, can be more significant. Research aimed at refining the presented calculation methodology for obtaining less conservative estimates through accounting for energy dissipation within the impactor is currently being undertaken by the authors and their colleagues. Insights into this component of energy dissipation can be found in Pathirana et al. [12.5]. In the interim, expressions presented in this chapter can be used to obtain conservative predictions for all impactor materials irrespective of whether crushing occurs on impact.

12.3 DETERMINATION OF PARTICIPATING MASS OF A PLATE ELEMENT

The recommended value of the participating mass (λm) for a given location along one of the diagonals can be estimated by the use of Equations (12.10a) or (12.10b), depending on the plate thickness:

$$\lambda m = \left(0.135 + 0.316 \times \frac{r_c}{r_d}\right)m \quad \text{for 1 mm thick plate} \tag{12.10a}$$

$$\lambda m = \left(0.067 + 0.039 \times \frac{r_c}{r_d}\right)m \quad \text{for 2 mm thick plate} \tag{12.10b}$$

where r_c is the radial offset of the point of strike from the centre of the square plate and r_d is half the length of the diagonal (refer Figure 12.2).

Applying either of these expressions for dealing with a plate that is thicker than 2 mm would give conservative estimates for the value of λm. Both equations can also be used for estimating λm for a given location along one of the perpendiculars by replacing r_c/r_d in the equations by d_c/d_d.

Equations (12.10a) and (12.10b) were derived from results recorded in impact experiments conducted by the authors (and colleagues) on square

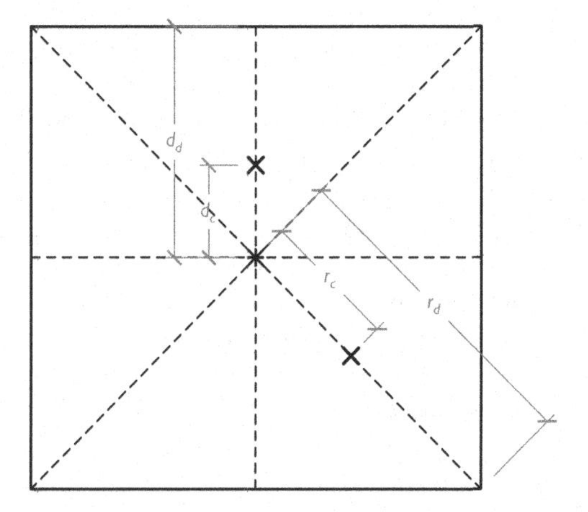

Figure 12.2 Impact location parameters.

plates of varying dimensions [12.4]. The impactors were solid steel spheres weighting between 110 grams and 535 grams. The incident velocity of impact was within 6.4 m/s. Thus, the recommended predictive relationships may only be applied with confidence to a limited range of impact scenarios.

An inexpensive procedure for the prediction of the participating mass (λm) of the panel involving non-destructive impact testing, which is generally applicable, is presented herein. First consider a hypothetical example in which the target is a lumped mass resting on a vertical spring. An impactor, which is another lumped mass (m), is dropped to strike the spring-supported lumped mass. The stiffness and type of supporting spring is immaterial in this illustration. The momentum transfer between the two lumped masses occurring on the impact is represented by the algebraic expression of Equation (12.11a) based on equal momentum principles. Rearranging Equation (12.11a) into (12.11b) provides the relationship for determining the value of λ once the velocities of motion of both lumped masses have been monitored:

$$mv_0 = -mv_1 + \lambda mv_2 \tag{12.11a}$$

$$\lambda = \frac{v_0 + v_1}{v_2} \tag{12.11b}$$

where, m is the mass of the impactor, v_0 and v_1 are the velocities of the impactor lumped mass immediately before and after the impact, respectively, and v_2 is the velocity of the spring-supported lumped mass immediately after the impact.

Although the hypothetical example as described appears trivial, this measurement methodology becomes very useful when the target is not a lumped mass, but is instead a plate-like element, as only a fraction of the mass of the plate (λm), known as the participating mass, is activated to respond to the strike. Importantly, the result varies with location of strike within the plate. The values of v_0 and v_1 can be inferred from a high-speed camera capture of the motion of the impactor. However, when dealing with a plate-like target such as an aluminium panel, the motion of the target cannot be measured at a single point. Strictly speaking, the response behaviour of the plate needs to be monitored over an array of points, which are distributed all over the plate. Immediately following an impact, the velocity measured at any point within the array is denoted as v_i, whereas the tributary mass is denoted as m_i. Applying the principles of equal momentum and equal energy gives Equation (12.12), which can be used to weight-average velocity values for determining the value of v_2, which is in turn used as input into Equation (12.11b) for finding the value of λ.

$$v_2 = \frac{\sum m_i v_i^2}{\sum m_i v_i} \tag{12.12}$$

An approximate value of v_2 can be found by inferring from the nodal velocity of the plate measured at the point of strike (v_n). The value of v_n can be inferred from the recording of an accelerometer that is attached to the plate. It has been found by the authors (through numerical simulation studies) that the value of v_2 is typically higher than the value of the velocity measured on the plate at the position of strike (v_n). The value of the v_n/v_2 has been found to typically vary within the range of 0.75 and 0.95, with the higher value representing an impact applied close to the corner of the plate. The numerical simulation technique is illustrated in Chapter 13 in detail. Taking v_2 to be simply equal to v_n would result in over-stating λ (in view of Equation (12.11b)). The predicted amount of permanent indentation into the panel would be conservative.

In summary, the use of Equation (12.11b) enables the value of λ, which is specific to a designated location within an aluminium plate to be determined once the values of v_0, v_1 and v_2 (which can be inferred from v_n) have been measured in a non-destructive experimental procedure as described.

12.4 VELOCITY OF HAIL IMPACT

This section puts the focus on hail impact, which deserves specific treatment, given the high terminal velocity of a large hailstone under free fall. Exacerbation of the motion of hail by wind is also considered. The resultant velocity of hail impact (U_R) combining the effects of free fall under gravity and horizontal wind velocity is defined by Equation (12.13). The velocity values so calculated using the expressions presented below are to be substituted into expressions presented in Section 12.2 for estimating the amount of permanent deformation (w_o) and the tensile strain in the aluminium (ε_s).

$$U_R = \sqrt{U_V^2 + U_H^2} \tag{12.13}$$

where U_V is the vertical velocity of freefall as defined by Equation (12.14) based on [12.7–12.9] for spherical specimens of hail.

$$U_V = \sqrt{\frac{4\rho_h D_h g}{3\rho_a C_D}} \tag{12.14}$$

where ρ_h is density of hailstone, D_h is diameter of the hailstone, ρ_a is density of air and C_D is the drag coefficient.

U_H is the horizontal velocity of hail under the influence of wind as defined by Equation (12.15) based on recommendation by Holmes [12.10]:

$$U_H = \frac{k_u U_w^2 t_u}{1 + k_u U_w t_u} \tag{12.15}$$

where U_w is the horizontal wind velocity. t_u is flight time of the hailstone and k_u is as defined by Equation (12.16):

$$k_u = \frac{3\rho_a C_D}{4\rho_h D_h} \tag{12.16}$$

The following assumptions may be made in the use of Equations (12.14) to (12.16), as per recommendations by Dieling et al. [12.9]; ρ_h = 917 kg/m^3, ρ_a = 1.225 kg/m^3 and C_D = 0.6 (noting that the range of values is 0.45–0.8). The value of t_u is typically in the range of 1–2 s.

The design of roofing panels should be based mainly on the vertical velocity of hail under free fall. When considering free-fall conditions (without incorporating the effects of wind), the velocity of impact increases monotonically with the size of the hailstone. When horizontal wind with a velocity of 36–48 m/s is considered to coexist with free-fall conditions, the resultant velocity of impact may be taken to remain roughly constant with the change in size of the hailstone. The combined design velocities for hailstones of all sizes in the vertical, and near vertical, directions relevant to impact on roof coverings have been found to be of the order of 30–40 m/s when the horizontal wind velocity is 36 m/s. The limits are increased to 40–50 m/s when the horizontal wind velocity is increased to 48 m/s.

The velocity of impact on a panel, which is installed vertically, should be resolved into the horizontal direction. The horizontal velocity of windborne hail is in the order of 15–25 m/s for D_h = 50 mm and 25–35 m/s for D_h = 25 mm.

The design velocities for hailstones of different sizes are presented in Figures 12.3(a) and 12.3(b) for U_H and U_R, respectively. U_V is plotted alongside U_R in Figure 12.3(b), based on t_u = 2 s for comparison.

12.5 WORKED EXAMPLES

12.5.1 Permanent Indentation into Aluminium Panel (Based on Use of Equation (12.10b))

Consider a 250 mm by 250 mm aluminium panel with thickness of 1 mm (alloy type: 5052-H34), which is subject to impact by a solid steel sphere weighing 110 grams with an impact velocity of 6.26 m/s. The aluminium panel weighs 158 grams, and has density of 2,700 kg/m^3 and characteristic yield strength of 138 MPa. The panel is fully fixed on all four edges. The value of COR for a steel impactor striking a smooth aluminium surface is taken as 0.5 in this example. An elaborate relationship for accurate

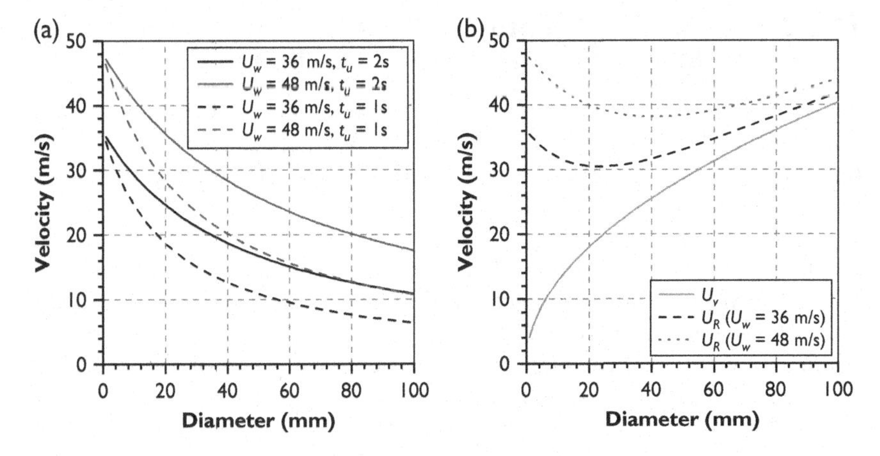

Figure 12.3 Velocity of hail impact: (a) horizontal velocity, (b) resultant velocity.

determination of COR can be found in Ref. [12.4]. Determine the maximum amount of permanent indentation into the panel at $r_c = 0.6r_d$. Check the likelihood of perforation of the panel given that the fracture strain for this type of aluminium alloy is 0.12 [12.6].

Solution

By the use of Equation (12.10a):

$$\lambda m = \left(0.135 + 0.316 \times \frac{r_c}{r_d}\right)m = (0.135 + 0.316(0.6))m = 0.325m$$

$$\lambda = 0.325$$

By the use of Equation (12.2):

$$M_c = \left(\frac{\lambda}{1+\lambda}\right)m = \left(\frac{0.325}{1+0.325}\right)(0.11) \approx 0.027 \text{ kg}$$

By the use of Equation (12.6b):

$$w_o = \sqrt{\frac{0.64 M_c v_0^2}{t_p f_y}(1 - \text{COR}^2)} = \sqrt{\frac{0.64(0.027)(6.26)^2}{0.001(138 \times 10^6)}(1 - 0.5^2)}$$

$$= 0.0019 \text{ m or } 1.9 \text{ mm}$$

By the use of Equation (12.9):

$$\varepsilon_{s,\max} = 2.65 \times 10^{-4}\, w_o^2 = 2.65 \times 10^{-4}\,(1.9)^2 \approx 0.001$$

The amount of indentation is visible but fracture is not predicted.

12.5.2 Permanent Indentation into Aluminium Panel (Based on Non-Destructive Testing)

Consider the same impact scenario as the previous example. Conduct a non-destructive (dropped ball) test instead of using Equation (12.10b) for obtaining the estimated value for λ. Velocities recorded from the dropped ball test are summarised as follows:

i. Incident velocity of impact (v_0) was 6.26 m/s (downward).
ii. The velocity of the impactor (v_1) at end of the restitution phase was 4.5 m/s (downward).
iii. The velocity of the plate at point of contact at end of the restitution phase was (v_n) 4.5 m/s (downward).

Solution
By the use of Equation (12.11b):

$$\lambda = \frac{v_0 + v_1}{v_2} = \frac{6.26 - 4.5}{4.5} = 0.39$$

By the use of Equation (12.2):

$$M_c = \left(\frac{\lambda}{1+\lambda}\right)m = \left(\frac{0.39}{1+0.39}\right)(0.11) \approx 0.031 \text{ kg}$$

By the use of Equation (12.6b):

$$w_o = \sqrt{\frac{0.64 M_c v_0^2}{t_p f_y}(1 - \text{COR}^2)} = \sqrt{\frac{0.64(0.031)(6.26)^2}{0.001(138 \times 10^6)}(1 - 0.5^2)}$$

$$\approx 0.002 \text{ m or } 2 \text{ mm}$$

which is comparable to the result obtained in the previous example.

12.5.3 Mass and Velocity of Hail Impact

Consider a hailstone 100 mm in size when idealised into a sphere. Calculate the resultant velocity of impact (U_R), taking into account the vertical velocity of free fall (U_V) and the horizontal velocity (U_H) in windborne conditions based on a wind velocity (U_w) of 36 m/s.

Solution
Equations (12.13) to (12.16) are employed to estimate the design velocities in the vertical and horizontal directions.

$$U_V = \sqrt{\frac{4\rho_h D_h g}{3\rho_a C_D}} = \sqrt{\frac{4(917)(0.1)(9.81)}{3(1.225)(0.6)}} = 40.4 \text{ m/s}$$

$$k_u = \frac{3\rho_a C_D}{4\rho_h D_h} = \frac{3(1.225)(0.6)}{4(917)(0.1)} = 0.006$$

$$U_H = \frac{k_u U_w^2 t_u}{1 + k_u U_w t_u} = \frac{0.006(36)^2(2)}{1 + 0.006(36)(2)} = 10.9 \text{ m/s}$$

$$U_R = \sqrt{U_V^2 + U_H^2} = \sqrt{40.4^2 + 10.9^2} = 41.8 \text{ m/s}$$

12.5.4 Permanent Deformation Generated by Hail Impact

Consider a 300 mm by 300 mm aluminium panel with a thickness of 2 mm (alloy type: 5052-H34) that has a density of 2700 kg/m^3, Poisson's ratio of 0.33 and characteristic yield strength of 220 MPa. Two impact scenarios are considered: (i) impact by a 50 mm diameter hailstone and (ii) impact by a 25 mm diameter hailstone. A wind velocity of 48 m/s and flight time of 2 s may be assumed. A COR of zero may be assumed. Determine the amount of deformation of the panel and the corresponding maximum tensile strain with both scenarios when the location of impact is at the centre of the plate, i.e. $r_c/r_d = 0$.

Solution

Only the horizontal component of the velocity, U_H, needs to be considered as the cladding is installed vertically. Based on Figure 12.3(a), when Uw = 48 m/s and t_u = 2 s, impact velocity is about 25 m/s and 35 m/s for hailstone diameter of 50 mm and 25 mm, respectively.

i. **D_h = 50 mm and v_0 = 25 m/s**

$$m = \rho_h \times \frac{4}{3}\pi\left(\frac{D_h}{2}\right)^3 = 917 \times \frac{4}{3}\pi\left(\frac{0.05}{2}\right)^3 = 0.06 \text{ kg}$$

By the use of Equation (12.10b):

$$\lambda m = \left(0.067 + 0.039 \times \frac{r_c}{r_d}\right)m = (0.067 + 0.039\ (0))m = 0.067\ m$$

$$\lambda = 0.067$$

$$M_c = \left(\frac{\lambda}{1+\lambda}\right)m = \left(\frac{0.067}{1 + 0.067}\right)(0.06) = 0.0038 \text{ kg}$$

By the use of Equation (12.6b):

$$w_o = \sqrt{\frac{0.64 M_c v_0^2}{t_p f_y}} = \sqrt{\frac{0.64(0.0038)(25)^2}{0.002(220 \times 10^6)}} = 0.0019 \text{ m or } 1.9 \text{ mm}$$

By the use of Equation (12.9):

$$\varepsilon_{s,\max} = 2.65 \times 10^{-4}\, w_o^2 = 2.65 \times 10^{-4}\,(1.9)^2 = 0.001$$

A fracture is not predicted, but the amount of permanent indentation may warrant replacement of the panel.

ii. **D_h= 25 mm and v_0 = 35 m/s**

$$m = \rho_h \times \frac{4}{3}\pi\left(\frac{D_h}{2}\right)^3 = 917 \times \frac{4}{3}\pi\left(\frac{0.025}{2}\right)^3 = 0.0075 \text{ kg}$$

By the use of Equation (12.10b):

$$\lambda m = \left(0.067 + 0.039 \times \frac{r_c}{r_d}\right)m = (0.067 + 0.039\,(0))m = 0.067\, m$$

$$\lambda = 0.067$$

$$M_c = \left(\frac{\lambda}{1+\lambda}\right)m = \left(\frac{0.067}{1+0.067}\right)(0.0075) = 0.00047 \text{ kg}$$

By the use of Equation (12.6b):

$$w_o = \sqrt{\frac{0.64 M_c v_0^2}{t_p f_y}} = \sqrt{\frac{0.64(0.00047)(35)^2}{0.002(220 \times 10^6)}} = 0.0009 \text{ m or } 0.9 \text{ mm}$$

By the use of Equation (12.9):

$$\varepsilon_{s,\max} = 2.65 \times 10^{-4}\, w_o^2 = 2.65 \times 10^{-4}\,(0.9)^2 = 0.00021$$

The first scenario featuring a larger-size hailstone generates a much higher level of deformation and tensile strain to the panel than the second scenario.

12.6 CLOSING REMARKS

The key parameter characterising damage to the panel is the amount of permanent indentation (w_o). Parameter w_o can be used to infer the maximum tensile strain ($\varepsilon_{s,\,max}$) generated by the impact. Expressions for estimating the value of key parameters, including w_o and $\varepsilon_{s,\,max}$, have been presented. An important feature of the presented analytical method is in incorporating the effects of the participating mass of the target on the amount of permanent indentation. The participating mass is in turn dependent on the location of impact, thickness of the plate and the mass of the impactor. Comparison of $\varepsilon_{s,\,max}$ with the limiting strain to cause fracture in aluminium can be used for estimating the likelihood of perforation of the panel, which can result in leaks during a storm. The vertical velocity of the hail impact associated with free fall and the wind-driven horizontal velocity, and their combinations, are considered. The resultant velocity of hail impact combining both phenomena is shown to be of the order of 30–50 m/s in conditions where the horizontal velocity of wind is in the range of 36–48 m/s. The resultant velocity should be taken as the velocity of impact when predicting damage to roofing, whereas horizontal velocity on its own should be taken when predicting damage to (vertically installed) claddings. The prediction of damage caused to aluminium panels is illustrated with worked examples for both roofing and cladding. The impact by idealised spherical objects and ice specimens have both been analysed.

12.7 APPENDIX

Equation (12.4) is re-written here as Equation (12.17):

$$\Delta E = \mathrm{KE}_0 - (\mathrm{KE}_1 + \mathrm{KE}_2) \tag{12.17}$$

The energy terms in Equation (12.17) are defined by Equations (12.18) to (12.20):

$$\mathrm{KE}_0 = \frac{1}{2} m v_0^2 \tag{12.18}$$

$$\mathrm{KE}_1 = \frac{1}{2} m v_1^2 \tag{12.19}$$

$$\mathrm{KE}_2 = \frac{1}{2} \lambda m v_2^2 \tag{12.20}$$

Substituting Equations (12.18) to (12.20) into Equation (12.17) gives Equation (12.21):

$$\Delta E = \frac{1}{2} m v_0^2 \left(1 - \left(\frac{v_1}{v_0} \right)^2 - \lambda \left(\frac{v_2}{v_0} \right)^2 \right) \tag{12.21}$$

Equations (3.1b) and (3.1c) are re-written here as Equations (12.22) and (12.23), respectively. Equation (12.23) can be re-arranged into Equation (12.24):

$$\frac{v_2}{v_0} = \frac{1 + \text{COR}}{1 + \lambda} \tag{12.22}$$

$$\text{COR} = \frac{v_1 + v_2}{v_0} \tag{12.23}$$

$$\frac{v_1}{v_0} = \text{COR} - \frac{v_2}{v_0} \tag{12.24}$$

Substituting Equations (12.22) and (12.24) into Equation (12.21) and algebraic manipulating gives Equation (12.25):

$$\Delta E = \frac{1}{2} m v_0^2 (1 - \text{COR}^2)\left(\frac{\lambda}{1 + \lambda}\right) \tag{12.25}$$

Substituting Equation (12.2) into Equation (12.25) gives Equation (12.26), which is identical to Equation (12.5):

$$\Delta E = \frac{1}{2} M_c v_0^2 (1 - \text{COR}^2) \tag{12.26}$$

REFERENCES

12.1 Duffey, T.A., 1967, *Large Deflection Dynamic Response of Clamped Circular Plates Subjected to Explosive Loading*, Sandia Corp., Albuquerque, N. Mex.

12.2 Calder, C.A. and Goldsmith, W., 1971, "Plastic deformation and perforation of thin plates resulting from projectile impact", *International Journal of Solids and Structures*, Vol. 7(7), pp. 863–881, doi:10.1016/0020-7683(71)90096-5

12.3 Mohotti, D., Ali, M., Ngo, T., Lu, J., Mendis, P., and Ruan, D., 2013, "Out-of-plane impact resistance of aluminium plates subjected to low velocity impacts", *Materials & Design*, Vol. 50, pp. 413–426, doi:10.1016/j.matdes.2013.03.023

12.4 Shi, S., Lam, N.T.K., Cui, Y., Zhang, L., Lu, G., and Gad, E.F., (in press), "Indentation into an aluminium panel by a rigid spherical object", *Thin-Walled Structures*.

12.5 Pathirana, M., Lam, N., Perera, S., Zhang, L., Ruan, D., and Gad, E., 2017, "Damage modelling of aluminium panels impacted by windborne debris", *Journal of Wind Engineering and Industrial Aerodynamics*, Vol. 165, pp. 1–12, doi:10.1016/j.jweia.2017.02.014

12.6 Herbin, A.H. and Barbato, M., 2012, "Fragility curves for building envelope components subject to windborne debris impact", *Journal of Wind Engineering and Industrial Aerodynamics*, Vol. 107–108, pp. 285–298, doi: 10.1016/j.jweia.2012.05.005
12.7 Matson, R.J. and Huggins, A.W., 1980, "The direct measurement of the sizes, shapes and kinematics of falling hailstones", *Journal Of Atmospheric Sciences*, Vol. 37(5), pp. 1107–1125, doi:10.1175/1520-0469(1980)037%3C1107:TDMOTS%3E2.0.CO;2
12.8 Paterson, D.A. and Sankaran, R., 1994, "Hail impact on building envelopes", *Journal of Wind Engineering and Industrial Aerodynamics*, Vol. 53(1), pp. 229–246, doi:10.1016/0167-6105(94)90028-0
12.9 Dieling, C., Smith, M., and Beruvides, M., 2020, "Review of impact factors of the velocity of large hailstones for laboratory hail impact testing consideration", *Geosciences*, Vol. 10(12), p. 500, doi:10.3390/geosciences10120500ho
12.10 Holmes, J., 2004, "Trajectories of spheres in strong winds with application to wind-borne debris", *Journal of Wind Engineering and Industrial Aerodynamics*, Vol. 92(1), pp. 9–22, doi:10.1016/j.jweia.2003.09.031

Chapter 13

Impact Dynamics Simulations

13.1 INTRODUCTION

The simulation of the response of a single-degree-freedom (SDOF) lumped mass system when subject to an idealised (short duration) transient action is first introduced. For a system responding within its linear elastic limit, the displacement time-history of the system responding to two, or more, of such short-duration impulses can be obtained by taking the sum (in the time domain) of the displacement generated by the individual short-duration impulses. This summation technique may then be used for dealing with transient actions of a finite duration which is pre-defined in the form of a forcing function. Towards the end of this chapter, the accuracy of predictions derived from this simulation methodology is validated by comparison against results derived from the classical method of solving for the response to rectangular- and triangular-shaped impulses. The principle of modal superposition is also employed for simulating the time-histories with multi-degree-of-freedom (MDOF) lumped mass systems. A cantilever element is used initially to illustrate the methodology. A more detailed illustration based on the example of a simply supported beam is also presented. Implementation of the calculation procedure including dynamic modal analysis on Microsoft Excel spreadsheets enables readers to develop a good mastery of structural dynamics computations. Results reported by a general-purpose structural analysis software on some simple building models can be checked independently, using the methodologies presented in this chapter, to develop confidence in the use of the software. Finally, the numerical simulation of the forcing function generated by the collision action (of a solid object on the surface of the target) is illustrated. The stiffness properties of materials of objects coming into contact are taken into account in the simulation. Full descriptions of every step in the simulation procedure based on the use of a worked example can be found towards the end of the chapter.

DOI: 10.1201/9781003133032-13

13.2 SIMULATIONS OF SINGLE-DEGREE-OF-FREEDOM LUMPED MASS SYSTEMS

Consider an initially stationary object, which is subject to a transient force propelling it into motion. The amount of impulse (J), which has the units force-time (i.e. kNs), transmitted to the object is equal to the integral of the applied transient force with respect to time (i.e. $\int F(t)dt$). The velocity of motion ($\dot{u}(\tau)$) at the expiry of the period of force application is then equal to J divided by the object mass (M), as expressed in Equation (13.1):

$$\dot{u}(\tau) = \frac{J}{M} = \frac{1}{M}\int_{t=\tau}^{t=\tau+t_\tau} F(t)dt \tag{13.1}$$

Equation (13.1) is valid, provided that the moving object is in "free-space," and is not subject to any external interference during the application of the force. A common type of example for illustrating dynamic simulations is shown in Figure 13.1, which features a pole supporting a lumped mass. The free-space assumption is strictly speaking not valid, as structural response of the pole may introduce interferences (via the structural force that results from displacement of the pole) affecting the motion behaviour of the lumped mass. The potential interference depends on the timing and duration of the application of the transient force. Equation (13.1) is therefore only valid when the applied transient force lasts for an infinitesimal duration (i.e. the force has a very short duration). As only a negligible amount of deflection occurs in the pole for a very short-duration transient force, the amount of resistance so generated from within the pole to interfere with the motion of the lumped mass is accordingly very small, and can be neglected (refer Figure 13.1 and the inset diagram showing the timing of the load application). The impulse depicted in the inset diagram is referred to in the rest of the chapter as a "short-duration impulse."

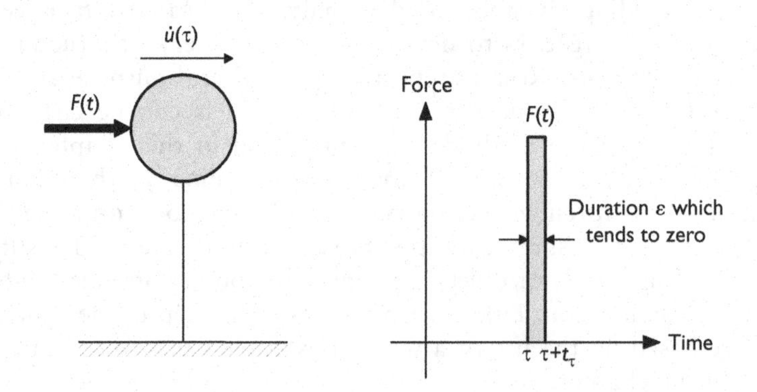

Figure 13.1 Single lumped mass system subject to the short-duration impulse.

If the short-duration impulse is equal to unity, as shown by Equation (13.2), the resulting velocity of motion is given by Equation (13.3):

$$\int_{t=\tau}^{t=\tau+t_\tau} F(t)\,dt = 1 \tag{13.2}$$

$$\dot{u}(\tau) = \frac{1}{M} \tag{13.3}$$

The short-duration impulse, which fulfils the requirements of Equation (13.2), is known as the *Dirac Delta* function. Predictions from Equation (13.3) can be scaled in a linear fashion to provide predictions for the velocity of motion of the targeted lumped mass for any given amount of impulse. Note that for a target, which is subject to one such short-duration impulse, then $v_2 = \dot{u}(\tau)$. The pole-like structure as depicted in Figure 13.1 is only one of many ways of representing a SDOF lumped mass system. There are other forms of SDOF systems (such as a trolley carrying a lumped mass) that behave in the same manner as the pole-like structure described. Thus, Equation (13.3) may be applied to give predictions for SDOF systems responding to any predefined short-duration impulse. Once the velocity of motion immediately following the application of the short-duration impulse (i.e. at $t = \tau$) is known, the displacement time-history of the SDOF system, $u(t)$, responding to the impulse is given by Equation (13.4) for $t \geq \tau$. Equation (13.4) may be re-written as Equation (13.5a) by making use of the relationship defined by Equation (13.6). Equation (13.5a) can also be written in the form of Equation (13.5b) by referring to Equation (13.1) and $v_2 = \dot{u}(\tau)$. The derivation of these expressions can be found in standard structural dynamics textbooks [13.1, 13.2]. The presentation of the derivation can also be found in the appendix of this chapter.

$$u(t) = \frac{J}{M\omega_n} e^{-\xi\omega_n(t-\tau)} \sin\omega_n(t-\tau) \tag{13.4}$$

$$u(t) = \frac{J}{M} h(t-\tau) \tag{13.5a}$$

$$u(t) = v_2 h(t-\tau) \tag{13.5b}$$

$$h(t-\tau) = \frac{1}{\omega_n} e^{-\xi\omega_n(t-\tau)} \sin\omega_n(t-\tau) \tag{13.6}$$

For a hard collision scenario in which the transmission of the impulse may be assumed to take place instantaneously, v_2 is defined by Equation (13.7), which was first introduced in Chapter 3 (refer to Equation (3.1b)) for estimating the velocity of a targeted lumped mass when collided upon by an object of mass m, and moving at velocity v_0.

$$v_2 = v_0\left(\frac{1 + \text{COR}}{1 + \lambda}\right) \tag{13.7}$$

Take an example where a short-duration impulse of 50 kNs with zero time lag (i.e $\tau = 0$) is applied to a SDOF lumped mass system. The lumped mass system weighs 100 tonnes, has a natural period of vibration of 0.5 s (which corresponds to a natural angular velocity of 12.57 rad/s) and viscous damping ratio (ξ) of 0.05. The displacement time-history, $u(t)$, is accordingly given by Equation (13.8a). In another example where a short-duration impulse of the same intensity is applied with a 0.5 s delay, the expression for $u(t)$ is given by Equation (13.8b):

$$u_1(t) = \frac{50}{100(12.57)} e^{-0.05(12.57)t} \sin 12.56t \quad \text{for} \quad \text{t} \geq 0 \text{ s} \tag{13.8a}$$

$$u_2(t) = \frac{50}{100(12.57)} e^{-0.05(12.57)(t-0.5)} \sin 12.56(t - 0.5) \quad \text{for} \quad \text{t} \geq 0.5 \text{ s} \tag{13.8b}$$

The expressions of Equations (13.8a) and (13.8b) so obtained can be used to present the displacement time-history of a SDOF system in the form of a graph, as shown in Figures 13.2(a) and 13.2(b), respectively.

The displacement time-history of a SDOF system responding within its linear elastic limit when subject to more than one short-duration impulse can be solved by the use of the presented expressions, and by employing the principles of superposition. For example, consider a case where two short-duration impulses, as depicted in Figures 13.2(a) and 13.2(b), are both imposed on the same SDOF system. The response displacement time-history can be derived by taking the arithmetic sum of Equations (13.8a) and (13.8b) at each time step, as shown by Equation (13.9):

$$u(t) = u_1(t) + u_2(t) \tag{13.9}$$

The displacement time-history generated by the application of the two short-duration pulses is shown in Figure 13.2(c) to illustrate the phenomenon of dynamic amplification.

If the time lag, τ, is changed from 0.5 s to 0.25 s, the displacement generated by the second impulse offsets that of the first impulse. The phenomenon of de-amplification is illustrated as shown in Figure 13.3.

A generalised expression for predicting the displacement time-history of a SDOF lumped mass system when subject to N number of impulses can be

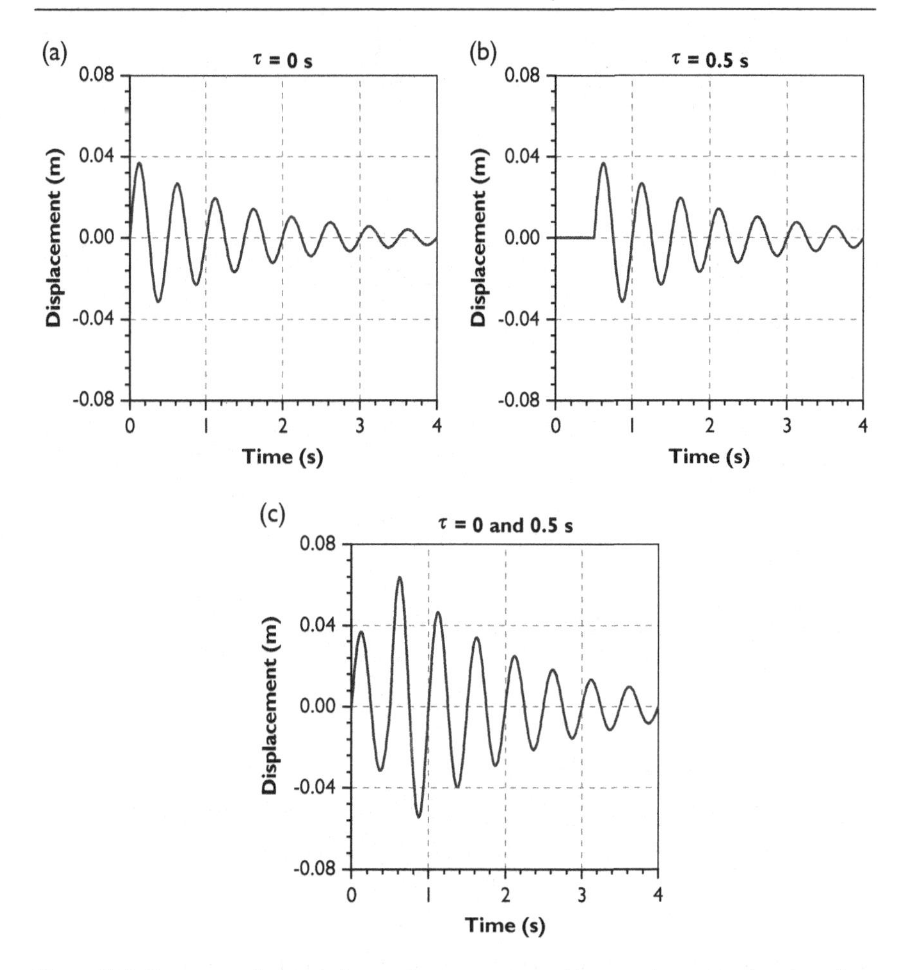

Figure 13.2 Response of a single lumped mass system subject to short-duration impulses.

derived by substituting Equation (13.4) into Equation (13.9), as shown in Equation (13.10):

$$u(t) = \frac{1}{M\omega_n} \sum_{i=1}^{N} \left[J_i e^{-\xi\omega_n (t-\tau_i)} \sin \omega_n (t - \tau_i) \right] \tag{13.10}$$

The use of Equation (13.10) for calculating the time-history of a response of a SDOF lumped mass system to multiple impulses is well established, and is known as the *Duhamel Integration Method.*

Implementation of the analysis procedure presented in this section for simulating the dynamic behaviour of a SDOF system in response to a predetermined forcing function is demonstrated by use of worked examples in Sections 13.5.1 to 13.5.3.

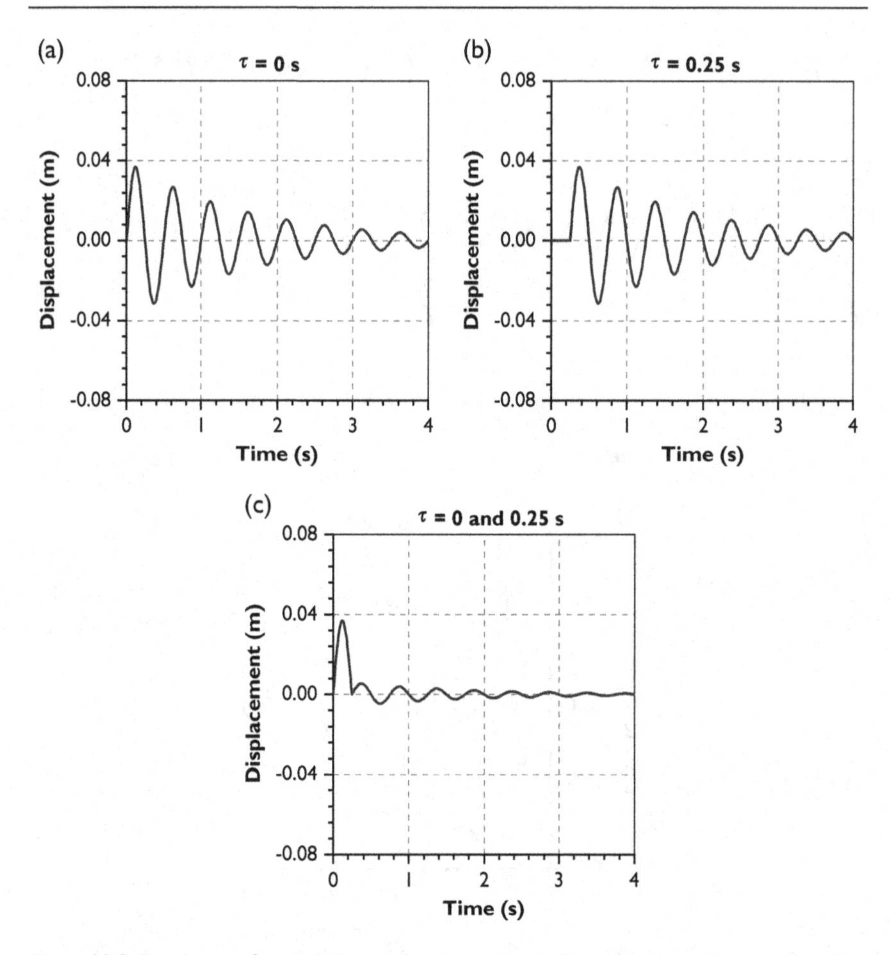

Figure 13.3 Response of a single lumped mass system subject to short-duration impulse.

13.3 SIMULATIONS OF MULTI-DEGREE-OF-FREEDOM LUMPED MASS SYSTEMS

The approach introduced herein for simulating the response of a MDOF lumped mass system is to make use of the principle of modal superposition, which is an established concept in structural dynamics. The procedure is broadly outlined in the following steps, which are subsequently described in detail:

1. Determination of the flexibility matrix and mass matrix of the MDOF system.
2. Determination of the mode shapes and natural periods for the different modes of vibration.

3. Solve the equation of dynamic equilibrium for the MDOF system to determine the generalised mass, generalised damping coefficient, generalised stiffness, excitation factor and participation factor for each mode of vibration.
4. Determine the displacement time-history for each mode of vibration using the *Duhamel Integration Method.*
5. Determine the overall displacement time-history of the MDOF system using modal superposition.

Step 1: Determination of the flexibility matrix and mass matrix of the MDOF system

A structure is first idealised into a finite number of lumped masses (or "nodes"). The analysis procedure can be commenced with the construction of the flexibility matrix of a vertical cantilever by employing static analyses of a single point load. First, a unit load, P_1, is applied at the uppermost node to result in the deflection of the cantilever element, as shown in Figure 13.4. The amount of deflection generated by the applied unit load is then calculated for each of the nodes in the element (e.g. $f_{3,1}$ is the displacement at node M_3 from the unit P_1 being applied at node M_1). Closed-form expressions for facilitating such calculations for common types of elements such as simply supported beams, fixed-end beams and cantilever beams are available. The calculated values are then entered into the first column of the flexibility matrix $[F]$ (i.e. $f_{1,1}$, $f_{2,1}$, $f_{3,1}$, and so on). Repeat the calculation and

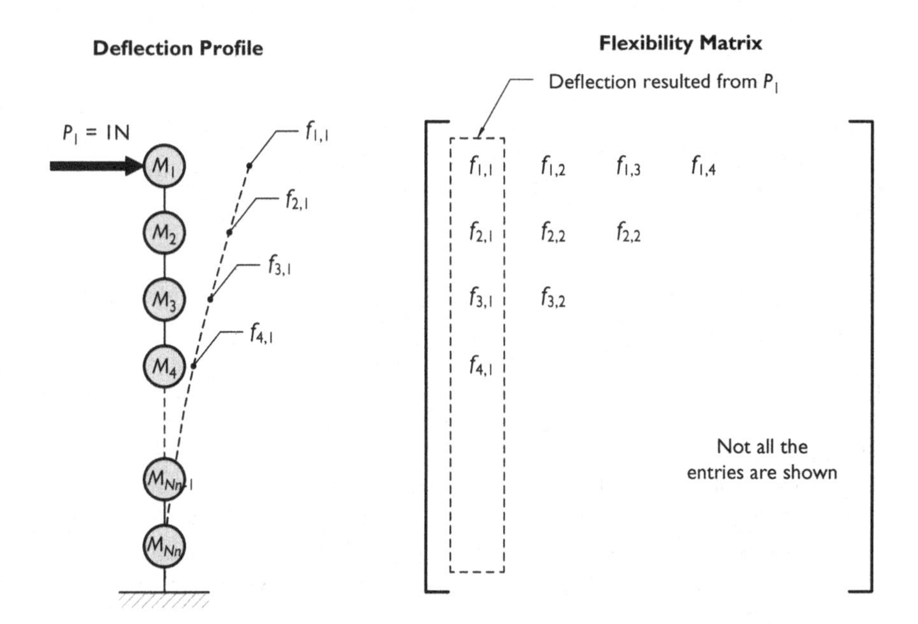

Figure 13.4 Beam deflection subjected to a unit load, $\boldsymbol{P_1}$.

cell entries for another unit load (P_2), which is applied at the next node down (i.e. second node counting from the top) for completing the entries to the second column of the [F] matrix (i.e. $f_{1,2}$, $f_{2,2}$, $f_{3,2}$, and so on). The rest of the [F] matrix is then populated by applying a unit load (one at a time) to the element at each of its nodes, from the top down. Construction of the first three columns of a [F] matrix is shown by Equation (13.11):

$$[F] = \begin{bmatrix} f_{1,1} & f_{1,2} & f_{1,3} & \cdots & f_{1,N_n} \\ f_{2,1} & f_{2,2} & f_{2,3} & \cdots & f_{2,N_n} \\ f_{3,1} & f_{3,2} & f_{3,3} & \cdots & f_{3,N_n} \\ \vdots & \vdots & \vdots & \ddots & \vdots \\ f_{N_n,1} & f_{N_n,2} & f_{N_n,3} & \cdots & f_{N_n,N_n} \end{bmatrix} \tag{13.11}$$

where subscript N_n is the total number of nodes.

The inversion of the flexibility matrix [F] results in the stiffness matrix [K], as shown by Equation (13.12):

$$[K] = [F]^{-1} \tag{13.12}$$

Meanwhile, the mass matrix can be constructed by entering the lumped mass value into the diagonal cells of the matrix, whereas all other cells have zero entries, as shown by Equation (13.13), where the subscript of each entry of M represents the node number:

$$[M] = \begin{bmatrix} M_1 & & & & \\ & M_2 & & & \\ & & M_3 & & \\ & & & \ddots & \\ & & & & M_{N_n} \end{bmatrix} \tag{13.13}$$

Step 2: Determination of the mode shapes and natural periods for the different modes of vibration

Once the mass matrix has been constructed, modal dynamic analysis may proceed. Step-by-step implementation of the mode shape iteration method of modal analysis is presented in Table 13.1. Once the [K] and [M] matrices are known, the standard iteration techniques may be applied to provide estimates for the modal frequencies (the eigenvalues ω_j, f_j and T_j) and the mode shape vectors (the eigenvectors $\{\phi\}_j$). Operational details of the method are illustrated by the use of worked examples, which are presented at the end of this chapter (Section 13.5).

Table 13.1 Step-by-step outline of modal analysis

Step	*Operation*	*Remarks*
1	Idealise structural member into line element with N_n number of equally spaced and equal mass nodes.	
2	Construct flexibility matrix $[F]$, which is then inversed into stiffness matrix $[K]$.	Refer to Figure 13.4 and Equations (13.11) and (13.12).
3	Construct mass matrix $[M]$.	Refer to Equation (13.13).
4	Determine dynamic matrix for the first mode of vibration. $[D]_1 = [F][M]$.	Not applicable to second mode and onwards.
5a	Define an arbitrary mode shape for the first iteration. $\{\phi\}_0 = \begin{Bmatrix} 1 \\ 1 \\ \vdots \\ 1 \end{Bmatrix}$	Number of elements in matrix $\{\phi\}_0 = N_n$.
5b	Conduct mode shape iteration by using the two following equations: $\{\phi_j\}'_c = [D]_j \{\phi_j\}_{c-1}$ $\{\phi_j\}_c = \frac{\{\phi_j\}'_c}{\phi'_{j,c,\max}}$ Calculate angular velocity for each iteration: $\omega_{j,c} = \sqrt{\frac{\{\phi_j\}_c^T [K] \{\phi_j\}_c}{\{\phi_j\}_c^T [M] \{\phi_j\}_c}}$ Repeat the iteration until the value of $\omega_{j,c}$ converges. It typically converges in less than 10 iterations. The first three iterations are shown as an example:	Subscript c represents the cth iteration $\{\phi_j\}'_c$ is the iterated mode shape before normalsation, and $\{\phi_j\}_c$ is the normalised mode shape. $\phi'_{j,c,\max}$ is the maximum value in the matrix $\{\phi_j\}'_c$.

(Continued)

TABLE 13.1 (Continued)

Step	*Operation*				*Remarks*
		First (c = 1)	Second (c = 2)	Third (c = 3)	
	$\{\phi_j\}'_c$	$[D]_j\{\phi\}_0$	$[D]_j\{\phi_j\}_1$	$[D]_j\{\phi_j\}_2$	
	$\{\phi_j\}_c$	$\frac{\{\phi_j\}'_1}{\phi'_{j,1,\max}}$	$\frac{\{\phi_j\}'_2}{\phi'_{j,2,\max}}$	$\frac{\{\phi_j\}'_3}{\phi'_{j,3,\max}}$	
	$\omega_{j,c}$	N/A	$\sqrt{\frac{\{\phi_j\}_1^T[K]\{\phi_j\}_1}{\{\phi_j\}_1^T[M]\{\phi_j\}_1}}$	$\sqrt{\frac{\{\phi_j\}_2^T[K]\{\phi_j\}_2}{\{\phi_j\}_2^T[M]\{\phi_j\}_2}}$	
5c	Calculate frequency and period. $f_j = \frac{\omega_j}{2\pi},\ T_j = \frac{1}{f_j}$				
6	Define the identity matrix. $[I] = \begin{bmatrix} 1 & & & & \\ & 1 & & & \\ & & 1 & & \\ & & & \ddots & \\ & & & & 1 \end{bmatrix}$				$[I]$ has size of $N_n \times N_n$.
7	Determine the sweeping matrix. $[S]_j = [I] - \frac{\{\phi_{j-1}\}\{\phi_{j-1}\}^T[M]}{\{\phi_{j-1}\}^T[M]\{\phi_{j-1}\}}$				Only applicable to second mode and onwards ($j \geq 2$).
8	Determine the dynamic matrix. $[D]_j = [D]_{j-1}[S]_j$				Only applicable to second mode and onwards ($j \geq 2$).
9	Repeat steps 5a–5c to determine mode shape and natural period.				
10	Repeat steps 7–9 for as many modes of vibration as required. For example, repeat up to j = 5 if 5 modes of vibration are considered.				

Step 3: Solve the equation of dynamic equilibrium for the MDOF system

The equation of dynamic equilibrium in the horizontal direction in a two-dimensional (2D) model can accordingly be expressed in terms of matrices and vectors, as shown by Equation (13.14), which is introduced in most textbooks on structural dynamics [13.1, 13.2].

$$[M]\{\ddot{u}(t)\} + [C]\{\dot{u}(t)\} + [K]\{u(t)\} = -\{r\}F(t) \tag{13.14}$$

where the three terms to the left-hand side of the equal sign represent the inertia, damping and elastic forces, respectively, whereas the term to the right-hand side of the equal sign is the product of the $\{r\}$ vector and the forcing function. Vector $\{r\}$ is used to define the location where the collision action is applied. For example, if the collision action is applied at the fifth lumped mass on the stick model (counted from the top as shown in the schematic diagram of Figure 13.5), the column vector $\{r\}$ is given by Equation (13.15), which shows that the only non-zero entry to the vector is on the fifth row. Equation (13.14) may also be re-written into the form of Equation (13.16) where the k-th lumped mass is subjected to the collision.

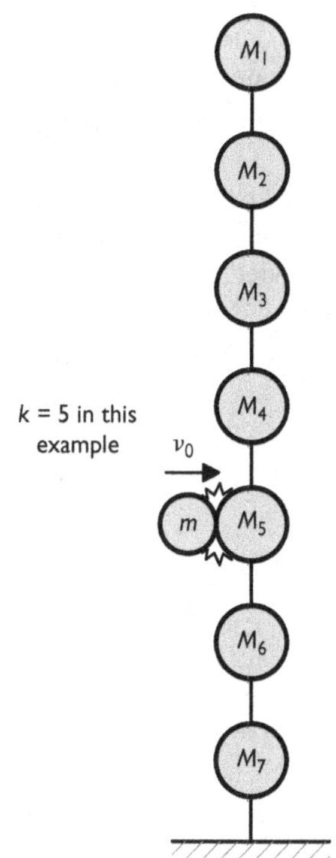

Figure 13.5 Multiple lumped mass stick model for impact response analyses.

$$\{r\} = \begin{Bmatrix} 0 \\ 0 \\ 0 \\ 0 \\ 1 \\ 0 \\ 0 \end{Bmatrix} \tag{13.15}$$

$$[M]\{\ddot{u}(t)\} + [C]\{\dot{u}(t)\} + [K]\{u(t)\} = -[M]\{r\}\frac{F(t)}{M_k} \tag{13.16}$$

where $F(t)$ is the forcing function of time, and M_k represents the kth lumped mass counting from the top down (e.g. $k = 5$ to represent the arrangement of Figure 13.5).

By the principle of modal superposition, the total response displacement time-history $\{u(t)\}$ of the stick model as described is resolved into n number of modes, as shown by Equation (13.17). The value of n is decided by the user depending on the significance of the contribution from each mode.

$$\{u(t)\} = \sum_{j=1}^{n} Y_j(t)\{\phi_j\} \tag{13.17}$$

where $Y_j(t)$ is the modal displacement for the j-th mode and $\{\phi_j\}$ is the corresponding mode shape vector, which defines the shape of deflection of the element for the j-th mode of vibration.

Consider the j-th mode of vibration, for example, the vector showing the displacement time-history of each node is represented by Equation (13.18a). Differentiating Equation (13.18a) once, and twice, with respect to t gives Equations (13.18b) and (13.18c), respectively:

$$\{u_j(t)\} = Y_j(t)\{\phi_j\} \tag{13.18a}$$

$$\{\dot{u}_j(t)\} = \dot{Y}_j(t)\{\phi_j\} \tag{13.18b}$$

$$\{\ddot{u}_j(t)\} = \ddot{Y}_j(t)\{\phi_j\} \tag{13.18c}$$

Substituting Equations (13.18a) to (13.18c) into Equation (13.16) and pre-multiplying both sides of the equation by $\{\phi_j\}^T$ gives Equation (13.19a), which can be re-written into the form of Equation (13.19b):

$$\begin{aligned}\{\phi_j\}^T[M]\{\phi_j\}\ddot{Y}_j(t) + \{\phi_j\}^T[C]\{\phi_j\}\dot{Y}_j(t) + \{\phi_j\}^T[K]\{\phi_j\}Y_j(t) \\ = -\{\phi_j\}^T[M]\{r\}\frac{F(t)}{M_k}\end{aligned} \tag{13.19a}$$

$$M_j \ddot{Y}_j(t) + C_j \dot{Y}_j(t) + K_j Y_j(t) = -L_j \frac{F(t)}{M_k} \tag{13.19b}$$

where M_j, C_j, K_j and L_j are the generalised mass, generalised damping coefficient, generalised stiffness and excitation factor, respectively, for the j-th mode, as defined by Equations (13.20a) to (13.20d):

$$M_j = \{\phi_j\}^T [M] \{\phi_j\} \tag{13.20a}$$

$$C_j = \{\phi_j\}^T [C] \{\phi_j\} \tag{13.20b}$$

$$K_j = \{\phi_j\}^T [K] \{\phi_j\} \tag{13.20c}$$

$$L_j = \{\phi_j\}^T [M] \{r\} \tag{13.20d}$$

Making the substitution that $C_j = 2\xi M_j \omega_j$ (as defined by Equation (2.4c) for a SDOF system, as presented in Chapter 2) into Equation (13.19b) and dividing every term by M_j gives Equation (13.21):

$$\ddot{Y}_j + 2\xi\omega_j \dot{Y}_j + \omega_j^2 Y_j = -\frac{L_j}{M_j} \frac{F(t)}{M_k} \tag{13.21}$$

where L_j/M_j is the participation factor for the j-th mode.

Step 4: Determine the displacement time-history for each mode of vibration

Given Equation (13.21) along with the findings from analysis of a SDOF lumped mass system as represented by Equations (13.5a) and (13.5b), the solution to Y_j can be expressed in the form of Equations (13.22a) and (13.22b), with $h_j(t)$ defined by Equation (13.23):

$$Y_j(t) = \frac{L_j}{M_j} \frac{J}{M_k} h_j(t - \tau) \tag{13.22a}$$

$$Y_j(t) = \frac{L_j}{M_j} v_k h_j(t - \tau) \tag{13.22b}$$

$$h_j(t - \tau) = \frac{1}{\omega_j} e^{-\xi\omega_j(t-\tau)} \sin \omega_j (t - \tau_i) \tag{13.23}$$

For a hard collision scenario, the amount of impulse that is transmitted to the target is defined by Equation (13.24a), which is in analogy to Equation (13.1) for dealing with a SDOF system. With a MDOF system, which is subjected to collision on one of its lumped masses at node k, the impulsive action that has been transmitted to the lumped mass through direct contact may be defined by Equation (13.24b). Equation (13.25) for estimating velocity (v_k) of the lumped mass at node k that is excited into motion by the collision is then derived by combining Equations (13.7), (13.24a) and (13.24b):

$$J = \lambda m v_2 \tag{13.24a}$$

$$J = M_k v_k \tag{13.24b}$$

$$v_k = v_0 \left(\frac{\lambda m}{M_k}\right)\left(\frac{1 + \text{COR}}{1 + \lambda}\right) \tag{13.25}$$

Step 5: Determine the overall displacement time-history of the MDOF system using modal superposition

The displacement time-history associated with each mode of vibration $u_j(t)$ may then be summed to give the total displacement time-history $x(t)$, as shown by Equation (13.26). Substituting Equation (13.22a) into Equation (13.18a) gives Equation (13.27a). The term J/M_k in Equation (13.27a) is replaced by v_k in Equation (13.27b):

$$\{x(t)\} = \sum_{j=1}^{n} \{u_j(t)\} \tag{13.26}$$

$$\begin{Bmatrix} x_1(t) \\ x_2(t) \\ x_3(t) \\ \vdots \\ x_{N_n}(t) \end{Bmatrix} = \sum_{j=1}^{n} \frac{L_j}{M_j} \frac{J}{M_k} h_j(t - \tau) \begin{Bmatrix} \phi_{1,j} \\ \phi_{2,j} \\ \phi_{3,j} \\ \vdots \\ \phi_{N_n,j} \end{Bmatrix} \tag{13.27a}$$

$$\begin{Bmatrix} x_1(t) \\ x_2(t) \\ x_3(t) \\ \vdots \\ x_{N_n}(t) \end{Bmatrix} = \sum_{j=1}^{n} \frac{L_j}{M_j} v_k h_j(t - \tau) \begin{Bmatrix} \phi_{1,j} \\ \phi_{2,j} \\ \phi_{3,j} \\ \vdots \\ \phi_{N_n,j} \end{Bmatrix} \tag{13.27b}$$

Equation (13.27a) gives the total displacement time-history for each discrete node when subjected to a single short-duration impulse. As for Equation (13.10), Equation (13.26a) can also be expanded by the use of the *Duhamel Integration Method*. The response to each of the individual discretised impulses can be summed, and modal contributions superposed, as shown in Equation (13.27c), to give the total displacement of the system responding to the collision.

$$\begin{Bmatrix} x_1(t) \\ x_2(t) \\ x_3(t) \\ \vdots \\ x_{N_n}(t) \end{Bmatrix} = \sum_{j=1}^{n} \sum_{i=1}^{N} \frac{L_j}{M_j} \frac{J_i}{M_k} h_j(t - \tau_i) \begin{Bmatrix} \phi_{1,j} \\ \phi_{2,j} \\ \phi_{3,j} \\ \vdots \\ \phi_{N_n,j} \end{Bmatrix} \tag{13.28}$$

where L_j is given by Equation (13.20d); M_j by Equation (13.20a); J_i/M_k can be replaced by v_k, which is defined by Equation (13.25); h_j is given by Equation (13.23) and mode shape vector by modal analysis as outlined in Table 13.1.

The utility of Equation (13.10) and Equation (13.27c) is to simulate the displacement time-history of a structural system responding to an impulse of finite duration. Such a finite duration impulse may be discretised into multiple short-duration impulses, which can then be input into Equation (13.27c) in which individual short-duration impulses are denoted by J_i. An example application is the collision of a vehicle on a barrier, which is connected to the side of the bridge deck on a bridge structure. The crumbling of the damaged vehicle on the barrier takes place over a period of time.

The advantage of Equation (13.27c) is that it is able to take any forcing function as input to simulate structural response behaviour. If the collision action delivers a short-duration impulse, Equation (13.27b) may be used instead. Step-by-step implementation of Equation (13.27b), or Equation (13.27c), in a spreadsheet program is presented in Table 13.2.

Implementation of the analysis procedure presented in this section for simulating the dynamic behaviour of a MDOF system in response to a pre-determined forcing function is demonstrated by use of worked examples in Sections 13.5.4 to 13.5.6.

Table 13.2 summarises the procedure for predicting the displacement of a SDOF, or MDOF, system responding to the impulse of a short, or finite, duration impulse. Calculations involved in dealing with the general case of a MDOF system, which is subject to a finite duration impulse are primarily based on the use of Equation (13.27c). The required operations for simulating system responses are summarised in Table 13.3.

13.4 SIMULATIONS OF CONTACT FORCING FUNCTION OF IMPACT

In Chapter 6, the fundamental principles concerning the modelling of collision actions localised at the point of contact were introduced for determining

Table 13.2 Step-by-step outline of simulating response of MDOF systems using the Duhamel Integration Method

Step	*Operation*	*Remarks*
1	Compute modal properties $\{\phi_j\}$ and T_j for as many modes as required.	Refer to Table 13.1.
2	Define $\{r\}$ based on the location subject to impulse.	
3	Calculate M_j and L_j.	Refer to Equations (13.20a) and (13.20d).
4	Declare time step interval (dt) and total duration (t_t) of simulated response. Time t ranges from 0 to t_t at an increment of dt.	Smaller dt gives higher resolution to the simulation.
	Steps 5 – 9	
Steps 5–9 differ for Equation (13.27b) and Equation (13.8). Apply Equation (13.27b) if (i) it is a hard impact scenario and (ii) input parameters m, v_0 and COR are available. Apply Equation (13.27c) if a digitised forcing function is available.		
	Equation (13.27b)	
5	Estimate generalised mass λm and λ.	Refer to Table 2.1.
6	Estimate v_k.	Refer to Equation (13.25).
7	Apply the following equation for each mode of vibration. $u_j(t) = \frac{1}{\omega_j} v_k e^{-\xi\omega_j t} \sin \omega_j t$	
8 – 9	N/A. Skip to step 10.	
	Equation (13.27c)	
5	Calculate total impulse J_t, i.e. area under the graph of the forcing function that the structure is subjected to.	
6	Declare the number of short-duration impulses N that the force time-history is discretised into. Time step between each impulse is then $d\tau = \frac{t_d}{N}$	t_d is the total duration of the forcing function. Higher N value gives better accuracy.
7	First impulse is taken to occur at $\tau_1 = 0.5d\tau$. For second impulse and onwards, $\tau_i = \tau_{i-1} + d\tau$. Search for the value of F_i at each time step τ_i from the forcing function.	

8	Determine the value of the short-duration impulse at each time step based on the force ratio. $J_i = \frac{F_i}{\sum_{i=1}^{N} F_i} \times J_t$	
9a	Apply the following equation for each short-duration impulse for each mode of vibration. $u_{i,j}(t) = \frac{J_i}{M_k \omega_j} e^{-\xi \omega_j (t-\tau_i)} \sin \omega_j (t - \tau_i)$	When $t < \tau$, $u_{i,j}(t) = 0$.
9b	Sum the response from each short-duration impulse. $u_j(t) = \sum_{i=1}^{N} u_{i,j}(t)$	Needs to be done for each mode of vibration.
	End of Steps 5 – 9	
10a	Apply the following equation for each mode of vibration. $x_{k,j}(t) = \frac{L_j}{M_j} \phi_{k,j} u_j(t)$	Applicable to all discretised nodes.
10b	Sum the responses from all modes of vibration. $x_k(t) = \sum_{j=1}^{n} x_{k,j}(t)$	Applicable to all discretised nodes.
11	Velocity and acceleration time-histories can also be computed by differentiating displacement time-histories twice. The following equations are to be applied at each time step. $v_k(t_i) = \frac{x_k(t_i) - x_k(t_{i-1})}{dt}$ $a_k(t_i) = \frac{v_k(t_i) - v_k(t_{i-1})}{dt}$	This step is optional. t_{i-1} and t_i represent two consecutive time steps, i.e. $t_i - t_{i-1} = dt$.

Table 13.3 Required operation for different systems and natures of impulse

System	*Impulse Duration*	*Required Operation*	*Remarks*
SDOF	Short	Equation (13.5a) or (13.5b)	–
	Finite	Table 13.2: step 4–9b	Omit subscript "j" and "k"
MDOF	Short	Table 13.2: step 1–4, 9a, 10a, 10b	Omit subscript "i"
	Finite	Table 13.2: full step	–

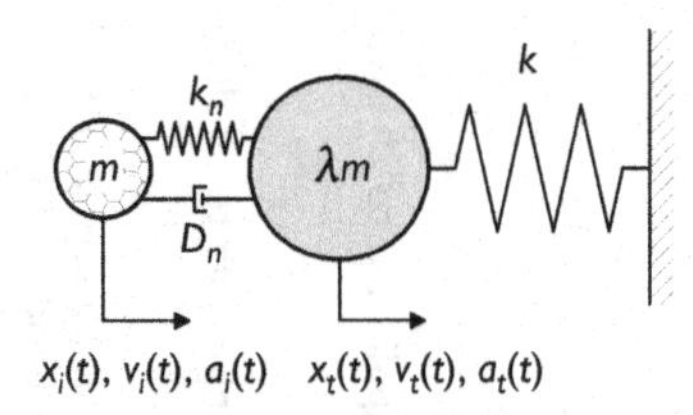

Figure 13.6 Two-degree-of-freedom lumped mass system.

maximum indentation and maximum contact force. This section follows from Chapter 6 to demonstrate the application of the Hunt and Crossley (H&C) model [13.3] in the form of a spring-connected lumped mass two-degree-of-freedom (2DOF) system, as illustrated by the schematic diagram of Figure 13.6.

Numerical simulation of the 2DOF system is aimed at determining the following:

i. Contact forcing function of time, $F(t)$; and
ii. Contact force–indentation, $F(t)$ vs δ, relationship.

A step-by-step outline of the operation details is listed in Table 13.4. Implementation of the simulation procedure in a Microsoft Excel spreadsheet is illustrated by a worked example at the end of the chapter (refer Section 13.5.6). This solution strategy was originally introduced in Sun et al. [13.4].

Implementation of the analysis procedure presented in this section for simulating the contact forcing function generated by a collision is demonstrated by use of a worked example in Section 13.5.7. A further example is presented in Section 13.5.8 to deal with a collision scenario in which the forcing function is not given. The solution, which is divided into three parts, involves integrating the simulation of the contact forcing function (in the pre-determined collision scenario) with the dynamic analysis of a MDOF system responding to the simulated forcing function.

Table 13.4 Step-by-step outline of the numerical simulation procedure to solve for forcing function of impact

Step	*Operation*	*Remarks*
1	Input parameters: v_o, m, λm, k, k_n, p and COR.	These parameters have been introduced in Chapter 6.
2	Calculate the value of D_n using Equation (6.11b): $D_n = (0.2p + 1.3)\left(\frac{1-\mathrm{COR}}{\mathrm{COR}}\right)\left(\frac{k_n}{\dot{\delta}_0}\right)$	$\dot{\delta}_0$ and v_0 can be used interchangeably in this context.
3	List all the variables that are functions of time. Time: t Velocity of the two lumped masses: $v_i(t)$ and $v_t(t)$ Displacement of the two lumped masses: $x_i(t)$ and $x_t(t)$ Relative displacement between the two lumped masses: $\delta(t)$ Relative velocity between the two lumped masses: $\dot{\delta}(t)$ Contact force between the two lumped masses: $F_c(t)$ Reaction force at the support of the rear lumped mass: $F_R(t)$ Accelerations of the two lumped masses: $a_i(t)$ and $a_t(t)$	
4	Declare time step interval dt.	Smaller dt gives higher resolution to the simulation.
5	Initiate all variables to zero at $t = 0$ and $i = 0$, except that $v_i(0) = v_0$.	i is a counter which starts at 0 and increases by 1 at every time step.
6	Apply the following equations at every time step from $i = 1$ onwards. $t(i) = t(i-1) + dt$ $v_i(i) = v_i(i-1) + a_i(i-1)dt$ $v_t(i) = v_t(i-1) + a_t(i-1)dt$ $x_i(i) = x_i(i-1) + v_i(i-1)dt$ $x_t(i) = x_t(i-1) + v_t(i-1)dt$ $\delta(i) = x_i(i) - x_t(i)$ $\dot{\delta}(i) = v_i(i) - v_t(i)$ $F_c(i) = k_n\delta(i)^p + D_n\delta(i)^p\dot{\delta}(i)$ $F_R(i) = kx_t(i)$ $a_i(i) = -\frac{F_c(i)}{m}$ $a_t(i) = \frac{F_c(i) - F_R(i)}{\lambda m}$	

13.5 WORKED EXAMPLES

13.5.1 Short-Duration Impulse Applied to a SDOF Lumped Mass System

Consider a SDOF lumped mass system with a mass of 10 tonnes, stiffness of 400 kN/m and is subject to an impulse of 50 kNs. A damping ratio of 5% is assumed. Simulate the displacement time-history of the system.

Solution

By the use of Equation (2.2e):

$$T = 2\pi\sqrt{\frac{M}{k}} = 2\pi\sqrt{\frac{10}{400}} = 1 \text{ s}$$

$$\omega_n = \frac{2\pi}{T} = \frac{2\pi}{1} = 6.3 \text{ rad/s}$$

From Equation (13.5a):

$$\begin{aligned} u(t) &= \frac{J}{M\omega_n} e^{-\xi\omega_n(t-\tau)} \sin \omega_n(t-\tau) \\ &= \frac{50}{10(6.3)} e^{-0.05(6.3)t} \sin 6.3t \\ &= 0.8e^{-0.315t} \sin 6.3t \end{aligned}$$

Plot $u(t) = 0.8e^{-0.315t} \sin 6.3t$ in Figure 13.7. Envelopes shown in Figure 13.7 represent the decay of amplitude with time.

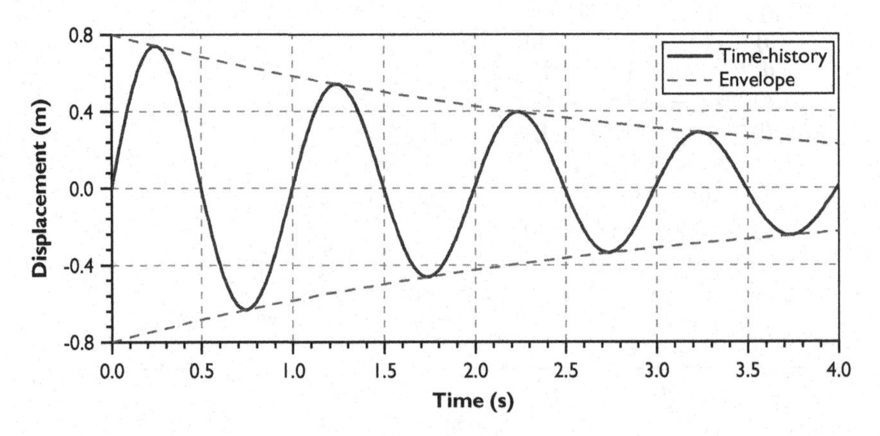

Figure 13.7 Displacement time history of a SDOF system subject to short-duration impulse.

13.5.2 Rectangular Impulse of Finite Duration Applied to a SDOF System

Consider a SDOF lumped mass system weighing 100 tonnes, with stiffness of 3,950 kN/m and a 5% viscous damping ratio. A rectangular-shaped impulse lasting for 0.5 seconds (as shown in Figure 13.8) is applied to excite the lumped mass into motion. Simulate the displacement time-history of the system responding to the impulse.

Solution

$$T = 2\pi\sqrt{\frac{M}{k}} = 2\pi\sqrt{\frac{100}{3950}} = 1 \text{ s}$$

$$\omega_n = \frac{2\pi}{T} = \frac{2\pi}{1} = 6.3 \text{ rad/s}$$

According to Table 13.3, steps 4 to 9b from Table 13.2 are applied for a SDOF system subject to an impulse of finite duration.

Step 4:
Take $dt = 0.03$ s and $t_t = 3$ s.

Step 5:

$$J_t = 100(0.5) = 50 \text{ kNs}$$

Step 6:
Discretise the forcing function into $N = 5$ short-duration impulses.

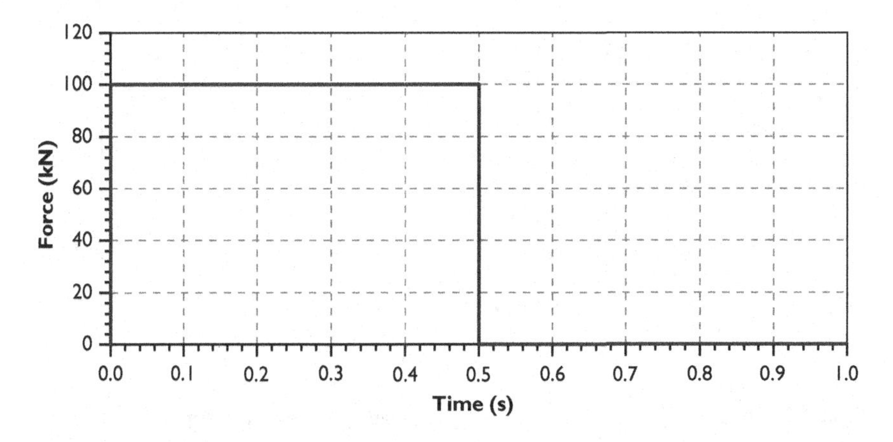

Figure 13.8 Rectangular-shaped impulse.

$$d\tau = \frac{t_d}{N} = \frac{0.5}{5} = 0.1 \text{ s}$$

Steps 7:
The first impulse occurs at $\tau_1 = 0.5d\tau = 0.5(0.1) = 0.05$ s. Other time steps are at increments of 0.1 s. The magnitude of F remains at 100 kN for all increments.

Step 8:
Determine the value of the individual short-duration impulses.

$$\sum_{i=1}^{N} F_i = 500 \text{ kN}$$

$$J_i = \frac{F_i}{\sum_{i=1}^{N} F_i} \times J_t = \frac{F_i}{500} \times 50 = 0.1F_i$$

Steps 7 and 8 are shown in Table 13.5 and illustrated graphically in Figure 13.9.

Table 13.5 Discretisation of a rectangular-shaped impulse

i	*1*	*2*	*3*	*4*	*5*
τ (s)	0.05	0.15	0.25	0.35	0.45
F (kN)	100	100	100	100	100
J (kNs)	10	10	10	10	10

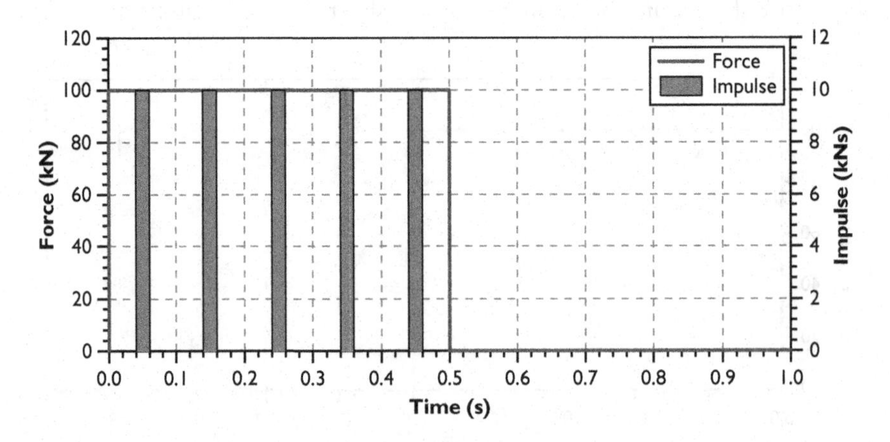

Figure 13.9 Resolving a rectangular impulse into five short-duration impulses.

Step 9:
Simulate the displacement time-history responding to each short-duration impulse.

$$u_i(t) = \frac{J_i}{M_k\omega_n} e^{-\xi\omega_n(t-\tau_i)} \sin\omega_n(t-\tau_i)$$

Note that when $t < \tau_i$, $u_i(t) = 0$.

Take the sum of responses to each short-duration impulse.

$$u(t) = \sum_{i=1}^{N=5} u_i(t) = u_1(t) + u_2(t) + u_3(t) + u_4(t) + u_5(t)$$

The first 10 time steps of simulation are shown in Table 13.6. All the values of $u_i(t)$ in Table 13.6 are presented in units of m. Note that the $u_4(t)$ and $u_5(t)$ entries in Table 13.6 are zeros since the table only covers a time period up to 0.27 s, which is prior to the $i = 4$ and $i = 5$ impulses being applied.

Plotting $u(t)$ versus t gives the displacement time-history of the responding system. It is shown that the displacement time-history so derived comes close to the prediction by Equations (13.28a) and (13.28b), which are the classical expressions that have been derived by calculus considering the dynamic equilibrium as presented in Chopra [13.1]. At time $t \le t_d$, the impulse is applied and at time $t > t_d$ the responding structure is undergoing natural vibration.

$$u(t) = u_{st}\left(1 - e^{-\xi\omega_n t}\cos\omega_n t\right) \quad \text{for } t \le t_d \quad (13.28a)$$

$$u(t) = u_{st} e^{-\xi\omega_n(t-t_d)}\left[\cos\omega_n(t-t_d) - e^{-\xi\omega_n t_d}\cos\omega_n t\right] \quad \text{for } t > t_d \quad (13.28b)$$

Table 13.6 Simulated displacement history for the first 10 time steps

t (s)	u_1 (t)	u_2 (t)	u_3 (t)	u_4 (t)	u_5 (t)	u (t)
0.000	0.000	0.000	0.000	0.000	0.000	0.000
0.030	0.000	0.000	0.000	0.000	0.000	0.000
0.060	0.001	0.000	0.000	0.000	0.000	0.001
0.090	0.004	0.000	0.000	0.000	0.000	0.004
0.120	0.007	0.000	0.000	0.000	0.000	0.007
0.150	0.009	0.000	0.000	0.000	0.000	0.009
0.180	0.011	0.003	0.000	0.000	0.000	0.014
0.210	0.013	0.006	0.000	0.000	0.000	0.019
0.240	0.014	0.008	0.000	0.000	0.000	0.022
0.270	0.015	0.010	0.002	0.000	0.000	0.027

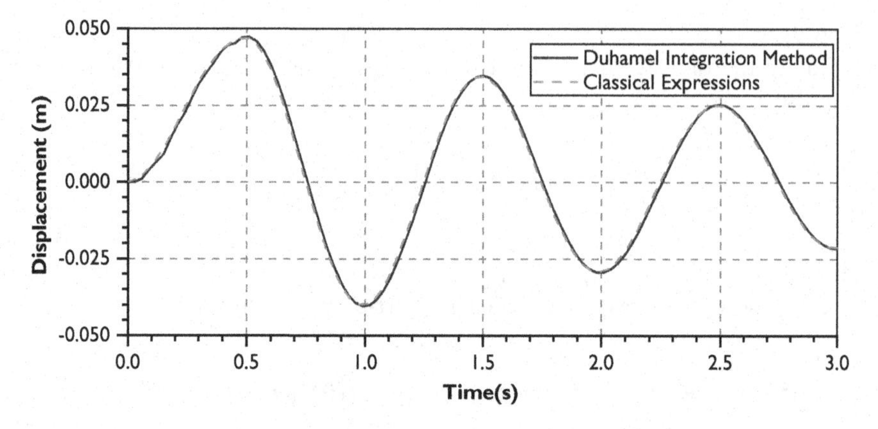

Figure 13.10 Simulation of response to a rectangular impulse of finite duration.

where u_{st} is the static displacement when subject to the maximum force of the forcing function in a static manner, as shown by Equation (13.29):

$$u_{st} = \frac{F_{\max}}{k} \tag{13.29}$$

Good consistencies between predictions from the simulation approach, and from the classical method, are shown in Figure 13.10.

13.5.3 Triangular Impulse of Finite Duration Applied to a SDOF System

Consider another SDOF system, which weighs 100 tonnes, has a stiffness of 3,950 kN/m and a viscous damping ratio of 1%. A triangular-shaped pulse lasting for 0.5 seconds, as shown in Figure 13.11, is applied to excite the

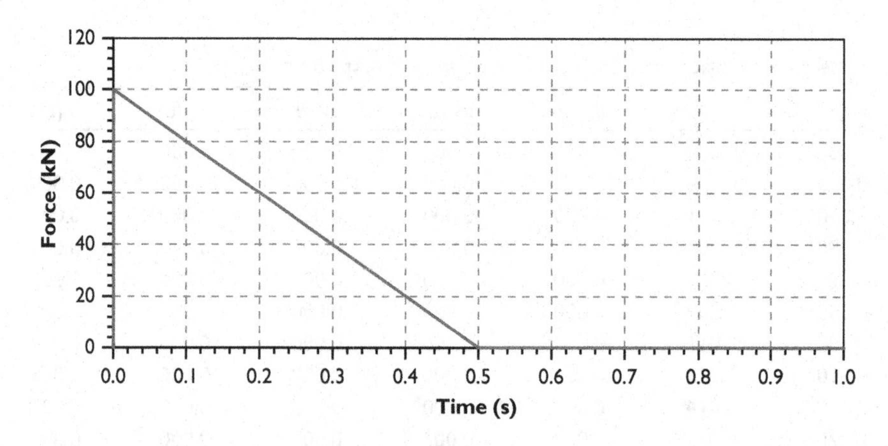

Figure 13.11 Triangular-shaped impulse.

lumped mass system. Simulate the displacement time-history of the responding system.

Solution

$$T = 2\pi\sqrt{\frac{M}{k}} = 2\pi\sqrt{\frac{100}{3950}} = 1 \text{ s}$$

$$\omega_n = \frac{2\pi}{T} = \frac{2\pi}{1} = 6.3 \text{ rad/s}$$

According to Table 13.3, steps 4 to 9b from Table 13.2 are applied.

Step 4:
Take $dt = 0.03$ s and $t_t = 3$ s.

Step 5:

$$J_t = \frac{1}{2}100(0.5) = 25 \text{ kNs}$$

Step 6:
Discretise the forcing function into $N = 5$ short-duration impulses.

$$d\tau = \frac{t_d}{N} = \frac{0.5}{5} = 0.1 \text{ s}$$

Step 7:
The first impulse is taken to occur at $\tau_1 = 0.5d\tau = 0.5(0.1) = 0.05$ s. Other time steps are at increments of 0.1 s. The value of F_i for all increments is read from Figure 13.11.

Step 8:
Determine the value of the individual short-duration impulses.

$$\sum_{i=1}^{N} F_i = 250 \text{ kN}$$

$$J_i = \frac{F_i}{\sum_{i=1}^{N} F_i} \times J_t = \frac{F_i}{250} \times 25 = 0.1F_i$$

Steps 7 and 8 are shown in Table 13.7, and illustrated graphically in Figure 13.12.

Table 13.7 Discretisation of a triangular-shaped impulse

i	1	2	3	4	5
τ (s)	0.05	0.15	0.25	0.35	0.45
F (kN)	90	70	50	30	10
J (kNs)	9	7	5	3	1

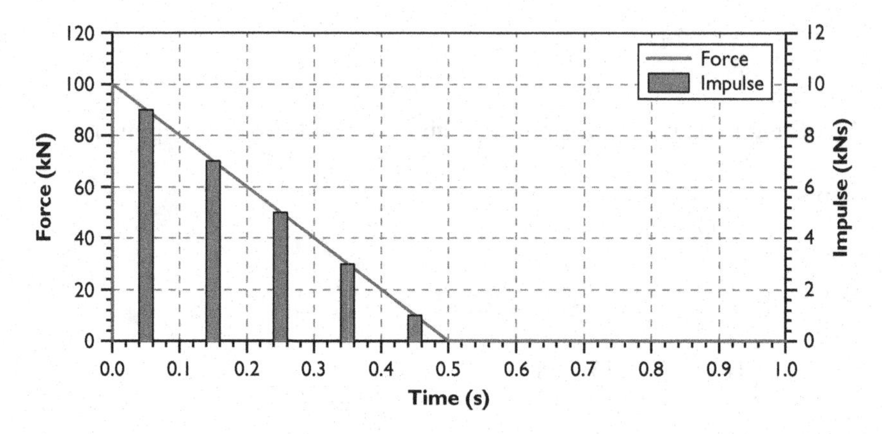

Figure 13.12 Resolving a triangular impulse into five short-duration impulses.

Step 9:
Simulate the displacement time-history in response to each short-duration impulse.

$$u_i(t) = \frac{J_i}{M_k \omega_n} e^{-\xi \omega_n (t-\tau_i)} \sin \omega_n (t - \tau_i)$$

Note that when $t < \tau_i$, $u_i(t) = 0$.

Take the sum of the response from each short-duration impulse.

$$u(t) = \sum_{i=1}^{N=5} u_i(t) = u_1(t) + u_2(t) + u_3(t) + u_4(t) + u_5(t)$$

The first 10 time steps of the simulation are shown in Table 13.8. The values of $u_i(t)$ for all i's are listed in Table 13.8, and are in units of m. Note that the $u_4(t)$ and $u_5(t)$ entries in Table 13.8 are zeros since the table only covers a time period up to 0.27 s, which is prior to the $i = 4$ and $i = 5$ impulses being applied.

Plotting $u(t)$ versus t gives the displacement time-history. Meanwhile, the classical solution to the displacement time-history in response to the triangular-shaped impulse is given by Equations (13.30a) and (13.30b).

Table 13.8 Simulated displacement history for the first 10 time steps

t (s)	u_1 (t)	u_2 (t)	u_3 (t)	u_4 (t)	u_5 (t)	u (t)
0.000	0.000	0.000	0.000	0.000	0.000	0.000
0.030	0.000	0.000	0.000	0.000	0.000	0.000
0.060	0.001	0.000	0.000	0.000	0.000	0.001
0.090	0.003	0.000	0.000	0.000	0.000	0.003
0.120	0.006	0.000	0.000	0.000	0.000	0.006
0.150	0.008	0.000	0.000	0.000	0.000	0.008
0.180	0.010	0.002	0.000	0.000	0.000	0.012
0.210	0.011	0.004	0.000	0.000	0.000	0.015
0.240	0.012	0.006	0.000	0.000	0.000	0.017
0.270	0.012	0.007	0.001	0.000	0.000	0.020

The slight decrease in the displacement amplitude as a result of the viscous damping is ignored in Equation (13.30a):

$$u(t) = u_{st}\left(1 - \frac{t}{t_d} - \cos\omega_n t + \frac{\sin\omega_n t}{\omega_n t_d}\right) \quad \text{for } t \leq t_d \tag{13.30a}$$

$$u(t) = e^{-\xi\omega_n(t-t_d)}\left[u(t_d)\cos\omega_n(t - t_d) + \frac{\dot{u}(t_d)}{\omega_n}\sin\omega_n(t - t_d)\right] \quad \text{for } t > t_d \tag{13.30b}$$

where $u(t_d)$ and $\dot{u}(t_d)$ are the displacement, and velocity, at $t = t_d$.

Good consistencies between predictions from the simulation approach, and from the classical method, of the system responding to the triangular-shaped impulse are shown in Figure 13.13.

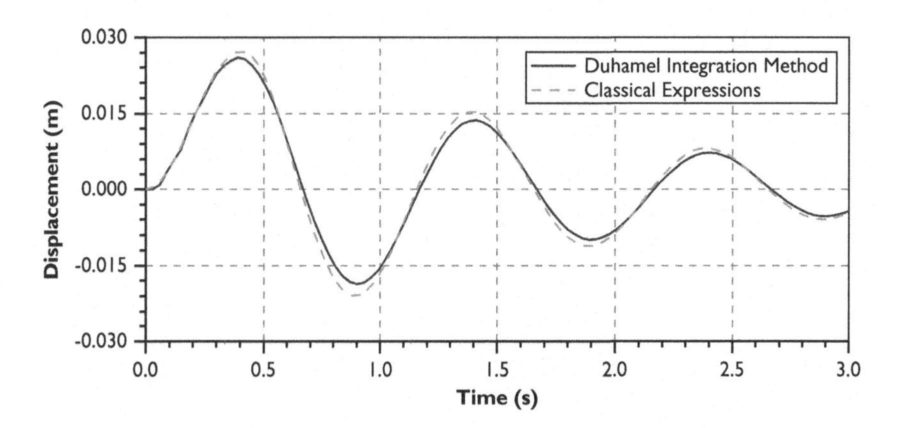

Figure 13.13 Simulation of response to a triangular impulse of finite duration.

13.5.4 Hard Collision Scenario on a Simply Supported Beam

A simply supported beam spanning 5 m is subject to the impact of a boulder weighing 170 kg, at a velocity of impact of 10 m/s at mid-span. A COR of 0.027 is assumed. The typical cross section of the beam is 0.25 m × 0.25 m and the Young's modulus is 200 GPa and density is 7,800 kg/m^3. Viscous damping ratio of 3% may be assumed. Simulate the displacement time-history at mid-span of the beam.

Solution

Total mass of beam $M = \rho BDL = 7800(0.25)(0.25)(5) = 2437.5$ kg

$$I = \frac{BD^3}{12} = \frac{0.25(0.25)^3}{12} = 3.255 \times 10^{-4}\ \text{m}^4$$

$$EI = 200 \times 10^9 (3.255 \times 10^{-4}) = 6.51 \times 10^7\ \text{Nm}$$

The simulation is carried out in steps as outlined in Table 13.2. Step 1 from Table 13.2 is to determine the modal properties of the beam as outlined in Table 13.1. Steps 1 to 10 (as shown in Table 13.1) are named with a decimal (as steps 1.1 to 1.10) in the following computation to avoid confusion with the step numbers from Table 13.2. Key parameters and matrices involved in the operation are presented in Table 13.9a to 13.9l.

Table 13.9a Offset of lumped masses from support on the left

k	*Offset (m)*
1	0.36
2	0.71
3	1.07
4	1.43
5	1.79
6	2.14
7	2.50
8	2.86
9	3.21
10	3.57
11	3.93
12	4.29
13	4.64

Table 13.9b Flexibility matrix [*F*] of the simply supported beam

a (m) / *x (m)*	*0.36*	*0.71*	*1.07*	⋯	*1.61*
0.36	2.8×10^{-9}	5.1×10^{-9}	6.8×10^{-9}	⋯	1.6×10^{-9}
0.71	5.1×10^{-9}	9.6×10^{-9}	13×10^{-9}	⋯	3.2×10^{-9}
1.07	6.8×10^{-9}	13×10^{-9}	18×10^{-9}	⋯	4.6×10^{-9}
⋮	⋮	⋮	⋮	⋱	⋮
4.64	1.6×10^{-9}	3.2×10^{-9}	4.6×10^{-9}	⋯	2.8×10^{-9}

Table 13.9c Stiffness matrix [*K*] of the simply supported beam

14.1×10^{9}	-13.6×10^{9}	5.9×10^{9}	⋯	1.05×10^{4}
-13.6×10^{9}	20.1×10^{9}	-15.2×10^{9}	⋯	-4.21×10^{4}
5.9×10^{9}	-15.2×10^{9}	20.5×10^{9}	⋯	15.8×10^{4}
⋮	⋮	⋮	⋱	⋮
1.05×10^{4}	-4.21×10^{4}	15.8×10^{4}	⋯	14×10^{9}

Table 13.9d Mass matrix [*M*] of the simply supported beam

187.5	0	0	⋯	0
0	187.5	0	⋯	0
0	0	187.5	⋯	0
⋮	⋮	⋮	⋱	⋮
0	0	0	⋯	187.5

Table 13.9e Dynamic matrix for the first mode of vibration $[D]_1$

5.3×10^{-7}	9.6×10^{-7}	12.7×10^{-7}	⋯	3.0×10^{-7}
9.6×10^{-7}	18.0×10^{-7}	24.4×10^{-7}	⋯	6.0×10^{-7}
12.7×10^{-7}	24.4×10^{-7}	34.0×10^{-7}	⋯	8.7×10^{-7}
⋮	⋮	⋮	⋱	⋮
3.0×10^{-7}	6.0×10^{-7}	8.7×10^{-7}	⋯	5.3×10^{-7}

Step 1.1:
The beam is idealised into a line element with 13 equally spaced lumped masses ($N_n = 13$), as shown in Figure 13.14. Values of their offset from the left-hand end of the beam are listed in Table 13.9a.

Step 1.2:
The deflection value (Δ) at offset x from the support on the left-hand end of a simply supported beam with length L is to be calculated. The beam is subject to point force, P, which is applied at a position as defined in

Table 13.9f Mode shape iterations

k	$\{\phi\}_0$	$\{\phi_I\}_I$	$\{\phi_I\}_2$	$\{\phi_I\}_3$	$\cdots$	$\{\phi_I\}_{10}$
1	1	0.226	0.223	0.223	$\cdots$	0.223
2	1	0.440	0.434	0.434	$\cdots$	0.434
3	1	0.629	0.624	0.623	$\cdots$	0.623
$\vdots$	$\vdots$	$\vdots$	$\vdots$	$\vdots$	$\ddots$	$\vdots$
13	1	0.226	0.223	0.223	$\cdots$	0.223
k		$\{\phi_I\}'_1$	$\{\phi_I\}'_2$	$\{\phi_I\}'_3$	$\cdots$	$\{\phi_I\}'_{10}$
1		1.48×10^{-5}	1.16×10^{-5}	1.15×10^{-5}	$\cdots$	1.15×10^{-5}
2		2.87×10^{-5}	2.25×10^{-5}	2.25×10^{-5}	$\cdots$	2.24×10^{-5}
3		4.11×10^{-5}	3.24×10^{-5}	3.23×10^{-5}	$\cdots$	3.23×10^{-5}
$\vdots$		$\vdots$	$\vdots$		$\ddots$	$\vdots$
13		1.48×10^{-5}	1.16×10^{-5}	1.15×10^{-5}	$\cdots$	1.15×10^{-5}
$\omega =$		139.11	139.02	139.02	$\cdots$	139.02

Table 13.9g Sweeping matrix for the second mode of vibration $[S]_2$

0.993	−0.014	−0.020	$\cdots$	−0.007
−0.014	0.973	−0.039	$\cdots$	−0.014
−0.020	−0.039	0.944	$\cdots$	−0.020
$\vdots$	$\vdots$	$\vdots$	$\ddots$	$\vdots$
−0.007	−0.014	−0.020	$\cdots$	0.993

Table 13.9h Dynamic matrix for the second mode of vibration $[D]_2$

1.62×10^{-7}	2.42×10^{-7}	2.46×10^{-7}	$\cdots$	-0.63×10^{-7}
2.42×10^{-7}	4.08×10^{-7}	4.40×10^{-7}	$\cdots$	-1.17×10^{-7}
2.46×10^{-7}	4.40×10^{-7}	5.28×10^{-7}	$\cdots$	-1.54×10^{-7}
$\vdots$	$\vdots$	$\vdots$	$\ddots$	$\vdots$
-0.63×10^{-7}	-1.17×10^{-7}	-1.54×10^{-7}	$\cdots$	1.62×10^{-7}

Figure 13.15. The values of Δ are given by Equations (13.31a) and (13.31b) based on principles of mechanics of materials [13.5]:

$$\Delta(x) = \frac{Pbx}{6EIL}(L^2 - b^2 - x^2) \qquad \text{for } x \leq a \tag{13.31a}$$

$$\Delta(x) = \frac{Pa(L - x)}{6EIL}(2Lx - a^2 - x^2) \quad \text{for } x > a \tag{13.31b}$$

Table 13.9i Mode shapes of the simply supported beam

k	$\{\phi_1\}$	$\{\phi_2\}$	$\{\phi_3\}$
1	0.223	−0.445	0.623
2	0.434	−0.802	0.975
3	0.623	−1.000	0.901
4	0.782	−1.000	0.434
5	0.901	−0.802	−0.223
6	0.975	−0.445	−0.782
7	1.000	0.000	−1.000
8	0.975	0.445	−0.782
9	0.901	0.802	−0.223
10	0.782	1.000	0.434
11	0.623	1.000	0.901
12	0.434	0.802	0.975
13	0.223	0.445	0.623

Table 13.9j Values of ω, f and T of a simply supported beam

Output	*Mode No. (j)*		
	1	*2*	*3*
ω (rad/s)	139	556	1251
f (Hz)	22	89	199
T (s)	0.045	0.011	0.005

Table 13.9k Values of M_j and L_j

Output	*Mode No. (j)*		
	1	*2*	*3*
M_j	1312.5	1380.9	1312.5
L_j	187.5	-8×10^{-8}	−187.5
L_j/M_j	0.14	-5.8×10^{-11}	−0.14

Table 13.9l First five time steps of the simulation (steps 7–10)

j =	*1*	*2*	*3*	*j =*	*1*	*2*	*3*	
ω_j =	*139*	*556*	*1251*	L_j/M_j =	*0.14*	-5.8×10^{-11}	*−0.14*	
				$\phi_{7,j}$ =	*1*	*0*	*−1*	
t (s)	u_1 *(t)*	u_2 *(t)*	u_3 *(t)*		$x_{7,1}$ *(t)*	$x_{7,2}$ *(t)*	$x_{7,3}$ *(t)*	x_7 *(t)*
0	0	0	0		0	0	0	0
0.0001	0.8	0.8	0.8		0.12	0.12	0	0.23
0.0002	1.6	1.6	1.6		0.23	0.23	0	0.46
0.0003	2.4	2.4	2.4		0.35	0.34	0	0.69
0.0004	3.3	3.1	3.2		0.47	0.44	0	0.91

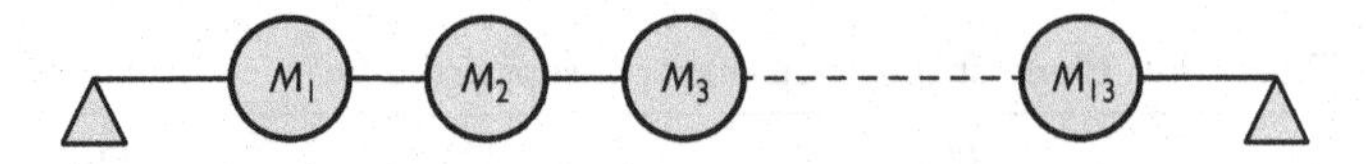

Figure 13.14 Simply supported beam idealised into MDOF system.

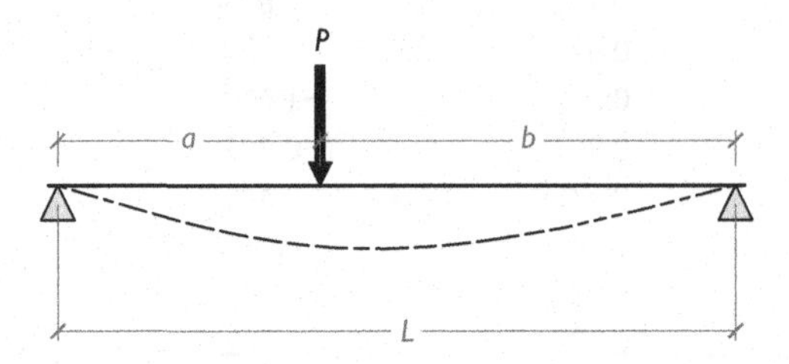

Figure 13.15 Simply supported beam subject to static force.

A unit load of $P = 1$ N is substituted into both Equations (13.31a) and (13.31b). Offset values as listed in Table 13.9a are substituted as a and x into both equations to construct the flexibility matrix $[F]$, as shown by Table 13.9b. For example, substitute $a = 0.36$ m and values of x ranging from 0.36 m to 4.64 m into Equations (13.31a) and (13.31b) to complete the first column of the matrix. The dimension of the flexibility matrix $[F]$ is 13 × 13.

Inversion of $[F]$ gives the stiffness matrix $[K]$, as shown in Table 13.9c, which is also of the same dimension.

Step 1.3:
Construct the mass matrix $[M]$, which is also of dimension 13 × 13, as shown in Table 13.9d. The diagonal entries to the matrix are the mass of each node as shown in the following:

$$\frac{M}{N_n} = \frac{2437.5}{13} = 187.5 \text{ kg}$$

Step 1.4:
Construct the dynamic matrix for the first mode of vibration as follows: $[D]_1 = [F][M]$.

The entries to $[D]_1$ are as listed in Table 13.9e.

Step 1.5:

i. Define an arbitrary mode shape for the first iteration, $\{\phi\}_0 = \{1\}$.
ii. Define $\{\phi_1\}'_1$ as follows: $\{\phi_1\}'_1 = [D]_1\{\phi\}_0$.
iii. Normalise the values of $\{\phi_1\}'_1$ to give $\{\phi_1\}_1$, i.e. divide each entry into $\{\phi_1\}'_1$ by the maximum value from $\{\phi_1\}'_1$ to give $\{\phi_1\}_1$.
iv. Estimate angular velocity by the use of the following equation:

$$\omega = \sqrt{\frac{\{\phi_1\}_1^T [K] \{\phi_1\}_1}{\{\phi_1\}_1^T [M] \{\phi_1\}_1}}$$

v. Repeat (ii)–(iv) for further iterations ($\{\phi_1\}_2, \{\phi_1\}_3 \ldots$) until convergence.

Ten iterations are used in this solution, as shown in Table 13.9f. The value of ω is shown to converge in the second iteration. Note that the mode shape iterations, as presented in Table 13.9f, are for the first mode of vibration ($j = 1$) only. Mode shape from the final iteration should be taken as the mode shape for the first mode of vibration, i.e. $\{\phi_1\} = \{\phi_1\}_{10}$.

Estimate the natural frequency and natural period of vibration using the converged value of ω.

$$f_1 = \frac{\omega_1}{2\pi} = \frac{139.02}{2\pi} = 22.1 \text{ Hz}$$

$$T_1 = \frac{1}{f_1} = \frac{1}{22.1} = 0.045 \text{ s}$$

Step 1.6:
Construct a 13 × 13 identity matrix.

$$[I] = \begin{bmatrix} 1 & & & & \\ & 1 & & & \\ & & 1 & & \\ & & & \ddots & \\ & & & & 1 \end{bmatrix}$$

Step 1.7:
Determine the sweeping matrix for the second mode of vibration by the following procedure:

$\{\phi_1\}$ is taken from the last column of Table 13.9f. $[S]_2$ so estimated is shown in Table 13.9g.

$$[S]_2 = [I] - \frac{\{\phi_1\}\{\phi_1\}^T [M]}{\{\phi_1\}^T [M] \{\phi_1\}}$$

Step 1.8:
Construct the dynamic matrix for the second mode of vibration by use of the following equation:

$$[D]_2 = [D]_1 [S]_2$$

Multiplying $[D]_1$ as presented in Table 13.9e by $[S]_2$ as presented in Table 13.9g gives $[D]_2$ as presented in Table 13.9h.

Step 1.9:
To determine mode shape and natural period of vibration for the second mode, repeat step 1.5 to construct a table similar to Table 13.9f. The only difference is the dynamic matrix, as $[D]_2$ is used instead of $[D]_1$. The first iteration then starts with $\{\phi_2\}'_1 = [D]_2\{\phi\}_0$.

Step 1.10:
Repeat steps 1.7 to 1.9 for the third mode of vibration ($j = 3$).

Results from steps 1.1 to 1.10 are shown in Table 13.9i, Figure 13.16 and Table 13.9j. Mode shapes as listed in Table 13.9i are plotted against the offset values from Table 13.9a to give results as shown in Figure 13.16. Values of ω, f and T are listed in Table 13.9j. Note that the results have been sorted based on the natural period of vibration, with the first mode having the highest T value. Note that the value of T decreases with increase in the modal number.

Once the modal properties are determined, the solution may proceed with step 2 from Table 13.2.

Step 2:
Define $\{r\}$ based on the location subject to impact. As the mid-span of the beam is subjected to impact, the seventh entry of $\{r\}$ is equal to 1, and the rest is equal to 0.

$$\{r\} = \begin{Bmatrix} 0 \\ 0 \\ 0 \\ 0 \\ 0 \\ 0 \\ 1 \\ 0 \\ 0 \\ 0 \\ 0 \\ 0 \\ 0 \end{Bmatrix}$$

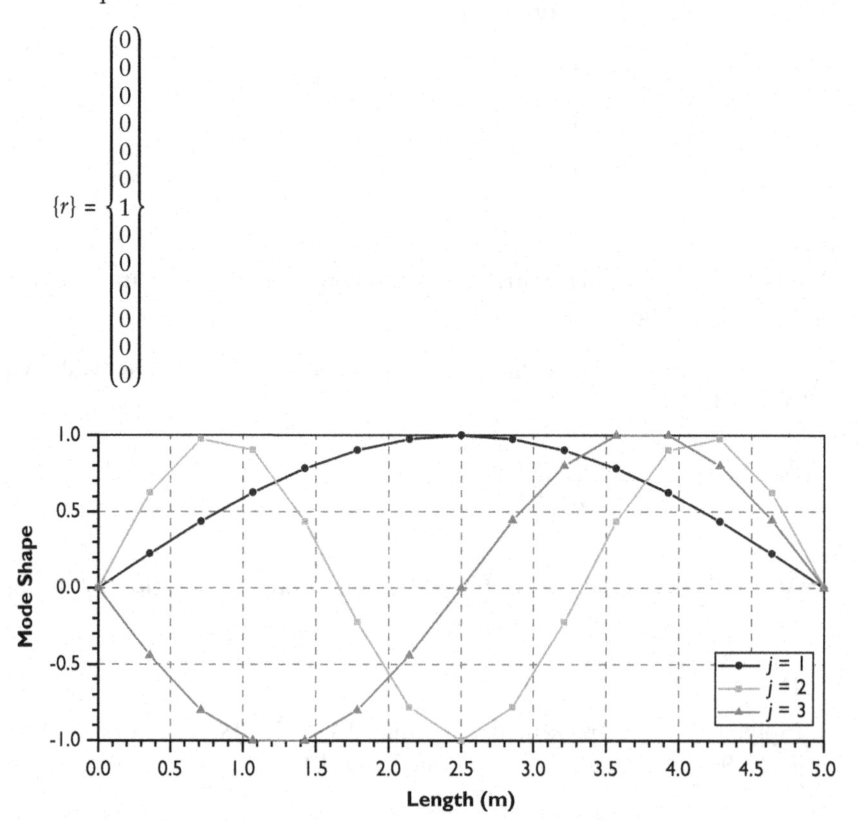

Figure 13.16 Mode shape of a simply supported beam.

Step 3:
By the use of Equations (13.20a) and (13.20d):

$$M_j = \{\phi_j\}^T [M] \{\phi_j\}$$

$$L_j = \{\phi_j\}^T [M] \{r\}$$

Substituting mode shapes into Table 13.9i gives estimates for M_j and L_j, as shown in Table 13.9k.

Step 4:
Declare time step interval $dt = 0.0001\ s$ and simulate response for $t_t = 0.05\ s$.

For steps 5 to 9, Equation (13.27b) is applied to simulate the response of the beam when subjected to a hard collision scenario where the values of parameters m, v_0 and COR are known.

Step 5:
According to Table 2.1, for a simply supported beam:

$$\lambda m = 0.5M = 0.5(2437.5) = 1218.75 \text{ kg}$$

$$\lambda = \frac{\lambda m}{m} = \frac{1218.75}{170} = 7.17$$

Step 6:
From step 1.3, $M_{k=7} = 187.5$ kg.

By the use of Equation (13.25):

$$v_k = v_0\left(\frac{\lambda m}{M_{k=7}}\right)\left(\frac{1 + \text{COR}}{1 + \lambda}\right) = 10\left(\frac{1218.75}{187.5}\right)\left(\frac{1 + 0.027}{1 + 7.17}\right) = 8.17 \text{ m/s}$$

Step 7:
Apply the following equation for each mode of vibration:

$$u_j(t) = \frac{1}{\omega_j} v_k e^{-\xi\omega_j t} \sin \omega_j t = \frac{1}{\omega_j}(8.17)e^{-0.03\omega_j t} \sin \omega_j t$$

Values of ω_j are taken from Table 13.9j and re-written into Table 13.9l. Sample calculated values of $u_j(t)$ are presented in the second to fourth columns of Table 13.9l.

Steps 8 and 9: N/A. Skip to step 10.

Step 10:
To simulate the mid-span response of the beam, take $k = 7$. Apply the following equation for each mode of vibration:

$$x_{7,j}(t) = \frac{L_j}{M_j}\phi_{7,j}u_j(t)$$

Values of L_j/M_j and $\phi_{7,j}$ are taken from Table 13.9k and 13.9i, respectively, and are re-written in Table 13.9l. Values of $u_j(t)$ are from the second to fourth columns of Table 13.9l as computed from step 7. Sample calculated values of $x_{7,j}(t)$ are shown in the $x_{7,1}(t)$, $x_{7,2}(t)$ and $x_{7,3}(t)$ columns of Table 13.9l.

Take the sum of responses associated with all three modes of vibration by the use of the following equation. Some sample results are shown in the last column of Table 13.9l. Note that all the displacement values in Table 13.9l have been multiplied by 1,000 to convert the unit from *m* to *mm*.

$$x_7(t) = \sum_{j=1}^{3} x_{7,j}(t)$$

Plotting $x_{7,1}(t)$, $x_{7,2}(t)$ and $x_{7,3}(t)$ against t gives the deflection time-histories which are associated with the first three modes of vibration (Figure 13.17(a)), whereas plotting $x_7(t)$ against t gives the total deflection time-history (Figure 13.17(b)). Note that these time-histories correspond to node no. 7, which is located at the mid-span of the beam. It is shown in Figure 13.17(a) that contributions from the second mode of vibration are negligible. This is

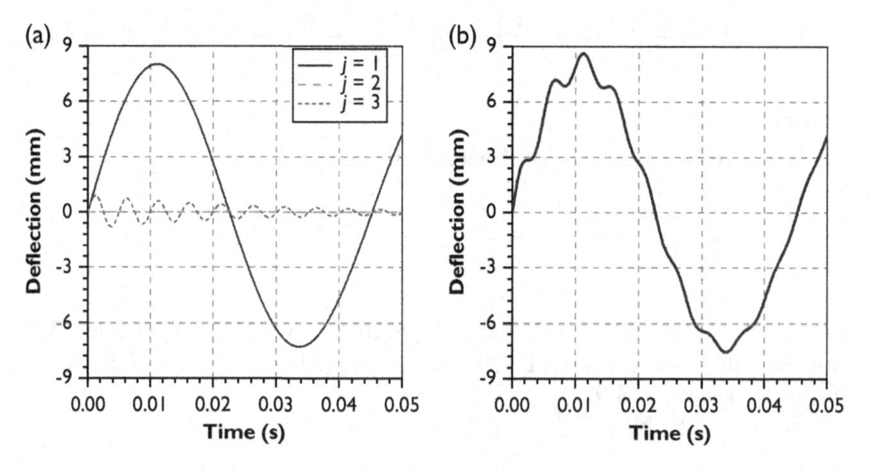

Figure 13.17 Simulated deflection time-history at mid-span of a simply supported beam: (a) deflection contributed by each mode, (b) total deflection.

because the mode shape of the second mode of vibration at node no. 7 is equal to 0 ($\phi_{7,2} = 0$), as shown in Table 13.9i and Figure 13.16.

The closed-form expression of Equation (3.7a), as derived in Chapter 3, has also been employed to estimate the maximum deflection of the simply supported beam for comparison purpose.

$$k = \frac{48EI}{L^3} = \frac{38(6.51 \times 10^7)}{5^3} = 25000 \text{ kN/m}$$

$$\Delta = \frac{mv_0}{\sqrt{mk}}\sqrt{\lambda\left(\frac{1 + \text{COR}}{1 + \lambda}\right)^2} = \frac{170(10)}{\sqrt{170(25000 \times 10^3)}}\sqrt{7.17\left(\frac{1 + 0.027}{1 + 7.17}\right)}$$
$$= 8.8 \text{ mm}$$

which is close to the maximum value from Figure 13.17(b) of 8.6 mm.

13.5.5 Impulsive Action Applied to a Cantilevered Wall

Consider a reinforced concrete barrier wall, which is fully fixed at the base that has a height of 1.5 m, width of 3 m ($B = 2h$) and thickness of 0.3 m. Its effective flexural of rigidity EI_{eff} is 93,200 kNm^2. The top of the barrier is subjected to a vehicular impact, which is defined by the impulse as shown in Figure 13.18. Simulate the deflection time-history at the top of the barrier.

Solution

The total mass of the beam is $M = \rho BDL = 2400(1.5)(3)(0.3) = 3240$ kg.

As for the example presented in Section 13.5.4, the modal properties of the wall are first determined by the use of steps as outlined in Table 13.1. Again, steps 1 to 10 from Table 13.1 are named as steps 1.1 to 1.10 in the following computation to differentiate the step numbers from Table 13.2.

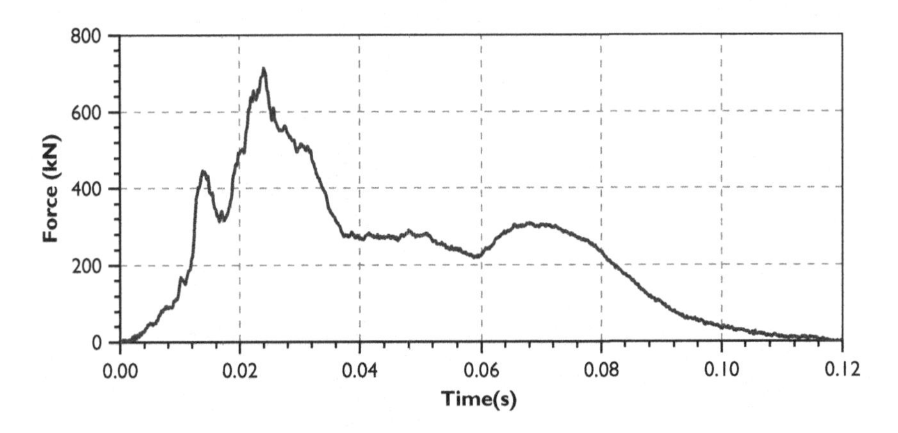

Figure 13.18 Forcing function representing a vehicular impact.

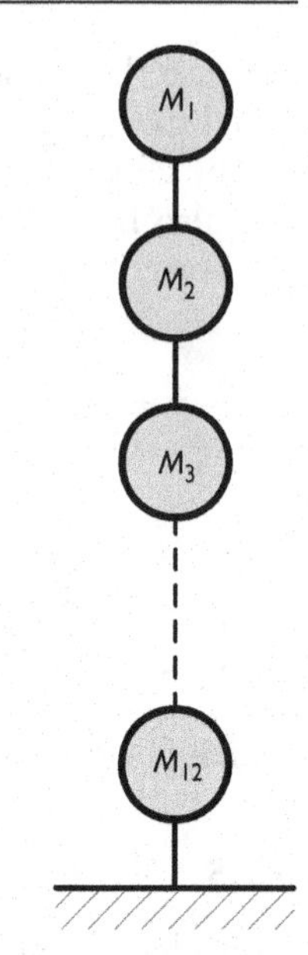

Figure 13.19 Cantilevered wall idealised into MDOF system.

Step 1.1:
The barrier is idealised into a line element with 12 equally spaced lumped masses (N_n = 12), as shown in Figure 13.19. Values of their offset from the top of the barrier are listed in Table 13.10a.

Step 1.2:
The deflection value (Δ) at offset x from the top of a cantilever of length L when subject to point force P is to be determined. The point force is applied

Table 13.10a Offset of lumped masses from the top end

k	*Offset (m)*
1	0
2	0.125
3	0.25
4	0.375
5	0.5
6	0.625
7	0.75
8	0.875
9	1
10	1.125
11	1.25
12	1.375

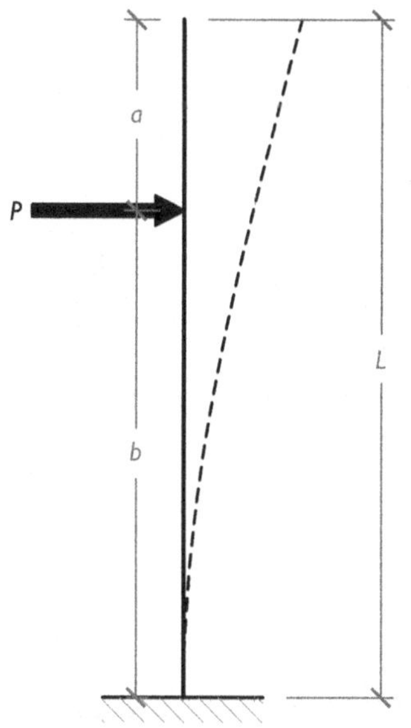

Figure 13.20 Cantilever subject to static force.

at a position as defined in Figure 13.20. Values of Δ are to be found by the use of Equations (13.32a) and (13.32b), which are based on principles of mechanics of materials [13.5]:

$$\Delta(x) = \frac{Pb^2}{6EI}(3L - 3x - b) \qquad \text{for } x \leq a \tag{13.32a}$$

$$\Delta(x) = \frac{P(L-x)^2}{6EI}(3b - L + x) \quad \text{for } x > a \tag{13.32b}$$

A unit load of $P = 1$ N is substituted into both Equations (13.32a) and (13.32b). The offset values as listed in Table 13.10a are substituted as a and x into both equations to construct the flexibility matrix [F], as shown by Table 13.10b. For instance, substitute $a = 0$ m and values of x ranging from

Table 13.10b Flexibility matrix [F] of the barrier

x (m) \ *a (m)*	*0*	*0.125*	*0.25*	⋯	*1.375*
0	12.1×10^{-9}	10.6×10^{-9}	9.1×10^{-9}	⋯	1.2×10^{-10}
0.125	10.6×10^{-9}	9.3×10^{-9}	8.0×10^{-9}	⋯	1.1×10^{-10}
0.25	9.1×10^{-9}	8.0×10^{-9}	7.0×10^{-9}	⋯	1.0×10^{-10}
⋮	⋮	⋮	⋮	⋱	⋮
1.375	1.2×10^{-10}	1.1×10^{-10}	1.0×10^{-10}	⋯	7.0×10^{-12}

Table 13.10c Stiffness matrix [K] of the barrier

7.7×10^{10}	-17.4×10^{10}	12.3×10^{10}	$\cdots$	-9.4×10^{5}
-17.4×10^{10}	47.1×10^{10}	-45.3×10^{10}	$\cdots$	5.6×10^{6}
12.3×10^{10}	-45.3×10^{10}	66.9×10^{10}	$\cdots$	-2.3×10^{7}
$\vdots$	$\vdots$	$\vdots$	$\ddots$	$\vdots$
-9.4×10^{5}	5.6×10^{6}	-2.3×10^{7}	$\cdots$	9.0×10^{11}

Table 13.10d Mass matrix [M] of the barrier

270	0	0	$\cdots$	0
0	270	0	$\cdots$	0
0	0	270	$\cdots$	0
$\vdots$	$\vdots$	$\vdots$	$\ddots$	$\vdots$
0	0	0	$\cdots$	270

0 m to 1.375 m into Equations (13.32a) and (13.32b) to complete entries to the first column of the matrix. The size of $[F]$ is 12 × 12.

The inversion of $[F]$ gives a stiffness matrix $[K]$, as shown in Table 13.10c.

Step 1.3:
Construct the mass matrix $[M]$ of dimensions 12 × 12, as shown in Table 13.10d. Diagonal entries to the matrix are populated with the mass of each node as shown in the following:

$$\frac{M}{N_n} = \frac{3240}{12} = 270 \text{ kg}$$

Step 1.4:
Construct the Dynamic Matrix for the first mode of vibration based on the relationship $[D]_1 = [F][M]$. The $[D]_1$ matrix is as shown in Table 13.10e.

Step 1.5

i. Define an arbitrary mode shape for the first iteration, $\{\phi\}_0 = \{1\}$.
ii. Define $\{\phi_1\}'_1$ as $\{\phi_1\}'_1 = [D]_1 \{\phi\}_0$.

Table 13.10e Dynamic matrix for the first mode of vibration $[D]_1$

3.26×10^{-6}	2.85×10^{-6}	2.45×10^{-6}	$\cdots$	3.30×10^{-8}
2.85×10^{-6}	2.51×10^{-6}	2.17×10^{-6}	$\cdots$	3.02×10^{-8}
2.45×10^{-6}	2.17×10^{-6}	1.89×10^{-6}	$\cdots$	2.74×10^{-8}
$\vdots$	$\vdots$	$\vdots$	$\ddots$	$\vdots$
3.30×10^{-8}	3.02×10^{-8}	2.74×10^{-8}	$\cdots$	1.89×10^{-9}

iii. Normalise entries to $\{\phi_1\}'_1$ to give $\{\phi_1\}_1$, i.e. divide values of the entries to $\{\phi_1\}'_1$ by the maximum value from $\{\phi_1\}'_1$ to give $\{\phi_1\}_1$.

iv. Estimate angular velocity by the use of the following equation:

$$\omega = \sqrt{\frac{\{\phi_1\}_1^T[K]\{\phi_1\}_1}{\{\phi_1\}_1^T[M]\{\phi_1\}_1}}$$

v. Repeat (ii) to (iv) for further iterations ($\{\phi_1\}_2$, $\{\phi_1\}_3$ ...) until convergence.

Ten iterations are used in this solution, as shown in Table 13.10f. The value of ω is shown to converge in the second iteration. Note that mode shape iterations of Table 13.10f are for the first mode of vibration ($j = 1$) only. The mode shape as calculated from the final iteration should be taken as the mode shape for the first mode of vibration, i.e. $\{\phi_1\} = \{\phi_1\}_{10}$.

Estimate natural frequency and natural period of vibration using the converged value of ω.

$$f_1 = \frac{\omega_1}{2\pi} = \frac{299.51}{2\pi} = 47.7 \text{ Hz}$$

$$T_1 = \frac{1}{f_1} = \frac{1}{22.1} = 0.021 \text{ s}$$

Table 13.10f Mode shape iterations

k	$\{\phi_0\}$	$\{\phi_1\}_1$	$\{\phi_1\}_2$	$\{\phi_1\}_3$	...	$\{\phi_1\}_{10}$
1	1	1.000	1.000	1.000	...	1.000
2	1	0.887	0.884	0.884	...	0.884
3	1	0.775	0.768	0.768	...	0.768
⋮	⋮	⋮	⋮	⋮	⋱	⋮
12	1	0.226	0.223	0.223	...	0.223
k		$\{\phi_1\}'_1$	$\{\phi_1\}'_2$	$\{\phi_1\}'_3$	...	$\{\phi_1\}'_{10}$
1		1.63×10^{-5}	1.13×10^{-5}	1.12×10^{-5}	...	1.11×10^{-5}
2		1.45×10^{-5}	1.00×10^{-5}	0.99×10^{-5}	...	0.99×10^{-5}
3		1.27×10^{-5}	0.87×10^{-5}	0.86×10^{-5}	...	0.86×10^{-5}
⋮		⋮	⋮		⋱	⋮
12		2.09×10^{-7}	1.30×10^{-7}	1.28×10^{-7}	...	1.28×10^{-7}
$\omega =$		300.70	299.51	299.51	...	299.51

Step 1.6:
Define a 12 × 12 identity matrix:

$$[I] = \begin{bmatrix} 1 & & & & \\ & 1 & & & \\ & & 1 & & \\ & & & \ddots & \\ & & & & 1 \end{bmatrix}$$

Step 1.7:
Determine the sweeping matrix for the second mode of vibration by the use of the following equation. The mode shape vector $\{\phi_1\}$ is taken from the last column of Table 13.10f. $[S]_2$ is estimated as shown in Table 13.10g.

$$[S]_2 = [I] - \frac{\{\phi_1\}\{\phi_1\}^T[M]}{\{\phi_1\}^T[M]\{\phi_1\}}$$

Step 1.8:
Construct the dynamic matrix for the second mode of vibration by the use of the following equation:

$$[D]_2 = [D]_1[S]_2$$

Multiplying $[D]_1$ from Table 13.10e by $[S]_2$ from Table 13.10g gives $[D]_2$, as shown in Table 13.10h.

Table 13.10g Sweeping matrix for the second mode of vibration $[S]_2$

0.714	−0.253	−0.220	⋯	−0.003
−0.253	0.776	−0.194	⋯	−0.003
−0.220	−0.194	0.831	⋯	−0.003
⋮	⋮	⋮	⋱	⋮
−0.003	−0.003	−0.003	⋯	1.000

Table 13.10h Dynamic matrix for the second mode of vibration $[D]_2$

1.62×10^{-7}	2.42×10^{-7}	2.46×10^{-7}	⋯	-0.63×10^{-7}
2.42×10^{-7}	4.08×10^{-7}	4.40×10^{-7}	⋯	-1.17×10^{-7}
2.46×10^{-7}	4.40×10^{-7}	5.28×10^{-7}	⋯	-1.54×10^{-7}
⋮	⋮	⋮	⋱	⋮
-0.63×10^{-7}	-1.17×10^{-7}	-1.54×10^{-7}	⋯	1.62×10^{-7}

Step 1.9:
To determine the mode shape and natural period for the second mode of vibration, repeat step 1.5 to construct a table similar to Table 13.10f. The only difference is the dynamic matrix, $[D]_2$, is used instead of $[D]_1$. The first iteration then starts with $\{\phi_2\}'_1 = [D]_2\{\phi\}_0$.

Step 1.10:
Repeat steps 1.7 to 1.9 for the third mode of vibration (j = 3).

Results derived from steps 1.1 to 1.10 are shown in Table 13.10i, Figure 13.21 and Table 13.10j. The mode shapes as listed in Table 13.10i

Table 13.10i Mode shapes of the barrier

k	$\{\phi_{j=1}\}$	$\{\phi_{j=2}\}$	$\{\phi_{j=3}\}$
1	1.000	−1.000	0.954
2	0.884	−0.536	0.141
3	0.768	−0.094	−0.517
4	0.654	0.297	−0.864
5	0.542	0.606	−0.817
6	0.435	0.805	−0.421
7	0.335	0.883	0.161
8	0.243	0.840	0.702
9	0.163	0.695	1.000
10	0.095	0.485	0.959
11	0.044	0.259	0.631
12	0.011	0.076	0.213

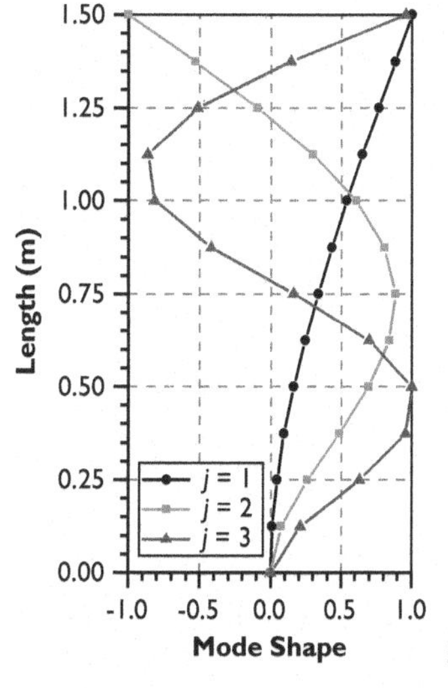

Figure 13.21 Mode shapes of a cantilevered wall.

Table 13.10j Values of ω, *f* and *T* of a cantilevered wall

Output	*Mode No. (j)*		
	1	2	*3*
ω (rad/s)	300	1884	5292
f (Hz)	48	300	842
T (s)	0.0210	0.0033	0.0012

are plotted against the offset values from Table 13.10a to give the mode shape profiles as presented in Figure 13.21. Values of ω, f and T are as listed in Table 13.10j.

Once the modal properties are determined, the solution may proceed with step 2 from Table 13.2.

Step 2:
Define the $\{r\}$ vector to indicate the location which is subject to collision. As the top of the barrier is subjected to the collision, the first entry of $\{r\}$ is equal to 1, and the rest are equal to 0.

$$\{r\} = \begin{Bmatrix} 1 \\ 0 \\ 0 \\ 0 \\ 0 \\ 0 \\ 0 \\ 0 \\ 0 \\ 0 \\ 0 \\ 0 \end{Bmatrix}$$

Step 3:
By the use of Equations (13.20a) and (13.20d):

$$M_j = \{\phi_j\}^T [M] \{\phi_j\}$$

$$L_j = \{\phi_j\}^T [M] \{r\}$$

Applying both equations by taking mode shapes from Table 13.10i gives the estimates shown in Table 13.10k.

Step 4:
Define the time-step interval $dt = 0.001\ s$ and simulate the response for $t_t = 0.1\ s$.

Table 13.10k Values of M_j and L_j

Output	*Mode No. (j)*		
	1	*2*	*3*
M_j	270	−270	258
L_j	943	1263	1531
L_j/M_j	0.286	−0.214	0.168

For steps 5 to 9, Equation (13.27c) is applied to incorporate the available forcing function (refer Figure 13.18).

Step 5:
The area under the graph of Figure 13.18 may be calculated by the use of the following equation:

$$J_t = \sum_{i=1}^{N_{td}} \left(\frac{F_i + F_{i+1}}{2} \right) \Delta t$$

It is found that $J_t = 26.7\ kNs$.

Step 6:
Discretise the forcing function into $N = 80$ short-duration impulses.

$$d\tau = \frac{t_d}{N} = \frac{0.12}{80} = 0.0015 \text{ s}$$

Step 7:
Take the first impulse to occur at $\tau_1 = 0.5d\tau = 0.5(0.0015) = 0.00075$ s. Other time steps are at increments of 0.0015 s. The value of F_i at each time step is read from Figure 13.18.

Step 8:
Determine the value of a short-duration impulse at each time step:

$$\sum_{i=1}^{N} F_i = 17880 \text{ kN}$$

$$J_i = \frac{F_i}{\sum_{i=1}^{N} F_i} \times J_t = \frac{F_i}{17880} \times 26.7 = 0.0015F_i$$

Steps 7 and 8 are shown in Table 13.10l and illustrated graphically in Figure 13.22.

Table 13.10l Impulse discretisation of an irregular impulse

i	1	2	3	...	80
τ (s)	0.00075	0.00225	0.00375	...	0.11925
F (kN)	4.88	14.58	25.63	...	1.70
J (kNs)	0.0073	0.0218	0.0384	...	0.0026

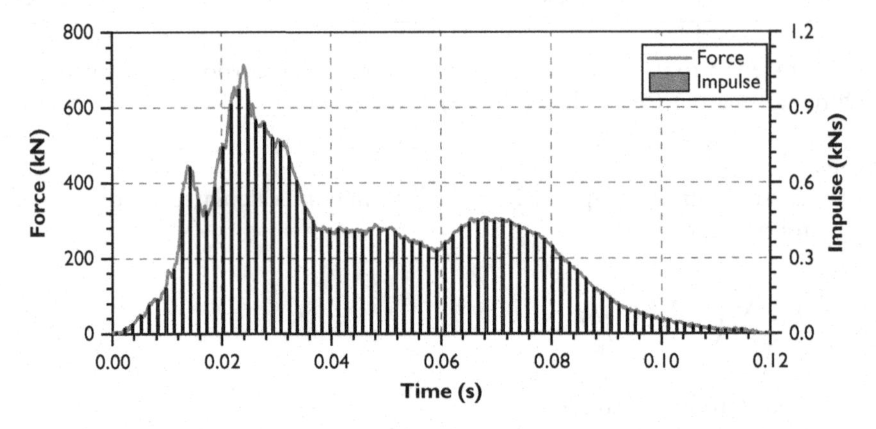

Figure 13.22 Resolving an irregular impulse into 80 short-duration impulses.

Step 9:
Simulate the displacement time-history for each short-duration impulse:

$$u_{i,j}(t) = \frac{J_i}{M_{k=5}\omega_j} e^{-\xi\omega_j(t-\tau_i)} \sin \omega_j (t - \tau_i)$$

For example, input parameters to specify the third short-duration impulse for the first mode of vibration are $J_3 = 0.0384$ kNs, $\tau_3 = 0.00375$ s and $\omega_1 = 300$ rad/s. As there are 80 short-duration impulses, the equations listed previously are to be applied 80 times for each mode of vibration. The results of the response are presented in a tabular format, as shown by Table 13.10m.

Table 13.10m First three time steps of $u_{i,j}$ (t) and u_j (t) (step 9)

	Mode 1, ω_1 = 300 rad/s				Mode 2, ω_2 = 1,884 rad/s				Mode 3, ω_3 = 5,292 rad/s			
t (s)	$u_{1,1}$	...	$u_{80,1}$	u_1	$u_{1,2}$	...	$u_{80,2}$	u_2	$u_{1,3}$	...	$u_{80,3}$	u_3
0	0	...	0	0	0	...	0	0	0	...	0	0
0.001	0.007	...	0	0.007	0.007	...	0	0.007	0.005	...	0	0.005
0.002	0.033	...	0	0.033	0.010	...	0	0.010	0.002	...	0	0.002

For each mode of vibration, the summation of the displacement to the individual short-duration impulses are shown as follows:

$$u_j(t) = \sum_{i=1}^{80} u_{i,j}(t) = u_{1,j}(t) + u_{2,j}(t) + u_{3,j}(t) + \cdots + u_{80,j}(t)$$

The summed time-histories are listed in the last column in Table 13.10m for each mode of vibration. Note that all the displacement values listed in Table 13.10m are in units of mm.

Step 10:
To simulate the response at the top of the barrier, take $k = 1$. Apply the following equation for each mode of vibration:

$$x_{1,j}(t) = \frac{L_j}{M_j}\phi_{1,j} u_j(t)$$

Values of L_j/M_j and $\phi_{1,j}$ are taken from Table 13.10k and 13.10i, respectively, and are re-written in Table 13.10n. Values of $u_j(t)$ are from the second to fourth columns of Table 13.10m, as computed in step 9. The sample calculated values of $x_{1,j}(t)$ are shown in the $x_{1,1}(t)$, $x_{1,2}(t)$ and $x_{1,3}(t)$ columns of Table 13.10n.

Take the sum of the responses from all three modes of vibration by the use of the equation following. Some sample results are shown in the last column of Table 13.10n. Note that all the displacement values in Table 13.10n are in units of mm.

$$x_1(t) = \sum_{j=1}^{3} x_{1,j}(t)$$

Plotting $x_{1,1}(t)$, $x_{1,2}(t)$ and $x_{1,3}(t)$ against t gives the deflection time-histories of the first three modes of vibration (Figure 13.23(a)), whereas plotting $x_1(t)$ against t gives the total deflection time-history (Figure 13.23(b)). Note that

Table 13.10n First five time steps of simulation of x_1 (t) (step 10)

j =	*1*	*2*	*3*	
L_j/M_j =	*0.286*	*−0.214*	*0.168*	
$\phi_{1,j}$ =	*1.000*	*−1.000*	*0.954*	
t (s)	$x_{1,1}$ (t)	$x_{1,2}$ (t)	$x_{1,3}$ (t)	x_1 (t)
0	0	0	0	0
0.0001	0.0019	0.0014	0.0008	0.0041
0.0002	0.0095	0.0022	0.0003	0.0119
0.0003	0.0334	0.0063	−0.0023	0.0374
0.0004	0.0702	0.0054	0.0038	0.0794

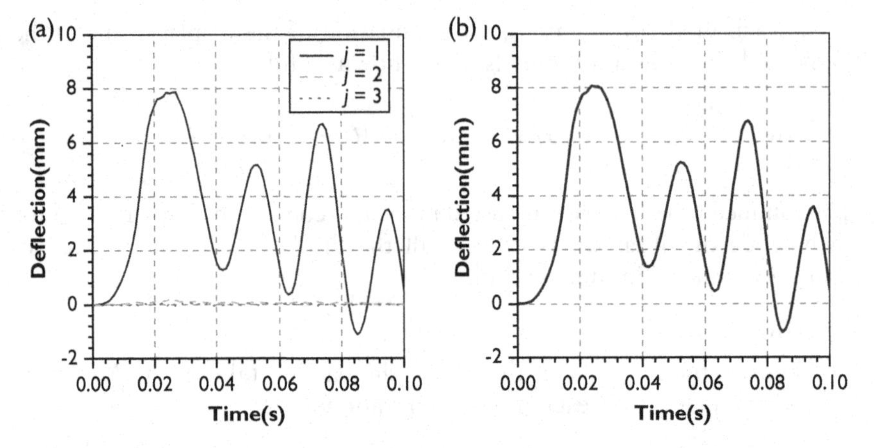

Figure 13.23 Simulated deflection time-history at the top of a barrier: (a) deflection contributed by each mode, (b) total deflection.

these time-histories correspond to node no. 1, which is located at the top of the barrier. The first mode of vibration is shown to dominate the response.

By repeating step 10 for all the other nodes, the deflection profile of the barrier at maximum deflection can be obtained, as shown in Figure 13.24. The only parameter that needs to be varied is $\phi_{k,j}$, which is readily available from Table 13.10i.

13.5.6 Collision Action on a Cantilever Wall

In the considered boulder impact scenario (first introduced in Section 8.5), a spherical granite boulder of 0.8 m in diameter (with density of 2,700 kg/m^3)

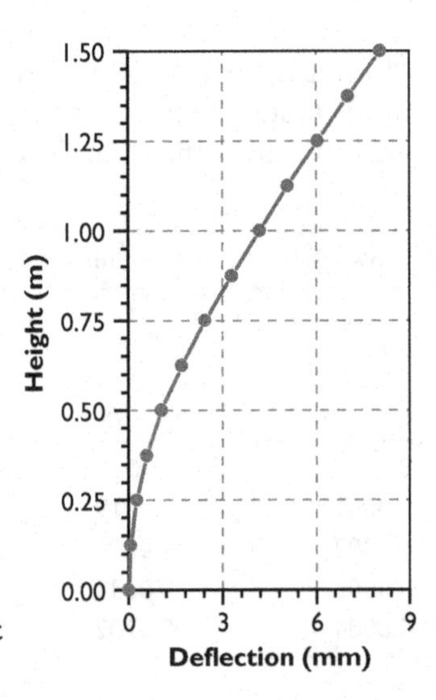

Figure 13.24 Deflection profile of the barrier at maximum deflection.

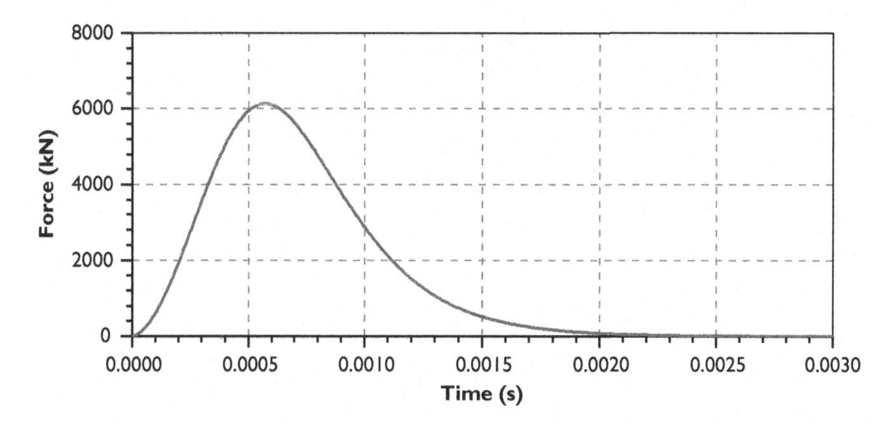

Figure 13.25 Contact force time-history delivered by a spherical granite boulder to a cantilever wall.

struck a RC barrier at an impact velocity of 7 m/s. The barrier was 3 metres tall, 6 metres long and 800 mm thick. The contact force time-history of the impact is as defined in Figure 13.25. Simulate the acceleration profile at the instance when maximum force occurs.

Solution

Step 1.1:
The barrier is idealised into a line element with 150 equally spaced lumped masses (N_n = 150), as shown in Figure 13.26. Values of their offset from the top of the barrier are listed in Table 13.11a.

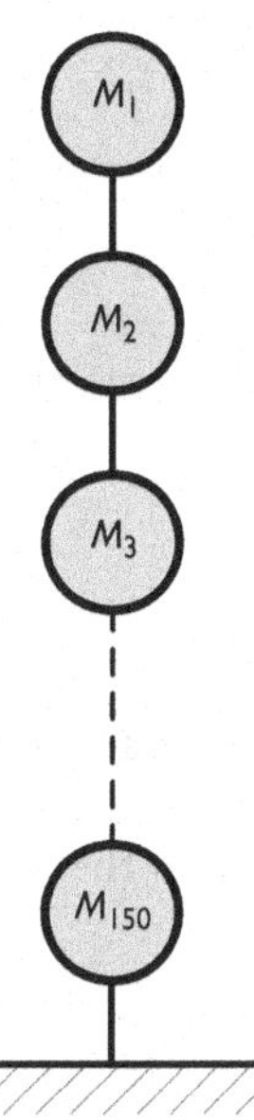

Figure 13.26 Cantilevered wall idealised into MDOF system.

Table 13.11a Offset of lumped masses from the top end

k	*Offset (m)*
1	0
2	0.02
3	0.04
4	0.06
⋮	⋮
147	2.94
148	2.96
149	2.98
150	3

Step 1.2:
Refer to step 1.2 from Section 13.5.5.

Step 1.3:
Construct a 150 × 150 mass matrix $[M]$ with diagonal entries based on the following:

$$\frac{m_w}{N_n} = \frac{34560}{150} = 230.4 \text{ kg}$$

Steps 1.4–1.10:
Refer to previous examples in Sections 13.5.4 and 13.5.5.

Results from modal analyses are presented in Figure 13.27 and Table 13.11b.

Step 2–10:
Refer to the previous example in Section 13.5.5.

Deflection time-histories can be simulated for each node (and there are in total 150 nodes, with a time-history for each node). Deflection time-history at the top of the wall is shown in Figure 13.28.

Step 11:
Each of the deflection time-histories can be converted into velocity and acceleration time-histories by applying the equations listed at each time step. An example for each of the velocity and acceleration time-histories are shown in Figure 13.29(a) and 13.29(b), respectively, for the top of the wall.

$$v_k(t_i) = \frac{x_k(t_i) - x_k(t_{i-1})}{dt}$$

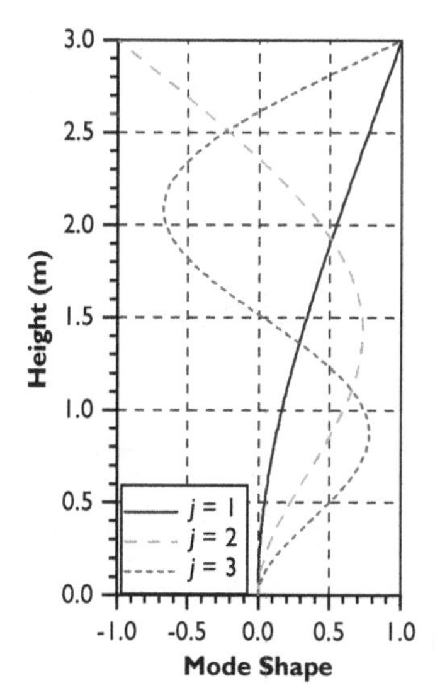

Figure 13.27 Mode shapes at the centreline of a cantilever wall.

Table 13.11b Values of ω, *f* and *T* of a cantilever wall

Output	*Mode No. (j)*		
	1	*2*	*3*
ω (rad/s)	222	1389	3890
f (Hz)	35	221	619
T (s)	0.0283	0.0045	0.0016

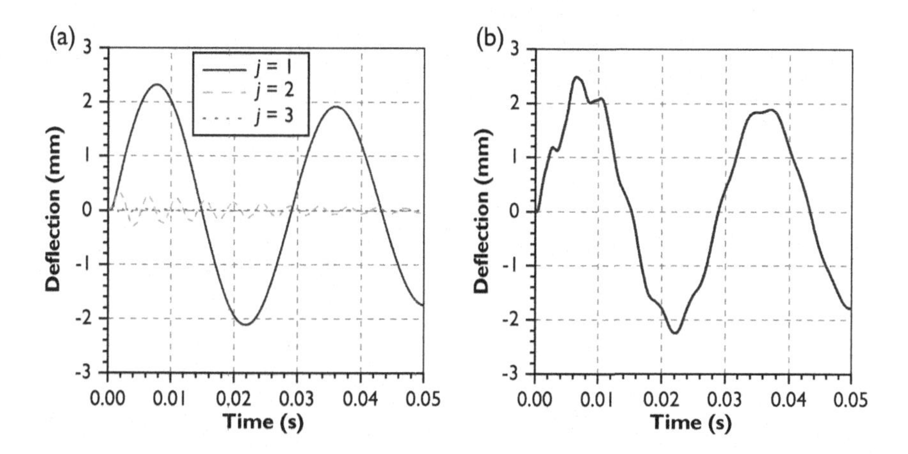

Figure 13.28 Simulated deflection time-history at the top of a cantilever wall: (a) deflection contributed by each mode, (b) total deflection.

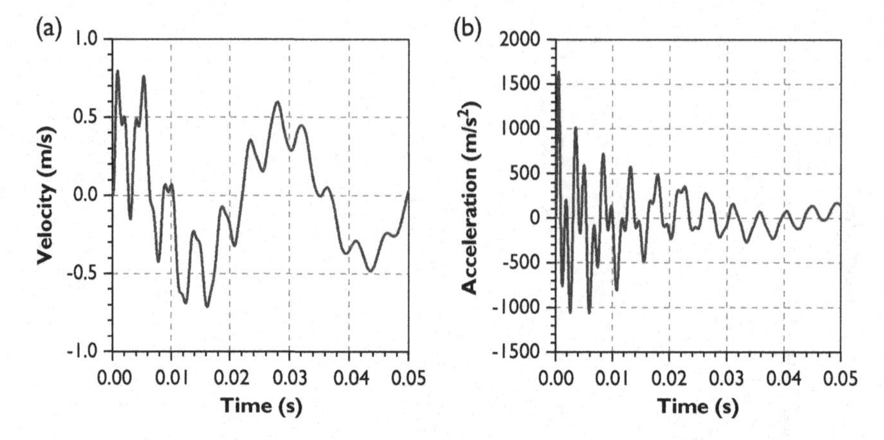

Figure 13.29 (a) Velocity and (b) acceleration time-histories at the top of a cantilever wall.

$$a_k(t_i) = \frac{v_k(t_i) - v_k(t_{i-1})}{dt}$$

It is found from the force time-history of Figure 13.25 that the maximum force, $F_{c,\,\max}$, occurred at $t = 0.57$ ms. The acceleration profile of the wall at this instance (as shown in Figure 13.30) is identical to that shown in Figure 8.13, which was used in the example of Section 8.5.

13.5.7 Simulation of Forcing Function at Contact

Consider the hailstorm scenario represented by the worked example of Section 6.7.1. Simulate the contact force time-history, and correlation between contact force and indentation.

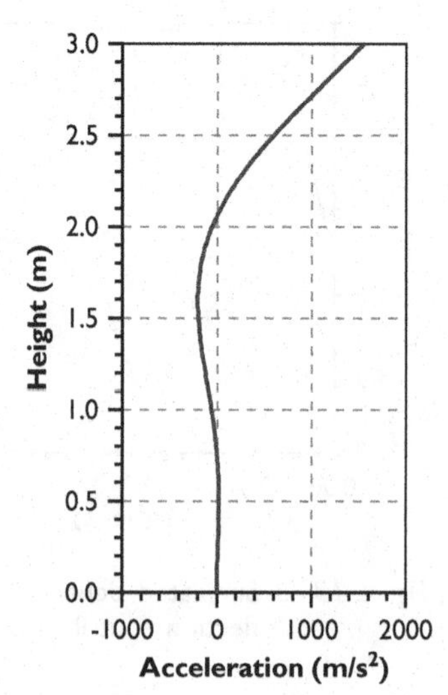

Figure 13.30 Acceleration profile of a cantilever wall.

Table 13.12 Input parameters for simulation of contact force time-history of hailstone impact

Parameters	*Values*
v_0	24.5 m/s
m	0.125 kg
λm	1 kg
k	50 kN/m
k_n	0.224 MN/m^p
p	1.508
COR	0.0245

Solution

The H&C model is employed in the simulation.

Step 1:
The input parameters are taken from Section 6.7.1 and are listed in Table 13.12.

Step 2:

$$\begin{aligned} D_n &= (0.2p + 1.3)\left(\frac{1-\text{COR}}{\text{COR}}\right)\left(\frac{k_n}{\dot{\delta}_0}\right) \\ &= (0.2(1.508) + 1.3)\left(\frac{1-0.0245}{0.0245}\right)\left(\frac{0.224 \times 10^6}{24.5}\right) \\ &= 5.83 \times 10^5 \end{aligned}$$

Step 3:
List all the variables as shown in the first row of Table 13.13.

Step 4:
Declare the time-step interval: $dt = 2 \times 10^{-6}$ s.

Step 5:
Initiate all variables to zero at $t = 0$ and $i = 0$, except that $v_1(0) = v_0 = 24.5$ m/s, as shown in the second row of Table 13.13.

Step 6:
Apply the following equations for $i = 1$ and onwards.

$$t(i) = t(i-1) + dt$$

$$v_i(i) = v_i(i-1) + a_i(i-1)dt$$

Table 13.13 First six time steps of simulation of contact force time-history

i	t	v_i	v_t	x_i	x_t	δ	$\dot{\delta}$	F_c	F_R	a_i	a_t
0	0	24.5	0	0	0	0	0	0	0	0	0
1	2×10^{-6}	24.5	0	4.9×10^{-5}	0	4.9×10^{-5}	24.5	4.6	0	−37	4.6
2	4×10^{-6}	24.4999	9.2×10^{-6}	9.8×10^{-5}	0	9.8×10^{-5}	24.4999	13.1	0	−105	13
3	6×10^{-6}	24.4997	3.5×10^{-5}	1.5×10^{-4}	1.8×10^{-11}	1.5×10^{-4}	24.4997	24.1	9.2×10^{-7}	−193	24
4	8×10^{-6}	24.4993	8.4×10^{-5}	2.0×10^{-4}	8.9×10^{-11}	2.0×10^{-4}	24.4993	37.2	4.5×10^{-6}	−297	37
5	10^{-5}	24.4987	1.6×10^{-4}	2.5×10^{-4}	2.6×10^{-10}	2.5×10^{-4}	24.4986	52.1	1.3×10^{-5}	−416	52

$$v_t(i) = v_t(i-1) + a_t(i-1)dt$$

$$x_i(i) = x_i(i-1) + v_i(i-1)dt$$

$$x_t(i) = x_t(i-1) + v_t(i-1)dt$$

$$\delta(i) = x_i(i) - x_t(i)$$

$$\dot{\delta}(i) = v_i(i) - v_t(i)$$

$$F_c(i) = k_n\delta(i)^p + D_n\delta(i)^p\dot{\delta}(i)$$

$$F_R(i) = kx_t(i)$$

$$a_i(i) = -\frac{F_c(i)}{m}$$

$$a_t(i) = \frac{F_c(i) - F_R(i)}{\lambda m}$$

Sample calculated values for up to step number $i = 5$ are as listed in Table 13.13.

F_c is plotted against t and δ to give contact force time-history of Figure 13.31(a) and contact force- indentation relationship of Figure 13.31(b), respectively.

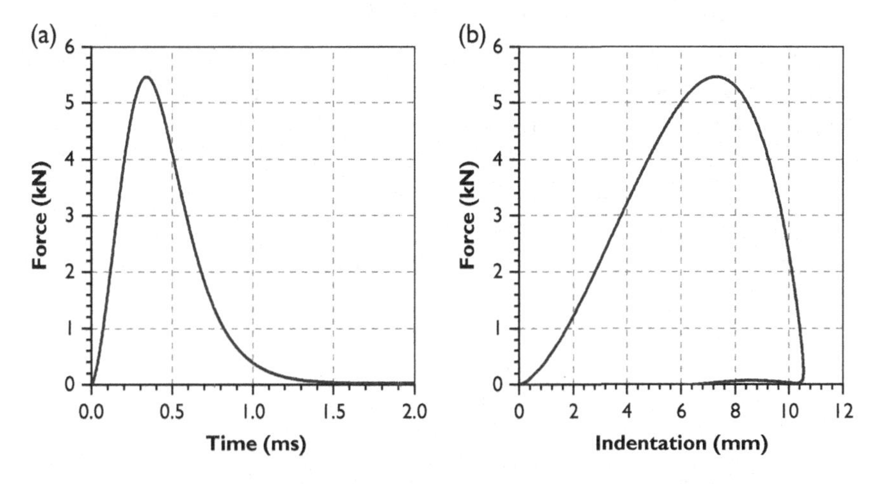

Figure 13.31 Simulations by the H&C model: (a) contact force time-history, (b) contact force versus indentation.

13.5.8 Collision Action on a Plate Simply Supported on Four Edges

Consider the impact scenario as described in Section 6.7.3: a 1 m diameter spherical granite boulder with mass of 1,400 kg and impact velocity of 10 m/s. The impact scenario is applied to a concrete plate element that is 6 m wide, 6 m long and 0.8 m thick, simply supported on all the edges. The contact forcing function is not given and needs to be simulated. The concrete has a characteristic cylinder compressive strength of 50 MPa. A Poisson's ratio of 0.2 is assumed for both the boulder and concrete. The *Young's* modulus of concrete and granite are taken as 32,000 MPa and 40,000 MPa, respectively. Simulate the dynamic response behaviour of the concrete plate, and present details at the instance when maximum contact force occurs.

Solution

There are three parts to the solution. Part 1 is about following the steps outlined in Table 13.4 to simulate the contact force time-history delivered by the impactor. The simulated contact force time-history is then carried over to Part 2 for simulating the response behaviour of the MDOF system by following the steps outlined in Table 13.2. In Part 3, the generalised mass for the considered localised action is calculated for comparing against the value used for input into the contact force simulation (in Part 1) to ensure consistency.

Part 1

The H&C model is employed in Part 1 of the simulation.

Total mass of plate: $M = \rho abt_p = 2400(6)(6)(0.8) = 69120$ kg

Based on the recommendations shown in Table 2.1, the generalised mass of the concrete plate, which is simply supported at the four edges, is initially taken as 25% of its total mass:

$$\lambda m = 0.25M = 0.25(69120) = 17280 \text{ kg}$$

The generalised stiffness of the plate may also be calculated based on the recommendations shown in Table 2.1:

$$D_p = \frac{Et_p^3}{12(1-\nu^2)} = \frac{32(0.8)^3}{12(1-0.2^2)} = 1422 \text{ MNm}$$

$$k = \frac{1}{4}D_p\pi^4 ab\left(\frac{1}{a^2}+\frac{1}{b^2}\right)^2 = \frac{1}{4}(1422)\pi^4(6)(6)\left(\frac{1}{6^2}+\frac{1}{6^2}\right)^2 = 3848 \text{ MN/m}$$

Other model parameters for the H&C model, namely k_n, p and COR, can be taken from Section 6.7.3. All the input parameters are listed in Table 13.14.

Table 13.14 Input parameters for simulation of contact force time-history of boulder impact assuming mass factor of 0.25

Parameters	*Values*
v_0	10 m/s
m	1,400 kg
λm	1,7280 kg
k	3,848 MN/m
k_n	3,174 MN/m^p
p	1.885
COR	0.0108

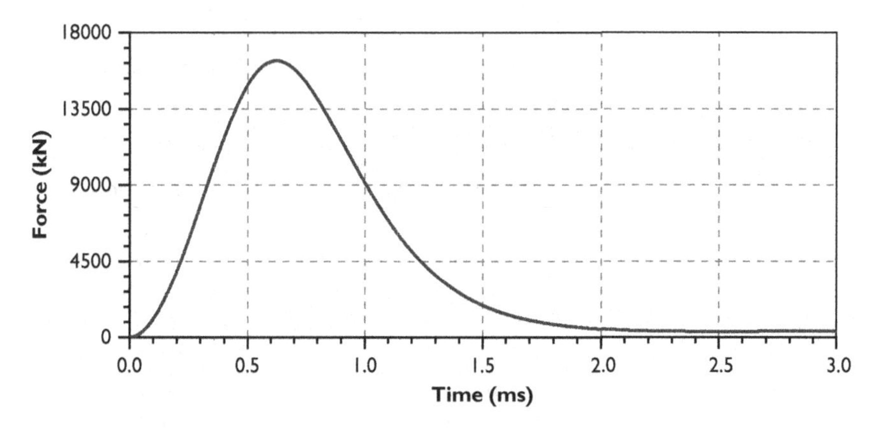

Figure 13.32 Simulated contact force time-history delivered by a spherical granite boulder to a plate element.

The contact force-time history is then simulated as shown in Figure 13.32. Refer to the previous example in Sections 13.5.7 for step-by-step implementation of the simulation.

Part 2

Steps employed in Section 13.5.5 may also be used to simulate the response of a plate member.

Step 1.1:

The concrete plate is idealised into an arrangement of 15 × 15 (totalling N_n = 225) equally spaced lumped masses (of identical masses), as shown by the plan view of Figure 13.33. Values of their offset from the bottom left-hand corner of the plate are as listed in Table 13.15a.

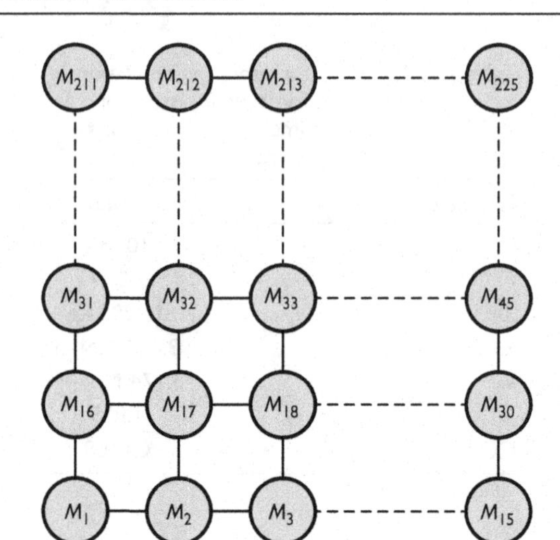

Figure 13.33 Rectangular plate simply supported on four edges idealised into MDOF system.

Table 13.15a Offset of lumped masses from bottom left-hand corner

k	*Offset x (m)*	*Offset y (m)*
1	0.375	0.375
2	0.750	0.375
3	1.125	0.375
⋮	⋮	⋮
15	5.625	0.375
16	0.375	0.750
17	0.750	0.750
18	1.125	0.750
⋮	⋮	⋮
223	4.875	5.625
224	5.250	5.625
225	5.625	5.625

Step 1.2:
The deflection value (Δ) at offset x and offset y when subject to point force P is to be calculated. The point force is applied at a position as defined in Figure 13.34. The value of Δ can be calculated by use of Equation (13.33) [13.6]:

$$\Delta(x, y) = \frac{4P}{\pi^4 D_p ab} \sum_m^\infty \sum_n^\infty \frac{\sin\left(\frac{m\pi x_1}{a}\right)\sin\left(\frac{n\pi y_1}{b}\right)}{\left[\left(\frac{m}{a}\right)^2 + \left(\frac{n}{b}\right)^2\right]^2} \sin\frac{m\pi x}{a} \sin\frac{n\pi y}{b} \tag{13.33}$$

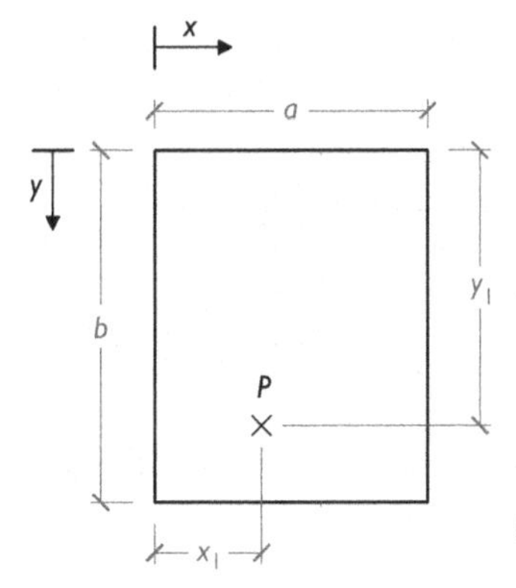

Figure 13.34 Rectangular plate simply supported on four edges subject to static force.

A unit load of $P = 1$ N is substituted into Equation (13.33). Offset x values as listed in Table 13.15a are substituted as a and x_1, and offset y values are substituted as b and y_1 into Equation (13.33) to construct the flexibility matrix [F]. In this example, the values of m and n are taken as 20. The dimension of [F] is 225 × 225. The inversion of [F] gives a stiffness matrix [K].

Step 1.3:
Construct a 225 × 225 mass matrix [M] with diagonal entries based on the following:

$$\frac{M}{N_n} = \frac{69120}{225} = 307.2 \text{ kg}$$

Steps 1.4 to 1.10:
Refer to the previous examples in Sections 13.5.4 and 13.5.5.

Results from modal analyses are presented in Figure 13.35 and Table 13.15b. Note that the plate spans in two directions. The mode shapes

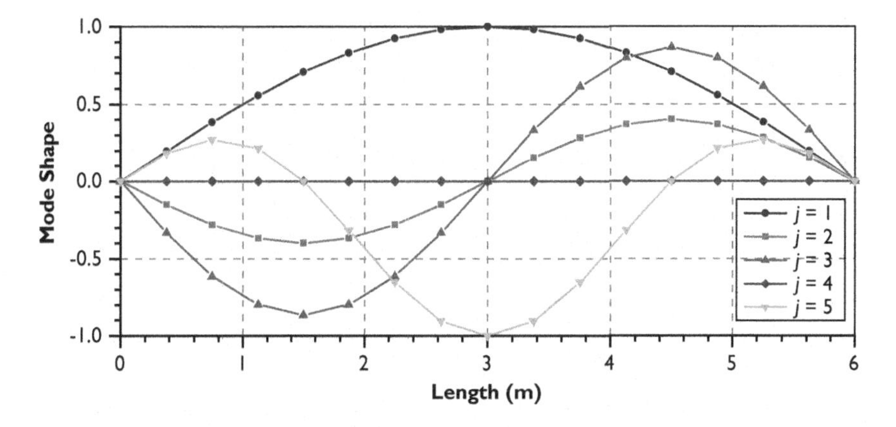

Figure 13.35 Mode shapes at the centreline of a concrete plate.

Table 13.15b Values of ω, *f* and *T* of a concrete plate

Output	*Mode No. (j)*				
	1	*2*	*3*	*4*	*5*
ω (rad/s)	387	968	968	1549	1936
f (Hz)	62	154	154	246	308
T (s)	0.0162	0.0065	0.0065	0.0041	0.0032

at the centreline are presented in one direction only in Figure 13.35 for illustration purposes.

Step 2 to 10:
Refer to the previous example in Section 13.5.5.

Deflection time-histories can be simulated for each node (and there are in total 225 nodes, with a time-history for each node). Deflection time-history for an example node (#113) at the centre of the plate is shown in Figure 13.36.

Step 11:
Each of the deflection time-histories can be converted into velocity and acceleration time-histories by applying the equations listed at each time step. An example for each of the velocity and acceleration time-histories are shown in Figure 13.37(a) and 13.37(b), respectively, for the centre of the plate.

$$v_k(t_i) = \frac{x_k(t_i) - x_k(t_{i-1})}{dt}$$

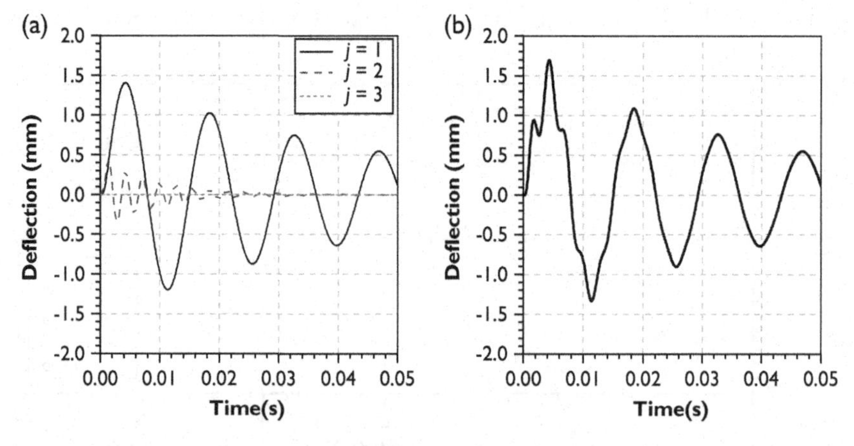

Figure 13.36 Simulated deflection time-history at the centre of a plate: (a) deflection contributed by each mode, (b) total deflection.

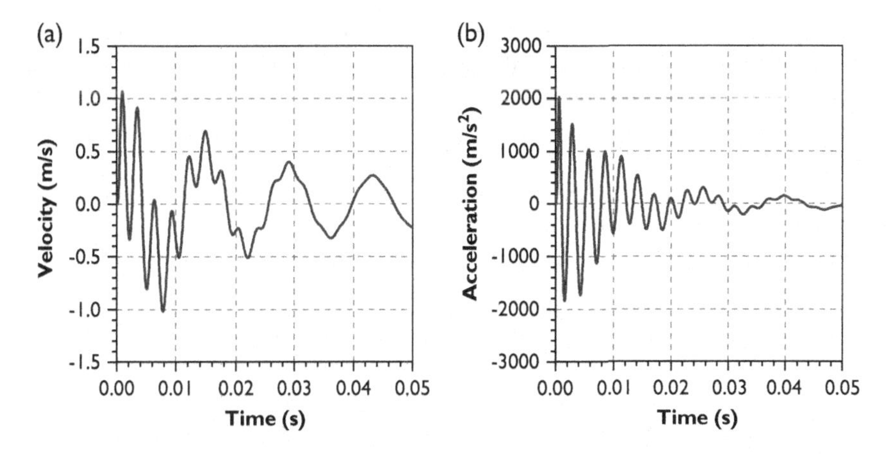

Figure 13.37 (a) Velocity and (b) acceleration time-histories at the centre of the plate.

$$a_k(t_i) = \frac{v_k(t_i) - v_k(t_{i-1})}{dt}$$

From Figure 13.32, it can be seen that the maximum force, $F_{c,\max}$, occurs at t = 0.62 ms. Acceleration profiles at three key locations at t = 0.62 ms are shown in Figure 13.38.

Part 3

Equation (12.12), which was introduced in Section 12.3 of Chapter 12 to weight-average velocity values on the surface of a plate for determining v_2, can be simplified as follows if all 225 discrete nodes have identical masses:

$$v_2 = \frac{\Sigma m_i v_i^2}{\Sigma m_i v_i} = \frac{\Sigma v_i^2}{\Sigma v_i}$$

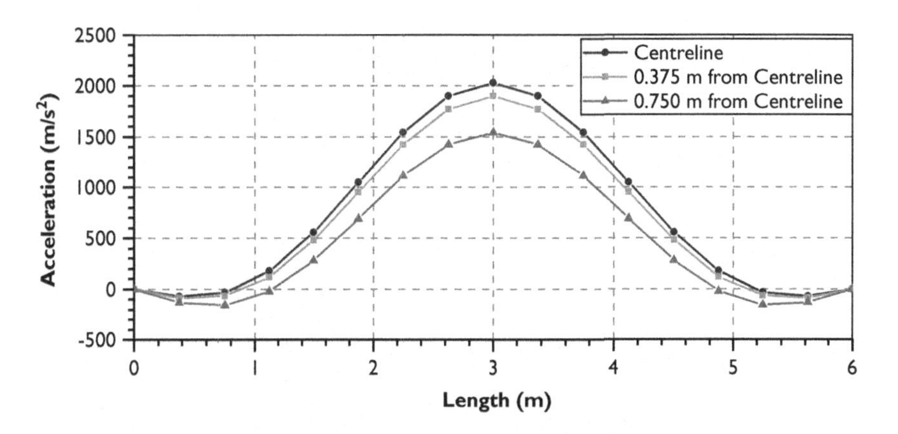

Figure 13.38 Acceleration profiles close to the centre of the plate.

To determine v_i, velocity time-histories for all 225 nodes are simulated in the same manner as the central node of the plate (Figure 13.37(a)). Peak velocity value in each time-history can be taken as the corresponding value for v_i. For instance, from Figure 13.37(a), velocity at the central node $v_{i=113} = 1.1$ m/s.

By the use of the simplified equation for weight-averaging velocity values as shown above: $v_2 = 0.56$ m/s.

From Table 13.14, COR = 0.0108. By the use of Equation (3.1c):

$$\text{COR} = \frac{v_1 + v_2}{v_0}$$

$$v_1 = \text{COR} \times v_0 - v_2 = 0.0108(10) - 0.56 = -0.45 \text{ m/s}$$

Multiplying both sides of Equation (12.11b) by m, and substituting the velocity values as shown above into the equation:

$$\lambda m = \left(\frac{v_0 + v_1}{v_2}\right) m = \left(\frac{10 - 0.45}{0.56}\right)(1400) = 23920 \text{ kg}$$

$$\text{Mass Factor} = \frac{\lambda m}{M} = \frac{23920}{69120} = 0.35$$

which is compared against the mass factor of 0.25 that was input into the contact force time-history analysis in Part 1 (refer Table 13.14 which shows $\lambda m = 17280$ kg). Thus, Part 1 is repeated with the value of the generalised mass updated to 23,920 kg while keeping the other parameters the same, as shown in Table 13.16.

The simulated contact force time-history based on input parameters of Table 13.16 is plotted alongside the original time-history (of Figure 13.32) in Figure 13.39. The second iteration of the contact force time-history

Table 13.16 Input parameters for simulation of contact force time-history of boulder impact assuming mass factor of 0.35

Parameters	*Values*
v_0	10 m/s
m	1,400 kg
λm	23,920 kg
k	3,848 MN/m
k_n	3,174 MN/m^p
p	1.885
COR	0.0108

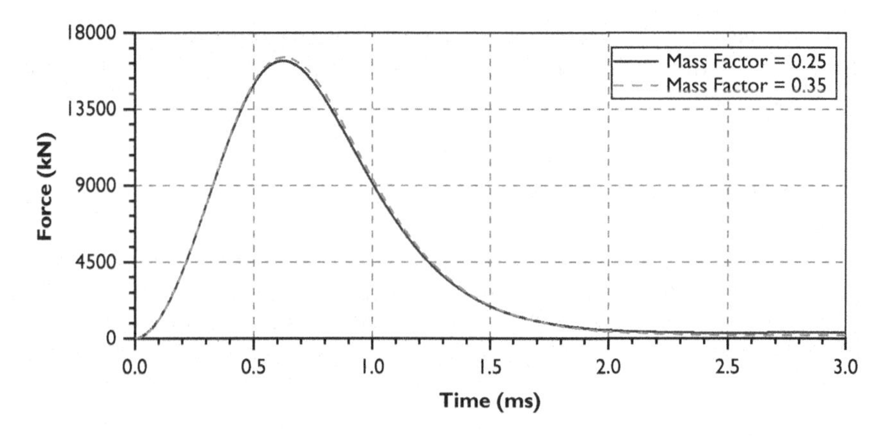

Figure 13.39 Comparison of contact force time-history for different mass factors.

corresponding to a mass factor of 0.35 is shown to be sufficiently close to the first iteration (where mass factor = 0.25). Thus, the plate response simulations as presented in Parts 2 and 3 need not be repeated. The analysis of the collision scenario may conclude at this point.

13.6 CLOSING REMARKS

In this chapter, the *Duhamel Integration Method* is first introduced to deal with transient actions of a finite duration such as a rectangular- or triangular-shaped impulse, which was applied to the lumped mass of a SDOF system. Good consistencies were demonstrated between predictions from the introduced technique and from established classical solutions add confidence to the use of the simulation methodology for dealing with a wide range of forcing functions of finite duration. The simulation methodology has also been extended to deal with a MDOF lumped mass model of a simply supported beam and a cantilevered wall. The simulated response of a simply supported beam is shown to reconcile with the closed-form expression from Chapter 3 for estimating maximum deflection. Although the beam element considered in the example was of a uniform cross section and with a uniform distribution of mass, the solution strategy is equally applicable to elements of non-uniform sections, and may well be applied to a wider form of structural elements such as a wall (or slab), frame, or a truss. A rational method of simulating a forcing function, which is generated by a collision scenario involving objects of known stiffness properties is also presented. This chapter concludes by demonstrating the implementation of the introduced simulation methodologies by going through a few worked examples. The final worked example deals with a collision scenario on a concrete plate. As the contact forcing function is not given in this example, the solution process involves integrating analytical techniques introduced in different parts of this chapter. The contact forcing function is first simulated; the response behaviour of the plate in response to the forcing function is then calculated.

13.7 APPENDIX

The solution to the equation of dynamic equilibrium is given by Equation (2.6) from Chapter 2, which is modified here into Equation (13.34). Note that ω_D is replaced by ω_n as $\omega_D \approx \omega_n$ for $\xi < 0.2$. As the impulse is assumed to commence at $t = \tau$, t in the equation is replaced by $t - \tau$. Differentiating Equation (13.34) with respect to t gives Equation (13.35):

$$u(t) = \Delta e^{-\xi\omega_n t} \sin(\omega_n(t-\tau) - \phi_p) \tag{13.34}$$

$$\dot{u}(t) = -\Delta\xi\omega_n e^{-\xi\omega_n(t-\tau)} \sin(\omega_n(t-\tau) - \phi_p) + \Delta e^{-\xi\omega_n(t-\tau)}\omega_n \cos(\omega_n(t-\tau) - \phi_p) \tag{13.35}$$

Both Equations (13.34) and (13.35) are valid for $t \geq \tau$. $u(t) = 0$ when $t < \tau$.

At $t = \tau$, $u(0) = 0$. Substituting the boundary condition into Equation (13.34) gives Equation (13.36), which is then re-arranged into Equation (13.37) to determine ϕ_p:

$$u(\tau) = \Delta e^0 \sin(-\phi_p) = 0 \tag{13.36}$$

$$\phi_p = 0 \tag{13.37}$$

At $t = \tau$, $\dot{u}(\tau)$ is defined by Equation (13.3), which is re-written here as Equation (13.38):

$$\dot{u}(\tau) = \frac{1}{M} \tag{13.38}$$

Substituting Equations (13.37) and (13.38) into Equation (13.35) gives Equation (13.39), which can be re-arranged into Equation (13.40) for determining Δ.

$$\dot{u}(\tau) = -\Delta\xi\omega_n e^0 \sin(0) + \Delta e^0 \omega_n \cos(0) = \frac{1}{M} \tag{13.39}$$

$$\Delta = \frac{1}{M\omega_n} \tag{13.40}$$

Substituting Equations (13.37) and (13.40) into Equation (13.34) gives Equation (13.41):

$$u(t) = \frac{1}{M\omega_n} e^{-\xi\omega_n t} \sin \omega_n (t - \tau) \tag{13.41}$$

If an impulse is of short duration, unit impulse of "1" in Equation (13.41) can be replaced by the magnitude of the impulse J. Equation (13.41) can then be written in the form of Equation (13.42), which is the same as Equation (13.4):

$$u(t) = \frac{J}{M\omega_n} e^{-\xi\omega_n t} \sin \omega_n (t - \tau) \tag{13.42}$$

REFERENCES

13.1 Chopra, A.K., 2012, *Dynamics of Structures - Theory and Applications to Earthquake Engineering*, 4th ed., Pentice Hall, Upper Saddle River, NJ.
13.2 Clough, R.W. and Penzien, J.P., 2003, *Dynamics of Structures*, 3rd ed., Computers & Structures, Inc., Berkeley, CA.
13.3 Hunt, K.H. and Crossley, F.R.E., 1975, "Coefficient of restitution interpreted as damping in vibroimpact", *Journal of Applied Mechanics*, Vol. 42(2), p. 440, doi:10.1115/1.3423596
13.4 Sun, J., Lam, N., Zhang, L., Ruan, D., and Gad, E., 2016, "Computer Simulation of Contact Forces Generated by Impact", *International Journal of Structural Stability and Dynamics*, Vol. 17(1), p. 1750005, doi:10.1142/S0219455417500055
13.5 Stephens, R.C., 1970, *Strength of Materials: theory and Examples*, Edward Arnold (Publishers), London.
13.6 Ugural, A.C., 2017, *Plates and Shells: theory and Analysis*, CRC Press, Boca Raton.

Chapter 14

Case Studies of Physical Experimentation

14.1 INTRODUCTION

In this final chapter of this book, more worked examples that are associated with analytical techniques introduced in this book across many chapters are presented. Each of the presented case studies of collision scenarios involved physical dynamic testing (mostly conducted by the authors and colleagues) along with calculations for giving estimates of the test results, and each case study consists of multiple tests based on varying the input parameters. The calculation procedure is illustrated in a step-by-step manner to guide the readers. Graphs are presented to show the comparison of the calculated results against experimental measurements from each test. The purpose of presenting authentic information from case studies involving dynamic testing is to: (i) enable estimates from analytical models presented in this book (and from elsewhere) to be benchmarked against reliable measured data, (ii) support the teaching of the subject to a cohort by presenting multiple versions of the same question in each case study and (iii) provide additional means for readers to practise applications of the learnt techniques.

The considered collision scenarios are: (i) a torpedo-shaped solid steel object colliding on a bare reinforced concrete (RC) barrier wall by use of a pendulum device, (ii) solid object impact on a free-standing rectangular RC blocks causing overturning, (iii)solid object impact on RC blocks causing sliding, (iv) impact of a fallen object on a RC beam, (v) impact testing of hailstones and windborne debris, (vi)granite boulder colliding on a concrete surface, (vii) a similar setup asscenario (i) but with the wall protected by a layer of gabion cushion placed infront of it and (viii) solid object impact on soil-embedded baffles.

14.2 BENDING OF A REINFORCED CONCRETE BARRIER WALL

This case study involves the use of the calculation techniques introduced in Chapters 2, 3 and 8. A series of impact tests on a cantilever RC wall was

DOI: 10.1201/9781003133032-14

Figure 14.1 Photograph of reinforced concrete wall experimental setup.

conducted by Yong et al. [14.1], as shown in Figure 14.1. The wall was supported on a 500 mm thick base slab, which was rigidly secured to the ground in order that only the cantilever wall, with a height of 1.5 m, length of 3.0 m and thickness of 0.23 m, and was subject to a flexural response from the collision action. In-situ compressive strength of concrete was 47 MPa. All the reinforcement in the RC wall was grade D500N reinforcing steel bars complying with specifications found in AS/NZS 4671 [14.2]. The vertical and horizontal reinforcement consisted of a layer of N20 bars at 200 mm spacing on both faces of the wall. The concrete cover was approximately 30 mm. The dimensions and reinforcement details of the wall are shown in Figure 14.2. The yield strength and modulus of elasticity of the reinforcement was 543 MPa and 194,000 MPa, respectively. Two torpedo-shaped steel impactors with a mass of 280 kg and 435 kg, respectively, were employed to collide with the wall at the top, and at the centreline horizontally. The impactor was released from different heights to achieve different impact velocities, as listed in Table 14.1. In each test, the value of COR was inferred from the measured velocities based on images recorded by a high-speed camera. A steel plate weighing 62.8 kg, which was attached to the wall in Test Nos. 1 to 8 to receive the impact (as to prevent localised damage at the point of impact in these tests), was removed for the remaining tests in which impact occurred to the bare concrete surface. Estimate the maximum deflection at the top of the wall in each test.

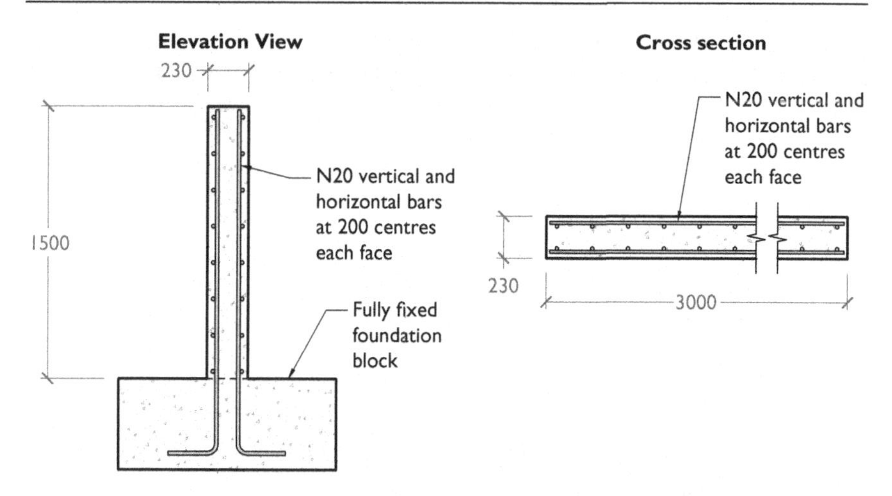

Figure 14.2 Dimensions and reinforcement details of reinforced concrete wall specimen.

Table 14.1 Details of reinforced concrete wall impact experiment

Test No.	*Impactor Mass (kg)*	*Impact Velocity (m/s)*	*COR*	*Steel Plate*
1	280	1.93	0.409	Yes
2	280	3.08	0.290	Yes
3	280	4.17	0.217	Yes
4	280	5.18	0.159	Yes
5	435	1.55	0.457	Yes
6	435	2.48	0.412	Yes
7	435	3.66	0.328	Yes
8	280	1.91	0.336	Yes
9	280	1.93	0.323	No
10	280	3.08	0.291	No
11	280	4.26	0.335	No
12	280	5.10	0.281	No

Solution

By conducting fibre-element analysis with the use of WHAM [14.3, 14.4], the moment-curvature relationship of the RC wall section was determined, as shown in Figure 14.3.

It is found from Figure 14.3 that $M_y = 402.5$ kNm and $\phi_y = 0.0232$ rad/m. By the use of Equation (8.16h):

$$E_c I_{\text{eff}} = \frac{M_y}{\phi_y} = \frac{402.5}{0.0232} = 17350 \text{ kNm}^2$$

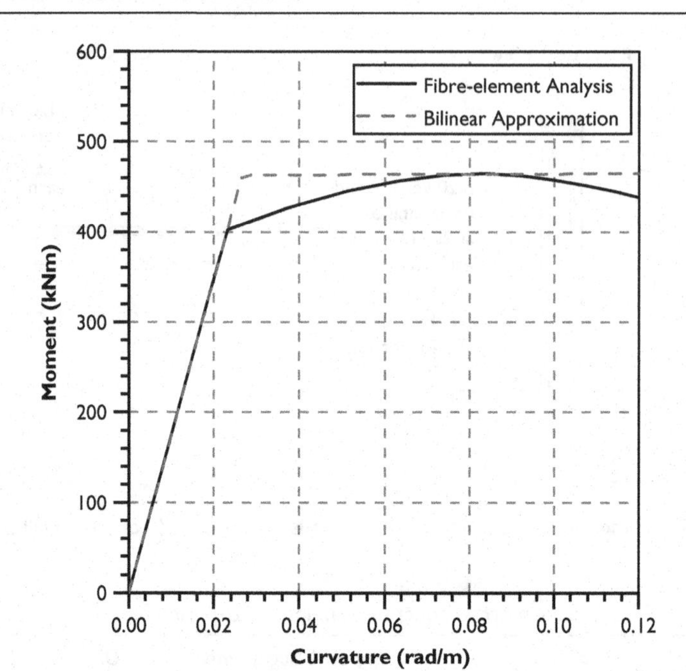

Figure 14.3 Moment-curvature relationship of RC wall.

Effective stiffness of the wall may then be calculated based on the relationship defined in Table 2.1 for a cantilevered member:

$$k_{\text{eff}} = \frac{3E_c I_{\text{eff}}}{H_w^3} = \frac{3(17350)}{1.5^3} = 15400 \text{ kN/m}$$

By the use of Equation (8.13a):

$$\Delta_y = \frac{\phi_y H_w^2}{3} = \frac{0.0232(1.5)^2}{3} \times 10^3 = 17.4 \text{ mm}$$

The yield strain of reinforcement may be estimated based on the information provided:

$$\varepsilon_{sy} = \frac{f_{sy}}{E_s} = \frac{543}{194000} = 0.0028$$

Every step of the calculation for obtaining estimates based on Test No. 1 (as defined in Table 14.1) is presented.

Total mass of stem wall:

$$m_w = \rho bDH_w = 2400(3)(0.23)(1.5) = 2484 \text{ kg}$$

The generalised mass of the wall may be taken as the sum of 25% of the total mass of the wall (as per the derivation in Chapter 2 and summary in Table 2.1) and the mass of the steel plate that received the impact:

$$\lambda m = 0.25m_w + 62.8 = 0.25(2484) + 62.8 = 683.8 \text{ kg}$$

$$\lambda = \frac{\lambda m}{m} = \frac{683.8}{280} = 2.44$$

By the use of Equation (3.7a):

$$\Delta = \frac{mv_0}{\sqrt{mk_{\text{eff}}}}\sqrt{\lambda\left(\frac{1 + \text{COR}}{1 + \lambda}\right)^2} = \frac{280(1.93)}{\sqrt{280(19000 \times 10^3)}}\sqrt{2.44\left(\frac{1 + 0.409}{1 + 2.44}\right)^2}$$
$$\times 10^3 = 4.74 \text{ mm}$$

The calculation procedure presented has also been repeated for the other tests, as listed in Table 14.1 (i.e. Test Nos. 2 to 12), to provide a series of estimates (Δ values) that can be compared against experimental measured values, as shown in Figure 14.4.

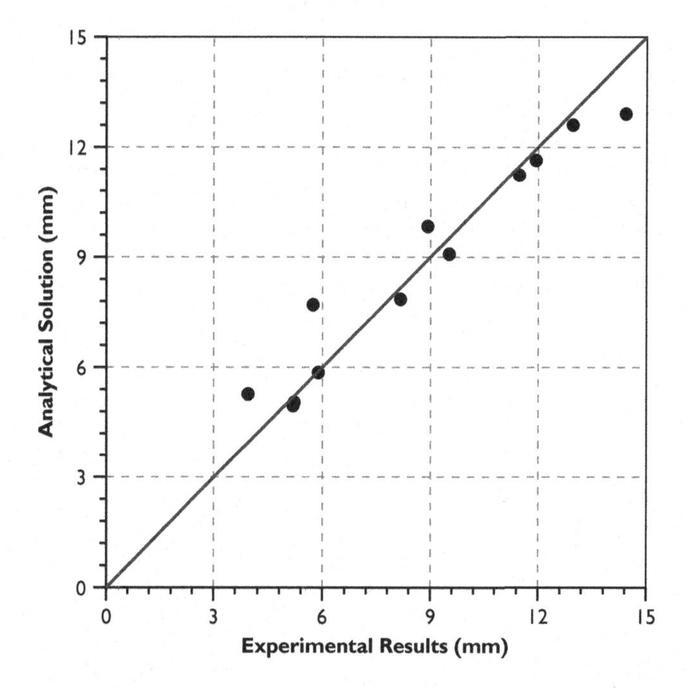

Figure 14.4 Deflection of RC wall estimated from analytical solution versus experimental results.

14.3 OVERTURNING OF FREE-STANDING TARGET

This case study involves the use of the calculation techniques introduced in Chapter 4. Consider a collision experiment carried out on free-standing concrete blocks, which was reported in Lam et al. [14.5], as shown in Figure 14.5. Three concrete blocks with height h of 0.4 m, 0.8 m and 1.2 m were used in the experiments. All three concrete blocks had cross-sectional dimensions of 0.2 m in thickness and 0.4 m in width. An impactor that had a mass of 5 kg was released from heights of 0.5 and 1.0 m to collide with the concrete block at the top. Values of COR as recorded from the tests were approximately 0.75 and 0.55 for release heights of 0.5 m and 1.0 m, respectively. Note that any sliding action was prevented by a stopper that was placed behind the block at the base in order that the motion generated by the impact was purely in the overturning mode. Estimate the horizontal displacement at the top of the block.

Solution

Consider an impact scenario where the impactor was released from a height of 0.5 m to strike a 0.4 m tall concrete block. The solution for this impact scenario is presented as follows.

$$v_0 = \sqrt{2gH} = \sqrt{2(9.81)(0.5)} = 3.13 \text{ m/s}$$

$$M = 2400(0.2)(0.4)(0.4) = 76.8 \text{ kg}$$

By the use of Equation (4.2):

$$I_\theta = M\left(\frac{h^2 + w^2}{3}\right) = 76.8\left(\frac{0.4^2 + 0.2^2}{3}\right) = 5.12 \text{ kgm}^2$$

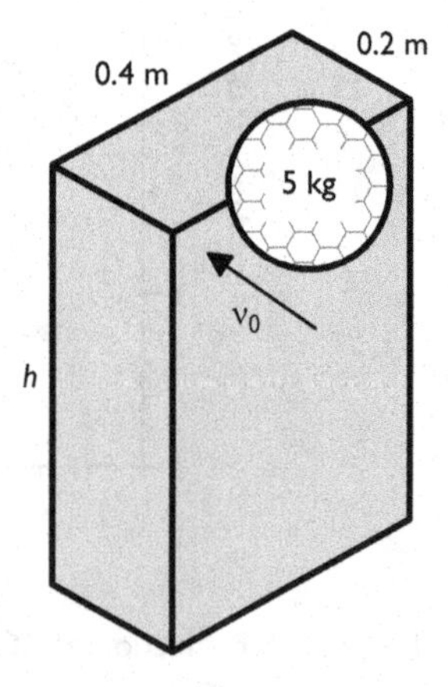

Figure 14.5 Impact test on overturning action of free-standing rectangular concrete blocks.

$$R = \sqrt{h^2 + w^2} = \sqrt{0.4^2 + 0.2^2} = 0.45 \text{ m}$$

By the use of Equation (4.6):

$$\kappa = \frac{I_\theta}{mhR} = \frac{5.12}{5(0.4)(0.45)} = 5.69$$

The calculated parameters are then substituted into Equation (4.15):

$$\begin{aligned}\Delta_{C.G.} &= \frac{mv_0^2}{2Mg}\frac{\kappa h}{R}\left(\frac{1 + COR}{1 + \kappa}\right)^2 \\ &= \frac{5(3.13)^2}{2(76.8)(9.81)}\left(\frac{5.69(0.4)}{0.45}\right)\left(\frac{1 + 0.75}{1 + 5.69}\right)^2 \\ &= 0.0113 \text{ m}\end{aligned}$$

By the use of Equations (4.16a)–(4.16c):

$$r = 0.5\sqrt{h^2 + w^2} = 0.5\sqrt{0.4^2 + 0.2^2} = 0.224 \text{ m}$$

$$\beta = \tan^{-1}\left(\frac{h}{w}\right) = \tan^{-1}\left(\frac{0.4}{0.2}\right) = 63.4°$$

$$\theta = \sin^{-1}\left(\frac{\frac{h}{2} + \Delta_{C.G.}}{r}\right) - \beta = \sin^{-1}\left(\frac{\frac{0.4}{2} + 0.0113}{0.224}\right) - 63.4 = 7.2°$$

Horizontal displacement may then be calculated using Equation (4.17):

$$\Delta = h \sin\theta = 0.4 \sin 7.2° = 0.0501 \text{ m} = 50.1 \text{ mm}$$

which can be compared against the experimental recorded value of 54 mm.

Other experimentally recorded results are shown in Figure 14.6, alongside curves to represent predictions obtained from the use of Equation (4.17). It is shown in Figure 14.6 that horizontal displacement decreases with an increase in height of the targeted block because of the increase in inertial resistance with increasing size of the block.

14.4 SLIDING OF FREE-STANDING TARGET

This case study involves the use of the calculation techniques introduced in Chapter 4. Consider a collision experiment that was carried out on a free-standing L-shaped concrete specimen as reported in Yong et al. [14.6]. The

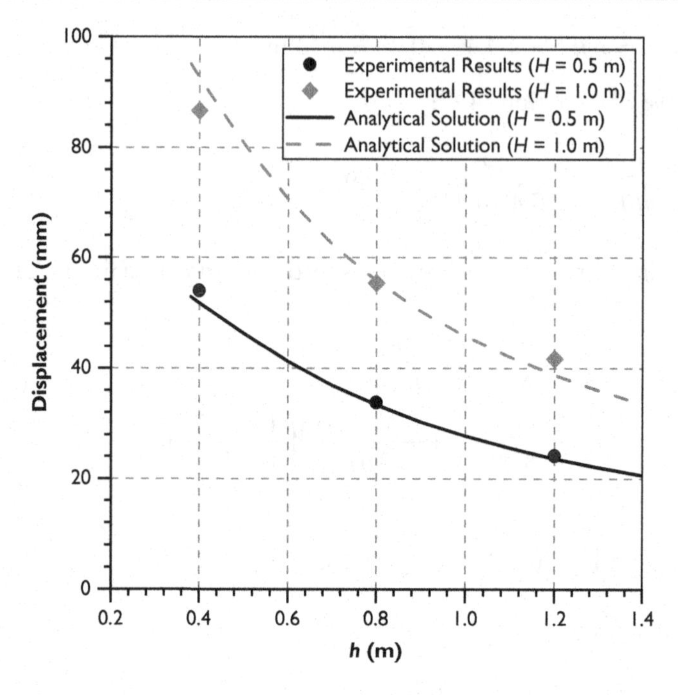

Figure 14.6 Horizontal displacement of rectangular blocks due to overturning action.

specimen had a mass of 127.8 kg. The bottom of the specimen was struck by a cast iron impactor, which had a mass of 5 kg, and was released from five different heights to achieve different impact velocities. Two laser displacement sensors (one at the top and one at the bottom) were employed to measure the sliding displacement of the specimen. The readings from both sensors have been found to be almost identical implying that the specimen only experienced sliding action (without any significant overturning nor bending actions). Velocity values inferred from the high-speed camera are listed in Table 14.2, including the incident velocity of impact (v_0), velocity of the impactor on rebound (v_1) and velocity of the target immediately following the impact (v_2). The coefficient of friction between the target specimen and the ground was 0.52. Estimate the amount of sliding displacement of the specimen.

Solution

Consider the case where $v_0 = 6.0$ m/s.

$$\lambda = \frac{127.8}{5} = 25.6$$

By the use of Equation (4.8):

Table 14.2 Velocities of impactor and target

v_0 (m/s)	v_1 (m/s)	v_2 (m/s)
2.5	0.88	0.13
3.7	1.26	0.19
4.4	1.36	0.22
5.5	1.60	0.28
6.0	1.70	0.30

$$\text{COR} = \frac{v_1 + v_2}{v_0} = \frac{1.7 + 0.3}{6} = 0.33$$

By the use of Equation (4.39):

$$\begin{aligned}\Delta &= \frac{v_0^2}{2\mu g}\left(\frac{1 + \text{COR}}{1 + \lambda}\right)^2 \\ &= \frac{6^2}{2(0.52)(9.81)}\left(\frac{1 + 0.33}{1 + 25.6}\right)^2 \\ &= 0.0088 \text{ m or } 8.8 \text{ mm}\end{aligned}$$

which can be compared against the experimental recorded value of 8.6 mm.

Estimates from calculations using Equation (4.39) based on other tests can also be compared against the recorded test results, as shown in Figure 14.7.

14.5 BENDING OF REINFORCED CONCRETE BEAMS

This case study involves the use of calculation techniques introduced in Chapters 2 and 5. A series of collision tests as reported in Fujikake et al. [14.7] was conducted on simply supported RC beams employing a drop hammer weighing 400 kg, and with a velocity of impact ranging from 1.7–6.9 m/s. The beam specimens were cast of concrete with a compression strength of 42 MPa, cross-sectional dimension of 0.15 m in width and 0.25 m in depth, and a span length of 1.4 m. The presentation of the test layout and the beam specimens are shown in Figure 14.8. Details of the experiments are listed in Table 14.3. The limiting force and deflection at the point of yield as inferred from the results of static tests carried out by Fujikake et al. [14.7] are listed in Table 14.4. Estimate the maximum deflection of the beam at the point of impact.

Solution

The solution for the impact scenario involving specimen S1616, which was subject to an impact with incident velocity of 1.72 m/s, is presented.

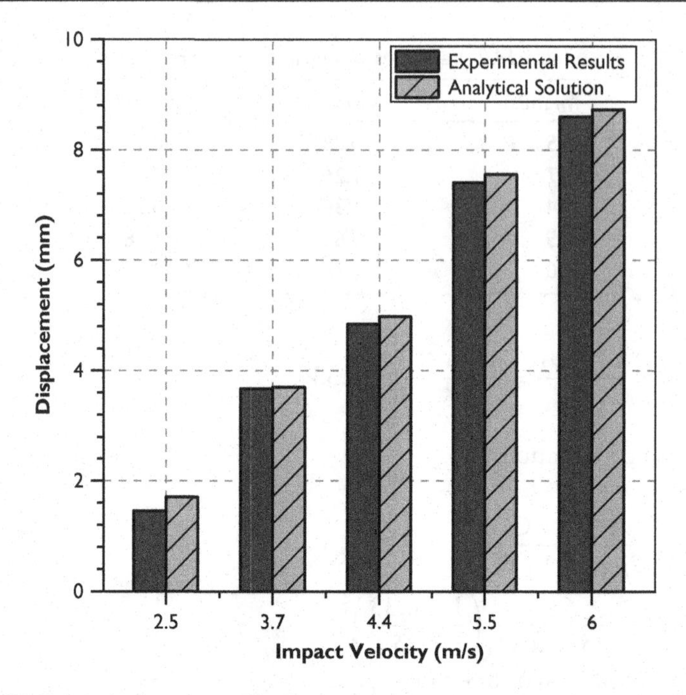

Figure 14.7 Sliding displacement of an L-shaped concrete specimen.

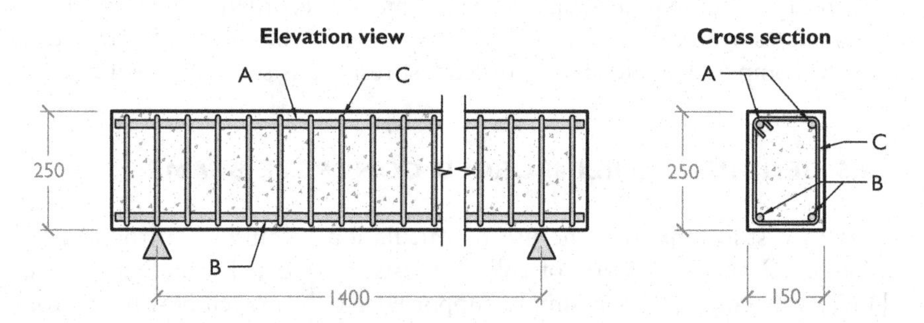

Figure 14.8 Layout of RC beam specimen.

Table 14.3 Test details of RC beam specimens

Specimen	*Reinforcement*			f_y *(MPa)*	v_0 *(m/s)*				
	A	*B*	*C*		*1.72*	*2.43*	*3.43*	*4.85*	*6.86*
S1616	2D16	2D16	D10@75	426	✓	✓	✓	✓	
S1322	2D13	2D22	D10@75	Top: 397 Bot: 418		✓	✓	✓	✓
S2222	2D22	2D22	D10@75	418		✓	✓	✓	✓

Table 14.4 Static tests results

Specimen	F_y (kN)	Δ_y (mm)
S1616	100	3.0
S1322	180	3.0
S2222	180	3.0

From Table 14.4, $F_y = 100$ kN and $\Delta_y = 3.0$ mm. Thus,

$$k_{\text{eff}} = \frac{F_y}{\Delta_y} = \frac{100}{3 \times 10^{-3}} = 33333 \text{ kN/m}$$

The generalised mass of the specimen may be calculated based on the use of the relationship as defined in Table 2.1 for a simply supported beam:

$$\lambda m = 0.5\rho AL = 0.5(2400)(0.15)(0.25)(1.4) = 63 \text{ kg}$$

$$\lambda = \frac{\lambda m}{m} = \frac{63}{400} = 0.16$$

As $\lambda < 1$, the impactor is assumed to be embedded into the target following the collision.

Thus, Equations (5.12a) and (5.12b) can be applied.

$$\Delta_s = \frac{mg}{k} = \frac{400(9.81)}{33333} = 0.12 \text{ mm}$$

$$\delta = \frac{mv_0}{\sqrt{mk_{\text{eff}}}}\sqrt{\frac{1}{1+\lambda}} = \frac{400(1.72)}{\sqrt{400(33333 \times 10^3)}}\sqrt{\frac{1}{1+0.16}}$$

$$= 0.0055 \text{ m or } 5.5 \text{ mm}$$

$$\Delta = \Delta_s + \sqrt{\Delta_s^2 + \delta^2} = 0.12 + \sqrt{0.12^2 + 5.5^2} = 5.6 \text{ mm}$$

As $\Delta > \Delta_y$, the beam is expected to have yielded. Thus, results obtained from use of Equation (5.12a) is not valid. Thus, Equation (5.17) is to be applied instead.

$$F_y - (1+\lambda)mg = 100 - (1+0.16)(400)(9.81) \times 10^{-3} = 95.45 \text{ kN}$$

$$\Delta = \frac{mv_o^2}{2\left(F_y - (1+\lambda)mg\right)}\left(\frac{1}{1+\lambda}\right) + \frac{F_y}{F_y - (1+\lambda)mg}\frac{\Delta_y}{2}$$

$$= \frac{400(1.72)^2}{2(95.45)}\left(\frac{1}{1+0.16}\right) + \frac{100}{95.45}\left(\frac{3}{2}\right)$$

$$= 6.9 \text{ mm}$$

Calculation procedures as presented have been repeated for other tests. Deflection values so estimated are compared with experimental results, as shown in Figure 14.9. The two anomalies shown in Figure 14.9 (where the estimated deflection values were close to 50 mm) were from tests in which specimens S1322 and S2222 were struck by the impactor at incident velocity of 6.86 m/s. The over-stating of the estimated deflection demand in three of the tested cases is believed to be the result of additional energy loss caused by the shattering of materials at the contact surface resulting from the high-intensity collision.

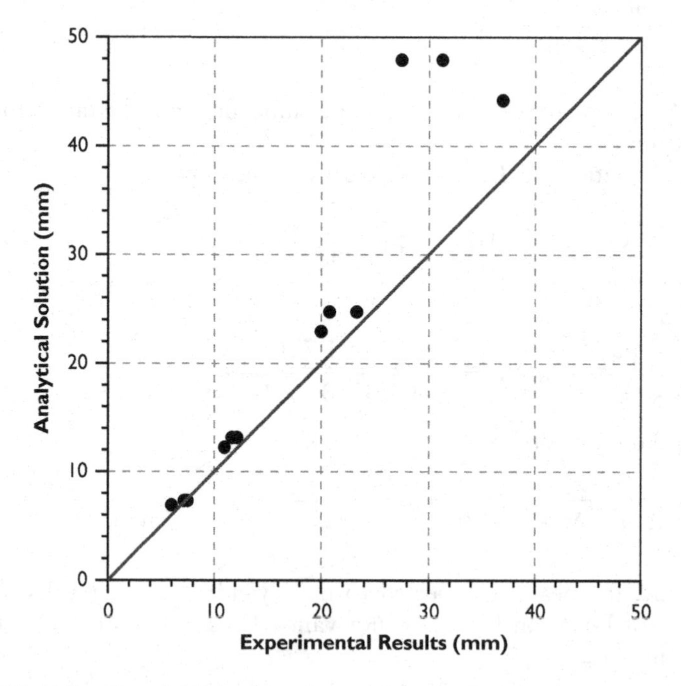

Figure 14.9 Deflection of RC beam estimated from analytical solution versus experimental results.

14.6 CONTACT FORCE GENERATED BY HAILSTONE AND WINDBORNE DEBRIS

This case study involves the use of the calculation techniques introduced in Chapter 6. A series of impact tests has been carried out with use of a gas gun (and setup), as reported in Sun et al. [14.8] and Perera et al. [14.9]. The tested debris materials include hail ice (weighing 59.5 g), concrete (weighing 306 g), birch wood (weighing 90 g) and brick (weighing 215 g). All the specimens were fired by a gas gun at velocities ranging from 7.9 to 29.1 m/s for hail ice specimens, 16.8 to 33.7 m/s for concrete specimens, 23.5 to 40.4 m/s for birch wood specimens and 9.5 to 38.6 m/s for brick specimens. A spring-connected target lumped mass weighing 2.164 kg was struck by the specimens. Estimate the amount of contact force delivered by the specimens.

Solution

The solution provided herein is for brick specimens with an impact velocity of 20 m/s.

$$\lambda = \frac{2.164}{0.215} = 10.06$$

By the use of Equation (6.10b):

$$M_c = \left(\frac{\lambda}{1+\lambda}\right)m = \left(\frac{10.06}{1+10.06}\right)(0.215) = 0.196 \text{ kg}$$

Equations (6.38a)–(6.38c) are used to determine the modelling parameters for brick debris impact.

$$k_n = 0.177v_0 + 2 = 0.177(20) + 2 = 5.54 \text{ MN/m}^p$$

$$p = 0.0136v_0 + 1.116 = 0.0136(20) + 1.116 = 1.39$$

$$\text{COR} = -0.0012v_0 + 0.094 = -0.0012(20) + 0.094 = 0.07$$

The calculated values for p and COR can be used for reading off the value of $C_{F_{c,\max}}$ by the use of a chart, as shown in Figure 6.11(b). With this example, it was found that $C_{F_{c,\max}} \approx 2.1$.

The peak contact force may then be estimated with the use of Equation (6.33):

$$F_{c,\max} = C_{F_{c,\max}} k_n \left[\left(\frac{p+1}{2k_n}\right)Mv_0^2\right]^{\frac{p}{p+1}}$$

$$= 2.1(5.54 \times 10^6)\left[\left(\frac{1.39+1}{2(5.54 \times 10^6)}\right)(0.196)(20)^2\right]^{\frac{1.39}{1.39+1}}$$

$$= 19.5 \text{ kN}$$

which can be compared against the experimental result of 19.2 kN.

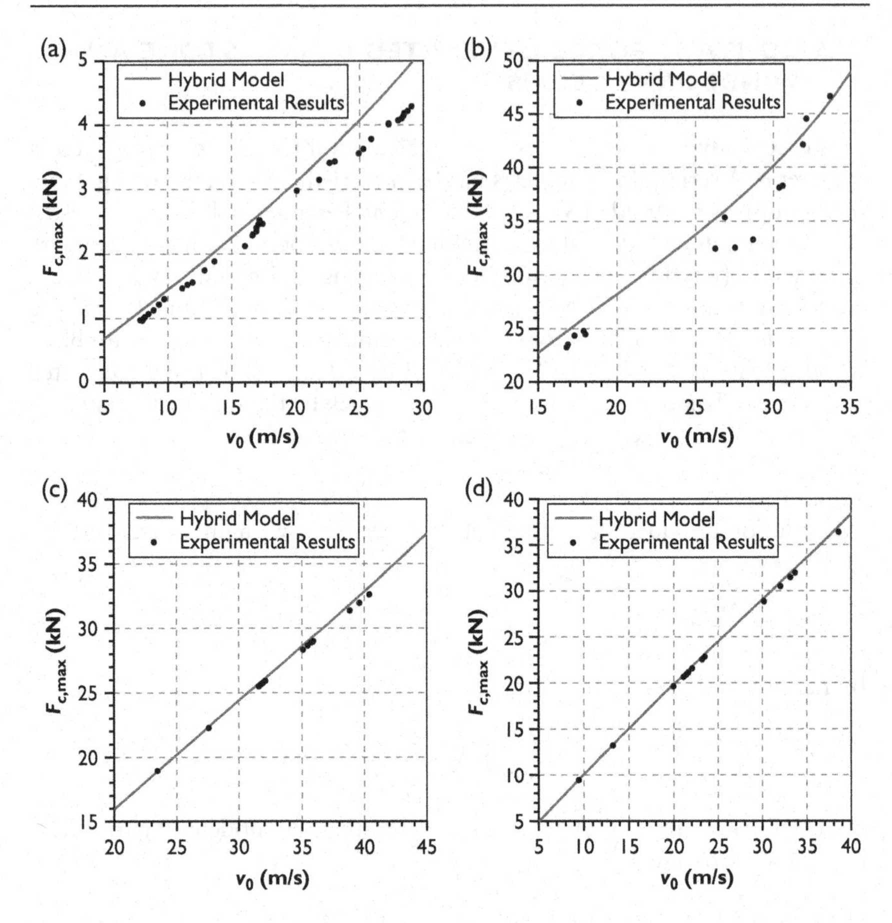

Figure 14.10 Contact force generated by (a) hail ice, (b) concrete, (c) birch wood and (d) brick.

The calculation procedure has been repeated for other test cases with different impact velocities. The results obtained from the calculations are compared against experimental measurements, as shown in Figures 14.10(a) to 14.10(d).

14.7 CONTACT FORCE GENERATED BY GRANITE BOULDER

This case study involves the use of the calculation techniques introduced in Chapter 6. Impact tests involving the use of a gas gun setup have been carried out with impactor specimens cored from granite boulders, and machined into a sphere, to fit in the apparatus as reported in Majeed et al. [14.10]. The tested granite spheres (as impactors) were 50 mm in diameter and 180 grams in mass. The spherical specimens were fired by a gas gun at velocities ranging from 9.5 to 26.3 m/s. Concrete specimens with dimensions of 90 mm × 90 mm × 60 mm (weighing 1.198 kg) were attached to a contact force measurement device, which was made up of a lumped mass of solid steel weighing 2.164 kg, and a helical spring that connected the lumped mass to a rigid support. Estimate the amount of contact force delivered by the impact.

Solution

The solution for this test case with an impact velocity of 26.3 m/s is presented.

$$\lambda = \frac{1.198 + 2.164}{0.18} = 18.68$$

By the use of Equation (6.10b):

$$M_c = \left(\frac{\lambda}{1+\lambda}\right)m = \left(\frac{18.68}{1 + 18.68}\right)(0.18) = 0.17 \text{ kg}$$

Equations (6.39a)–(6.39c) may be used for determining the modelling parameters for the granite as the impactor.

$$k_{n100} = 84.273v_0 + 160.86 = 84.273(26.3) + 160.86 = 2377.24 \text{ MN/m}^p$$

$$k_{nD} = k_{n100}\sqrt{\frac{D}{100}} = 2377.24\sqrt{\frac{50}{100}} = 1681 \text{ MN/m}^p$$

$$p = 1.21 + 0.0175v_0 + 0.0005D = 1.21 + 0.0175(26.3) + 0.0005(50)$$
$$= 1.695$$

$$\text{KE}_0 = \frac{1}{2}mv_0^2 = \frac{1}{2}(0.18)(26.3)^2 = 62.25 \text{ J} = 0.06225 \text{ kJ}$$

$$\text{COR} = 0.068\text{KE}_0^{-0.433} = 0.068(0.06225)^{-0.433} = 0.226$$

The calculated values for p and COR can be used for reading off the value of $C_{F_{c,\max}}$ by use of a chart, as shown in Figure 6.11(b). With this example, it was found that $C_{F_{c,\max}} \approx 1.18$.

The peak contact force may then be estimated by the use of Equation (6.33):

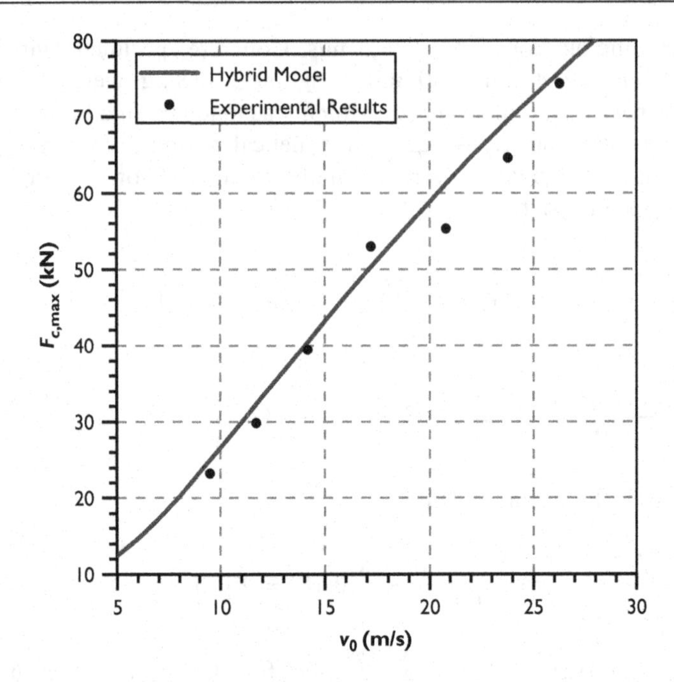

Figure 14.11 Contact force generated by granite specimens.

$$F_{c,\max} = C_{F_{c,\max}} k_n \left[\left(\frac{p+1}{2k_n} \right) M v_0^2 \right]^{\frac{p}{p+1}}$$

$$= 1.18(1681 \times 10^6) \left[\left(\frac{1.695 + 1}{2(1681 \times 10^6)} \right) (0.17)(26.3)^2 \right]^{\frac{1.695}{1.695+1}}$$

$$= 75.6 \text{ kN}$$

which can be compared against the experimentally measured result of 74.3 kN.

The calculation procedure was repeated for tests with different velocities and specimens. Results from the calculation can be compared against experimental measurements, as shown in Figure 14.11.

14.8 BENDING OF CUSHIONED REINFORCED CONCRETE STEM WALL

This case study involves the use of the calculation techniques introduced in Chapter 7. Consider the same RC wall as presented in Section 14.2. A layer of gabion cushion was placed in front of the wall to receive the impact of

Figure 14.12 Photograph of cushioned reinforced concrete wall experimental setup.

three impactors weighing 280 kg, 435 kg and 1,020 kg, respectively. Detailed descriptions of the experimental setup can be found in Perera et al. [14.11]. The cushion layer consisted of cubical gabions with side dimensions of 500 mm. The gabions, which were filled with crushed gravels, were arranged in a 3 × 3 array, as shown in Figure 14.12. Each gabion had a density of 1,520 kg/m^3 and a modulus of elasticity of 3,000 kPa, as measured from unconfined compression tests. The internal angle of friction of the fill materials is taken as 40° for crushed gravels. Estimate the maximum deflection of the wall when subject to the impact scenarios listed in Table 14.5.

A photograph of a gabion cushion placed in front of a reinforced concrete wall showing the experimental setup is described in the text.

Solution

Every step of the calculation based on Test No. 1 (as defined in Table 14.5) is presented.

Equations (7.1a)–(7.1c) are used to estimate the added mass and the mass ratio.

Table 14.5 Details of cushioned reinforced concrete wall impact experiment

Test No.	*Impactor Mass (kg)*	*Impact Velocity (m/s)*	*r (m)*
1	280	3.13	0.15
2	280	5.24	0.15
3	435	2.51	0.15
4	435	3.37	0.15

$$R_1 = R_i + e \tan 20° = 0.15 + 0.5 \tan 20° = 0.33 \text{ m}$$

$$m_g = \frac{\pi}{3}(R_1^2 + R_1 R_i + R_1^2) e \rho_g = \frac{\pi}{3}(0.33^2 + 0.33(0.15) + 0.15^2)(0.5)(1520)$$
$$= 145.3 \text{ kg}$$

Referring to the previous worked example (Section 14.2), the sum of the generalised mass of the wall and steel plate was found to be 683.8 kg.

$$\lambda m = 683.8 + m_g = 683.8 + 145.3 = 829.1 \text{ kg}$$

$$\lambda = \frac{\lambda m}{m} = \frac{829.1}{280} = 3$$

Equation (7.5) can be used for determining the maximum contact force ($F_{c,\max}$).

$$\text{KE}_0 = \frac{1}{2} m v_0^2 = \frac{1}{2}(280)(3.13)^2 = 1372 \text{ J}$$

$$\begin{aligned} F_{c,\max} &= 1.82 e^{-0.5} R_i^{0.7} M_E^{0.4} \tan \phi_k (\text{KE}_0)^{0.6} \\ &= 1.82(0.5)^{-0.5}(0.15)^{0.7}(3000 \times 10^3)^{0.4} \tan 40°(1372)^{0.6} \times 10^{-3} \\ &= 17 \text{ kN} \end{aligned}$$

The stiffness of the gabion cushion may then be estimated by the use of Equation (7.4).

$$k_n = \frac{F_{c,\max}^2}{112\text{KE}_0}\left(\frac{1+\lambda}{\lambda}\right) = \frac{(17 \times 10^3)^2}{112(1372)}\left(\frac{1+3}{3}\right) = 2510 \text{ N/m}$$

Equations (7.8a) and (7.8b) are then used for calculating the natural period of the cushion and the target, respectively.

$$T_m = 2\pi\sqrt{\frac{m}{k_n}} = 2\pi\sqrt{\frac{280}{2510}} = 2.1 \text{ s}$$

From Section 14.2, k_{eff} = 15,400 kN/m.

$$T_M = 2\pi\sqrt{\frac{\lambda m}{k}} = 2\pi\sqrt{\frac{829.1}{15400 \times 10^3}} = 0.046 \text{ s}$$

$$\frac{T_m}{T_M} = \frac{2.1}{0.046} = 46$$

By the use of the calculated values of λ and T_m/T_M, and with the aid of the design chart of Figure 7.5, the γ value of 0.29 can be identified.

The maximum deflection generated by the cushioned collision action may then be calculated using Equation (7.7), as shown in the following:

$$\Delta = \frac{mv_0}{\sqrt{mk_{eff}}}\sqrt{\frac{1}{1+\lambda}} \times \gamma = \frac{280(3.13)}{\sqrt{280(15400 \times 10^3)}}\sqrt{\frac{1}{1+3}} \times 0.29$$
$$= 0.0019 \text{ m or } 1.9 \text{ mm}$$

The calculation procedures presented have been repeated for other tests, the details of which are listed in Table 14.5 (i.e. Test Nos. 2 to 4) to provide estimates of Δ values for comparison against experimental measured values (as shown in Figure 14.13).

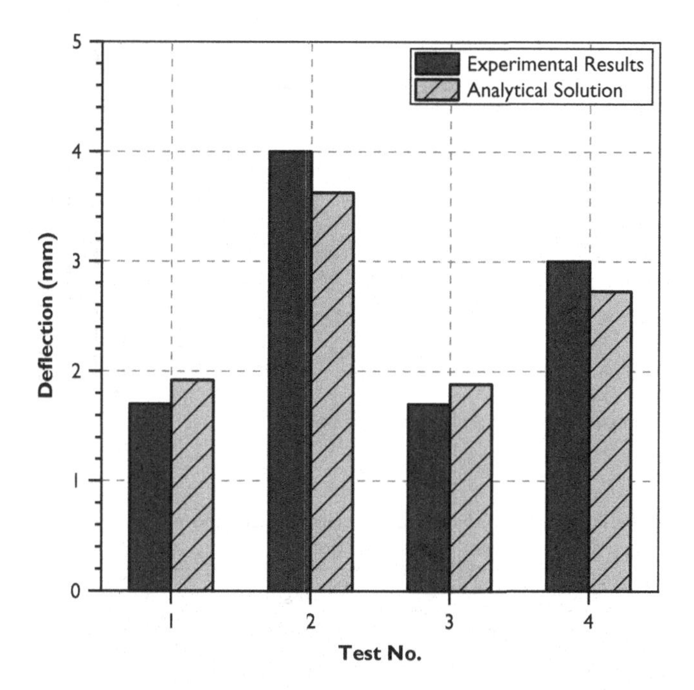

Figure 14.13 Deflection of cushioned RC wall: analytical estimates versus experimental results.

14.9 BENDING OF SOIL-EMBEDDED BAFFLES

This case study involves the use of the calculation techniques introduced in Chapter 11. Consider a collision experiment carried out on soil-embedded baffles, as reported in Perera and Lam [14.12]. As shown in Figure 14.14, the impactor had a mass of 28 kg and a varying impact velocity, v_0. Three steel sections that were manufactured to AS/NZS 1163 [14.13] were used as the baffles, which includes two square hollow sections (SHS) and a circular hollow section (CHS), as listed in Table 14.6, alongside the values of mass per unit length and a second moment of area. The baffles were embedded into compacted sandy gravel with a η_h value of 80 MN/m^3 in accordance with the sensitivity study carried out by Perera and Lam [14.12]. The embedded segment of the baffle is cast in concrete, forming a jacket of square cross section and of side lengths measuring 85 mm. Two embedment lengths (L_e), 350 and 500 mm, were considered for each of the sections.

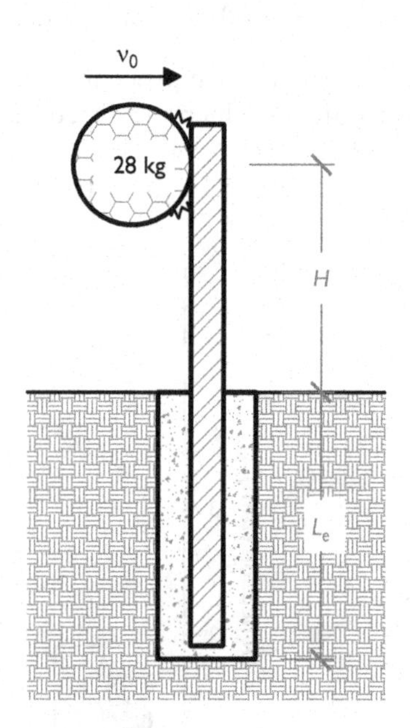

Figure 14.14 Impact tests on soil-embedded baffles.

Table 14.6 Mass per unit length and second moment of area of steel sections

Section Type	*Mass (kg/m)*	*I (mm⁴)*
30 × 1.6 SHS	1.38	23,100
35 × 2.5 SHS	2.42	52,900
33.4 × 32 CHS	2.42	36,000

Table 14.7 Test details of soil-embedded baffles

Test No.	L_e *(m)*	*H (m)*	*Section Type*	v_0 *(m/s)*
1	0.35	0.65	30 × 1.6 SHS	1.4
2				3.0
3				3.3
4			35 × 2.5 SHS	2.2
5				3.3
6				3.8
7			33.4 × 32 CHS	3.4
8	0.5	0.5	30 × 1.6 SHS	2.2
9				3.2
10			35 × 2.5 SHS	2.2
11				3.2
12				3.8
13			33.4 × 32 CHS	3.4

The Young's modulus of the steel sections and concrete jacket are 210 GPa and 27,000 MPa, respectively. A cracked effective stiffness ratio of 0.1 may be assumed for concrete. Thirteen tests were carried out, as listed in Table 14.7. Estimate the maximum amount of displacement of the baffle when collided upon by the impactor.

Solution

Every step of the calculation based on Test No. 1 (as defined in Table 14.7) is presented.

$$EI_{pole} = 210000(23100) \times 10^{-9} = 4.85 \text{ kNm}^2$$

$$I_{concrete} = \frac{85 \times 85^3 - 30 \times 30^3}{12} = 4.28 \times 10^6 \text{ mm}^4$$

$$EI_{concrete} = 0.1(27000)(4.28 \times 10^6) \times 10^{-9} = 11.56 \text{ kNm}^2$$

$$EI_{combined} = EI_{pole} + EI_{concrete} = 4.85 + 11.56 = 16.41 \text{ kNm}^2$$

Calculations based on the use of Equation (11.2) are shown as follows:

$$T_s = \sqrt[5]{\frac{EI_{combined}}{\eta_h}} = \sqrt[5]{\frac{16.41}{80 \times 10^3}} = 0.183 \text{ m}$$

By reading off the design chart of Figure 11.3: $[L_e/T_s]_{\text{lim}} = 1.9$ approximately for $T_s = 0.183$ m. The embedment length (L_e) is at least equal to 0.35 m (=1.9 × 0.183 m) in order to ensure flexible behaviour of the soil-embedded pole.

By the use of Equation (11.1):

$$C = 0.45[L_e/T_s]_{\text{lim}}\, T_s = 0.45(1.9)(0.183) = 0.156 \text{ m}$$

By the use of Equation (11.6b):

$$k_e = \left[\frac{(H + C)^3 - H^3}{3EI_{\text{combined}}} + \frac{H^3}{3EI_{\text{pole}}}\right]^{-1} = \left[\frac{(0.5 + 0.156)^3 - 0.5^3}{3(16.41)} + \frac{0.5^3}{3(4.85)}\right]^{-1}$$
$$= 84.8 \text{ kN/m}$$

The pole is considered a cantilevered member with a height $H + C = 0.656$ m. Thus, the total mass is:

$$m_{\text{pole}} = 0.656(1.38) = 0.9 \text{ kg}$$

The generalised mass of a cantilever member is calculated as follows as per the recommendations presented in Table 2.1 (of Chapter 2):

$$\lambda m = 0.25 m_{\text{pole}} = 0.25(0.9) = 0.23 \text{ kg}$$

$$\lambda = \frac{\lambda m}{m} = \frac{0.23}{28} = 0.008$$

By the use of Equation (11.7):

$$\Delta = \frac{m v_o}{\sqrt{m k_e}} \sqrt{\frac{1}{1 + \lambda}} = \frac{28(1.4)}{\sqrt{28(84.8 \times 10^3)}} \sqrt{\frac{1}{1 + 0.008}}$$
$$= 0.0253 \text{ m or } 25.3 \text{ mm}$$

The calculation procedures presented have been repeated for other tests, the details of which are listed in Table 14.7 (i.e. Test Nos. 2 to 13) to provide estimates of Δ values for comparison against experimental measured values (as shown in Figure 14.15).

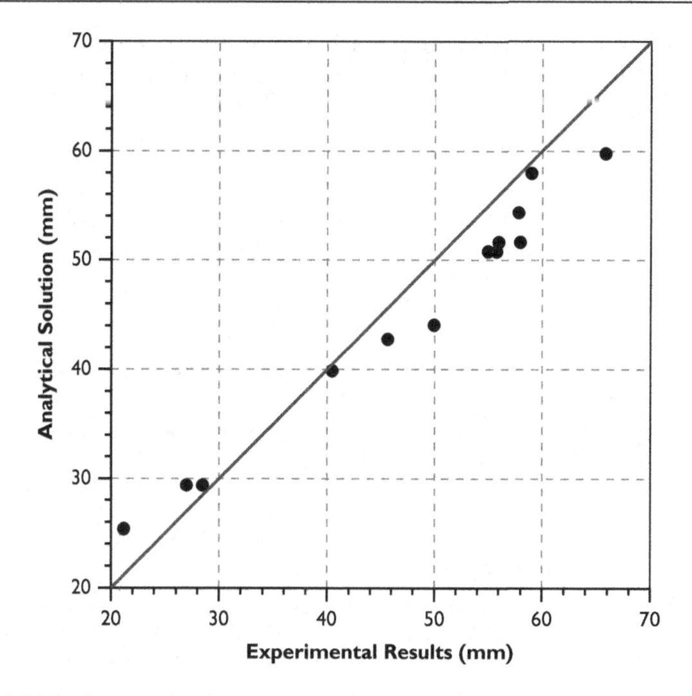

Figure 14.15 Deflection of soil-embedded baffle estimated from analytical solution versus experimental results.

REFERENCES

14.1 Yong, A.C.Y., Lam, N.T.K., Menegon, S.J., and Gad, E.F., 2020, "Experimental and analytical assessment of the flexural behaviour of cantilevered RC walls subjected to impact actions", *Journal of Structural Engineering*, Vol. 146(4), p. 04020034, doi:10.1061/(ASCE)ST.1943-541X.0002578

14.2 Standards Australia and Standards New Zealand, 2019, *AS/NZS 4671:2019 Steel for the reinforcement of concrete*, Standards Australia Limited and Standards New Zealand, Sydney and Wellington.

14.3 Menegon, S.J., Wilson, J.L., Lam, N.T.K., and McBean, P., 2017, "Seismic design and detailing of reinforced concrete wall buildings", *Concrete in Australia*, Vol. 43(4), pp. 55–58.

14.4 Menegon, S.J., 2019, "WHAM: a user-friendly and transparent non-linear analysis program for RC walls and building cores", *Available from*: downloads.menegon.com.au/1/20190901

14.5 Lam, N.T.K., Yong, A.C.Y., Lam, C., Kwan, J.S.H., Perera, J.S., Disfani, M.M., and Gad, E., 2018, "Displacement-based approach for the assessment of overturning stability of rectangular rigid barriers subjected to point impact", *Journal of Engineering Mechanics*, Vol. 144(2), pp. 04017161-1 –04017161-15, doi:10.1061/(ASCE)EM.1943-7889.0001383

14.6 Yong, A.C.Y., Lam, C., Lam, N.T.K., Perera, J.S., and Kwan, J.S.H., 2019, "Analytical solution for estimating sliding displacement of rigid barriers subjected to boulder impact", *Journal of Engineering Mechanics*, Vol. 145(3), p. 04019006, doi:10.1061/(ASCE)EM.1943-7889.0001576

14.7 Fujikake, K., Li, B., and Soeun, S., 2009, "Impact response of reinforced concrete beam and its analytical evaluation", *Journal of Structural Engineering*, Vol. 135(8), pp. 938–950, doi:10.1061/(ASCE)ST.1943-541X.0000039

14.8 Sun, J., Lam, N., Zhang, L., Ruan, D., and Gad, E., 2015, "Contact forces generated by hailstone impact", *International Journal of Impact Engineering*, Vol. 84, pp. 145–158, doi:10.1016/j.ijimpeng.2015.05.015

14.9 Perera, S., Lam, N., Pathirana, M., Zhang, L., Ruan, D., and Gad, E., 2016, "Deterministic solutions for contact force generated by impact of windborne debris", *International Journal of Impact Engineering*, Vol. 91, pp. 126–141, doi:10.1016/j.ijimpeng.2016.01.002

14.10 Majeed, Z.Z.A., Lam, N.T.K., Lam, C., Gad, E., and Kwan, J.S.H., 2019, "Contact force generated by impact of boulder on concrete surface", *International Journal of Impact Engineering*, Vol. 132, p. 103324, doi: 10.1016/j.ijimpeng.2019.103324

14.11 Perera, J.S., Lam, N., Disfani, M.M., and Gad, E., 2021, "Experimental and analytical investigation of a RC wall with a Gabion cushion subjected to boulder impact", *International Journal of Impact Engineering*, Vol. 151, p. 103823, doi:10.1016/j.ijimpeng.2021.103823

14.12 Perera, J.S., and Lam, N., 2022, "Soil-embedded steel baffle with concrete footing responding to collision by a fallen or flying object", *International Journal of Geomechanics*, Vol. 22(3), p. 04021311, doi:10.1061/(ASCE)GM.1943-5622.0002299

14.13 Standards Australia, 2016, *AS/NZS 1163 Cold-formed structural steel hollow sections*, Standard Australia Limited, New South Wales, Australia and Wellington, New Zealand.

Index

Page numbers in **bold** indicate tables; page numbers in *italics* indicate figures

Printed in the United States
by Baker & Taylor Publisher Services